The SHEEP BOOK *for* Smallholders

Tim Tyne

HomeFarmer

The Sheep Book For Smallholders
Published by The Good Life Press Ltd., 2009, 2012, and by
Home Farmer, an imprint of The Good Life Press Ltd., in 2015.
This book contains some material that first appeared in
Country Smallholding magazine.

ISBN 978190487 1644

A catalogue record for this book is available from the
British Library.

Published by
Home Farmer
PO Box 536
Preston
PR2 9ZY

www.homefarmer.co.uk

Visit Tim and Dot Tyne at
www.viableselfsufficiency.co.uk

All photographs by Tim and Dot Tyne unless credited otherwise.
Design and layout by Ruth Tott

Printed in Scotland by Bell and Bain Ltd, Glasgow

The **SHEEP BOOK** *for Smallholders*

Some few years ago I attended a talk given by David Henderson, author of *The Veterinary Book for Sheep Farmers.* At the end of the evening I asked him if he would be so kind as to sign my copy of his book, which he did. He also added a concise message, comprising only two words, which I reproduce below for your benefit:

"Good Shepherding!"

Author's Note:

Firstly, this is not a veterinary manual, nor is it a miserable catalogue of the numerous ways in which a sheep may choose to die. Although information on various diseases has been included where appropriate, the emphasis throughout is on practical stockmanship. Secondly, I make no apology for my apparently random use of both metric and imperial measurements. I simply wrote down whichever seemed easiest to visualise at the time, in the hope that other people would be able to visualise it too. For readers who are not conversant with both systems, I've included conversion tables in the appendices.

ACKNOWLEDGEMENTS

John Thorley OBE, FRAgS, Lucy Tyne MRCVS (one of my sisters), Daphne Tyne (my mother), Katie Thear, Mike & Janice Randle of Bryn Bychan, John Hughes of Llanllechid, Helen Garnham of Innovis Ltd (photo, p.60), Val Grainger "The Woolly Shepherd" (photo, p.201), Andy Nickless (photos, p.261, 264, 280), Sam Boon of Signet (photos, p.72, 74), Andrew Forgrave / Daily Post (photo p.156), Andrew Todd of Brintons Carpets, Geraint Evans of Tithebarn Ltd (photo p.91), Ian Wilkinson of Cotswold Seeds (photos, p.9, 212, 213, 214, 215, 221, 222, 226, 233), Gwynedd Council Trading Standards Dept., Farmers Guardian (photo, p.13), Jill Tyrer (photos p.6, 32), Richard Wilson, Mark Hardy of High Weald Dairy (photos, p.203, 204, 205), Simon Donovan (photo, p.34), Helen Arthan (photo, p.31), and all the breed society secretaries and others who responded to my requests for information and pictures.

I'm also indebted to all the shepherds and farmers I've worked with over the years who, wittingly or unwittingly, have taught me so much, either by direct example or by simply leaving me alone to do things my own way, and put my own ideas into practice. And to Diane Cowgill, former editor of Country Smallholding magazine, without whose encouragement I'd probably never have started writing in the first place.

And of course, special thanks to my wife, Dot, who has put a tremendous amount of effort into this book too, and to my children Iestyn, Llinos and Rhian, who've put up with their dad spending a whole summer shut in his office (and being very grumpy).

CONTENTS

FOREWORD

The Sheep Book for Smallholders by Tim Tyne is an absolute must for anyone with an interest in sheep and of particular value to those who already have a few or on the other hand who aspire to reach the dizzy heights of having some to care for.

The first prerequisite of anyone who wishes to become a shepherd or flock-master in any sense is that they have to be enthusiastic about this remarkable animal which comes in all shapes and sizes, fits all sorts of purposes and has the ability to make even the most patient person totally exasperated and then come back for more!

Tim Tyne is clearly such a person – the book reflects a passion and a depth of knowledge about his beloved sheep which he portrays in an easy flowing style which cannot help but be a must for the bookshelf of all sheep farming families, large and small and be of special interest for people getting started or who wish to expand their knowledge about this marvellous, often undervalued species.

The book briefly traces the history of the sheep and its relationship with man, makes the links to the number of breeds, the importance of the sheep dog – and how to train him or her, how the partnership between sheep, man and dog has stood the test of centuries and explains why it is such an important part of the present and future.

Tim takes us through the entire shepherding calendar with plenty of clear descriptions of how to carry out many of the tasks associated with being a good shepherd, and takes us all the way through getting started, handling, weaning, shearing, fattening, touching on marketing, slaughtering and butchery as well as the other products of milk and wool.

As such it is a timeless introduction to sheep keeping which will be enjoyed by everyone with an interest in sheep or who wishes to expand their knowledge.

It would be impossible to summarize it properly in this Foreword, suffice to say I'm certain it will be read from cover to cover with rapt attention – it has something for everyone – even one or two things which might form the basis of an animated discussion or two!

John Thorley OBE, FRAgS

Since time immemorial the sheep has fed man, clothed him, and accompanied him on his travels to all four corners of the Earth.

A BRIEF HISTORY OF SHEEP AND MAN

The early domestication of sheep, some 8,000 years ago in Asia, possibly represented the single most significant step that the human race has ever made. It was sheep that enabled us to give up our precarious existence as hunter-gatherers, settle down in one place, and become farmers. All civilisation stems from this, and sheep are, in some way or another, intrinsically tied up with almost every momentous episode in the history of mankind since that time. Their importance is freely acknowledged through mythology, folklore and religion. Shepherding must surely rank as one of the oldest professions (second only to prostitution, perhaps?), and ancient cultures, pre-dating Christianity by several thousand years, venerated the role of the shepherd. One of the first gods of the ancient Egyptians was the ram headed Khnum, and the Sumerians had deities devoted solely to the protection of their extensive flocks. The New Testament's Lamb of God is a relatively recent concept built upon older (deeper?) beliefs, making use of an existing socially acceptable icon. Greek mythology adopts some similar ideas, and the sacrificed golden ram of Heracles is honoured by the constellation of Aries. The Chinese zodiac also honours sheep, although the early star gazers could scarcely have predicted that China would one day have the largest sheep population in the world! The sacrificial lamb appears throughout Celtic and pagan folklore too, with records of burnt offerings being made as recently as the late 1800s in the Isle of Man, and possibly in parts of Scotland.

The first domestic sheep was the Mouflon. These animals would have provided the early shepherds with meat and hides and, presumably, milk. Wool was still unheard of, as the Mouflon had a hairy coat. At some point, however, someone must have realised that soft fibres plucked from the body of a living animal could be used to produce warm garments, while allowing the animal to continue to grow and breed and produce milk. By the time sheep arrived in Britain in around 4000 BC they would very definitely have had a fleece

Wild Mouflon

rather than hair, although the art of spinning and weaving didn't really catch on here for another 2000 years. However, by the end of the Bronze Age the spindle and loom had become important items in the economy of the times. From being a convenient way of providing a meat meal without having to go hunting, the sheep had evolved into a true multi-purpose animal, combining both commerce and subsistence. Wool, by virtue of the fact it doesn't go "off", became a tradeable commodity that could be exported, and, together with milk, superseded meat and hides as the principal products. There are still more dairy sheep in the world than there are dairy cows.

By the beginning of the Iron Age the little mouflon derived sheep had accompanied settlers to all parts of Britain, and even to the far flung outposts of the Western and Northern Isles, where their descendants can easily be recognised to this day in the shape of the Soay breed. The Iron Age represented a period of turmoil and conflict, and it is clear that the inhabitants of Britain were arranging themselves into tribal groups, with settlements, farms and forts located within walled enclosures for security. It appears from archaeological findings that fleeces were spun at home by the families of the smaller farmers, and a large proportion of the resultant yarn was given to the hierarchical leader as a form of rent. Here, in the overlord's establishment, the yarn was woven into cloth, and, by the end of the Iron Age, the regional chieftains had established a thriving export trade in woollen fabrics to mainland Europe.

Meanwhile, woollen clothing had become a symbol of wealth in the great civilisations of Greece and the Roman Empire. Considerable advancements in animal husbandry led to the rapid development of new breeds of sheep – indeed, the Romans are credited with the creation of an early type of merino. When the Romans invaded Britain in AD 43 they brought with them improved breeds of sheep with white fleeces that required shearing (not plucking, as had previously been the norm). These large, longwoolled Roman sheep gave rise to the ancestors of distinctive breeds such as the Cotswold and the Lincoln Longwool. Further north they were crossed with the existing local types (which were very similar to today's Soay), producing yet more regional variants.

TIP! When you reach an introspective time of life, and begin to feel that you ought to put your affairs in order before departing this mortal existence, remember to request that you be buried with a lock of wool in your hand. The Lord is a shepherd also, and understands that flocks must be tended on the Sabbath – when you produce the token of your trade at the pearly gates you'll be forgiven for all your absences from regular worship!

The Romans' stay in Britain wasn't lengthy – a mere four hundred years or so – and it didn't take long for the ordered lifestyle of the Roman occupation to descend into disarray when they left. There followed the dark ages. Apart from the areas where the Roman sheep had been dominant, the local breeds retained most of the characteristics of the older "native" types, being rather small and wiry, with poor quality coloured fleeces. These sheep were eventually displaced and pushed northwards by the improved breeds of the Saxon invaders. In the meantime a similar influx was occurring in the north, as Viking raiders settled with their own flocks of multi-horned sheep. The Soay sheep were already in existence long before the Vikings turned up, and, apart from giving the Island (and hence the breed) a name, the invaders seem to have had very little impact there. Elsewhere the Viking influence is more marked, and can be seen in the Shetland, Orkney, Hebridean, Manx Loaghtan, Boreray and Icelandic breeds of sheep. Together with some other Scandinavian breeds these form the group known as the "Northern short tailed". The little Herdwick sheep of the Lake District also has its origins in Scandinavia. Similar types to those that became restricted to offshore islands would have been found over much of mainland Scotland, and the ancestors of many of our modern hill breeds may have been created at the interface between the Saxon and the Northern races.

The British sheep industry continued to operate in a fragmented manner throughout the Anglo-Saxon period, although the emergence of the manorial system that governed both land use and social hierarchy enabled some progress to be made. By the time of the Norman invasion, Britain had once again established itself as an exporter of woollen products, albeit in a rather disorganised fashion. The immense

Mediaeval spinning and weaving

scale and potential of the sheep industry was clear from the Doomesday survey, but it was only the monks – the Cistercians in particular – who had the organisational infrastructure to exploit this resource. The retreating Romans had, some 600 years previously, relocated their textile industry to Belgium, and, whether by chance or not, the region continued to be noted for its expertise in weaving and dyeing. The Cistercians had strong links with Flanders, and it didn't take long for a highly efficient trade to develop between the powerful new abbeys in England and their industrial contacts abroad. The Cistercian abbeys extended their ranges, running vast flocks of sheep, and, as other monastic orders followed their example, the wealth of mediaeval England became dominated by (and dependant upon) wool. Many of the consequences of this trade – the taxes imposed by royalty (and subsequent leaders), the banking system created by the Knights Templar, the large areas of enclosed grazing and the increasing industrialisation of manufacturing processes – have a significant impact on our lives and landscapes to this day.

Wool continued to be the principal output of the British sheep flocks until rising population in the 18th century increased the demand for meat. Ironically, it was the invention of two items – the Flying Shuttle and the Spinning Jenny – used to mechanise the process of manufacturing finished cloth from raw wool that sparked the industrial revolution which ultimately resulted in the increase in urban populations which, due to their density and location, were unable to feed themselves. The British flockmasters turned their attention to mutton production, and the raw materials required by the textile industry were imported from abroad. Initially supplies came in from mainland Europe, but, when Britain's relationship with France made this impossible, the mill owners looked to the new colonies in Australia to satisfy their requirements. British sheep breeds were exported "down under" to form the genetic basis of a rapidly developing industry, and, as if history were repeating itself, another continent began to grow wealthy on the back of the humble sheep.

SHEEP FARMING IN THE UK

The British sheep industry (or at least a large part of it) is described as "stratified". It is quite unique in the way it works, and we like to think of it as the UK's "traditional" sheep management system – indeed, when new legislation descends upon us from Europe,

as it all too frequently does, and threatens the core structure of our industry, exponents often bewail the damage being done to our shepherding heritage. It's true that the stratified system is based upon time-honoured regional husbandry techniques, but, in its current form, it is dominated by relatively few breeds. Therefore, one can only assume that the evolution of this industry structure has brought about the demise of less competitive local types, despite the traditional nature of each individual sector of the system. Having said that, older breeds do re-emerge from time to time as key players in the game, following changes in market requirements or shifts in provincial land use. Alongside this, a number of breeds (the Lleyn being the chief example) have shown themselves capable of satisfying almost every criteria within the whole of the system, by-passing the stratification altogether.

How It Works

As its name suggests, the stratified system operates as a series of layers or steps. At the top we have the true mountain breeds, which, in the harsh environment in which they've evolved, are bred pure. All the female offspring are retained, and older "draft" ewes are sold downhill. Male lambs are also sold to lowland farmers for fattening, and ewe lambs spend their first winter away on kinder pasture before returning to their birthplace to take the place of the draft ewes. Many of these flocks are "hefted", making them virtually irreplaceable. The loss of hefted flocks during the 2001 foot and mouth outbreak had a knock-on effect felt by the whole industry.

If we follow the progress of the draft ewes downhill, we find them on marginal land farms, being mated to "crossing" sires. Male lambs may be sold as stores, or fattened on farm,

Suffolk ewe & lamb

depending on the individual circumstances – it's irrelevant really, for here the principal output is the first cross female offspring. These, in several variations (depending on the specific parentage) are the most popular commercial ewes in the UK. They're sold further downhill, in their thousands each year, to join productive lowland flocks.

On the lowland farms the crossbred ewes, displaying both the maternal ability of their dams and the prolificacy of their paternal breed, are put to rams of terminal sire breeds to produce the final end product in the system – quality prime lamb.

Core Breeds Of The Stratified System

At present, the predominant hill breeds at the top of the tree are the Swaledale and the Scottish Blackface, and the principal crossing sire is the Bluefaced Leicester. The results of this union are the North Country Mule and the Scotch Mule, the UK's two most popular commercial ewes. The Texel is the most frequently used terminal sire, with the Suffolk and Charollais jostling one another for second place. This handful of breeds forms the

backbone of our industry today. But of course there's more to it than that, and if we spread the net a little wider we find several other competitive runners in the downhill race: The Welsh mountain ewes, in particular the speckle faced types, also produce a popular mule when mated to the Bluefaced Leicester, as does the Cheviot. Of the more traditional breeds both the Exmoor Horn and the Llanwenog are proving themselves to be useful mule mothers.

The Bluefaced Leicester itself has a rival in the shape of the Border Leicester. The crossbred progeny by the Border Leicester ram are generally known as Halfbreds, and there are halfbred parallels to most of the popular types of mule. On the Blackface and the Swaledale the Border Leicester gives us the Greyface. The whiter faced Welsh mountain breeds produce the Welsh Halfbred (probably the closest commercial rival to the Mule) and the Cheviot ewe is used to produce the Scotch Halfbred. The offspring of a Clun Forest ewe (a breed not dissimilar to the Llanwenog) mated to a Border Leicester ram is known as an English Halfbred. And if we cast our net a little wider still we find another couple of crossing sires, the Wensleydale and the Teeswater, being used on Dalesbred ewes to produce the Masham. So, although there's a dominant thread holding the stratified system together, on closer inspection the whole thing appears to be a fairly complex web. And that's only part of the story!

Integral Sectors

The system as I've explained it above wouldn't work without a number of other key players in the game:

Right at the top, at the very, very top, over and above the commercial flocks of mountain ewes, are the men who breed the Blackface and Swaledale tups. Many of these flocks have been handed on from generation to generation, and the evocative names by which they're known have become synonymous with the history and development of the breeds they represent. We may smile knowingly, and mutter about pre-arranged deals, when we see tups from famous flocks selling for tens of thousands of pounds, but really we shouldn't begrudge these shepherds their moment of glory – they are, after all, at the apex of the industry.

Another important figure is the store lamb finisher. Generally the upland farms, and quite a few of the marginal land holdings too, haven't the quality of grazing required to rear male lambs to slaughter weights, and besides, as we've seen, their principal outputs are female. Most of the male lambs, of both the pure hill flocks and the mule producing flocks, are sold as stores for fattening elsewhere. In many cases the buyers are arable farmers who sow large acreages of stubble turnips after harvest, and fatten lambs over the winter. Dairy farmers also play a key role in taking lambs off the hills in winter, although in this case it's the female lambs. They return to their holding of birth in the spring.

The crossing sires are often bred in nucleus flocks on the farms of the mule breeders, although many are found in specialist flocks on smaller holdings, as are the terminal sires. In fact, it is in the breeding of terminal sires that the commercially minded smallholder plays a significant role in the stratification of the industry.

Outputs

Currently there are just over 14 million breeding sheep in the UK, producing some 325,000 tonnes of meat (34.3% of EU production) and 32,000 tonnes of greasy wool. For comparison, Australia produces 692,000 tonnes of sheepmeat and 474,800 tonnes of wool from a breeding population of 79 million ewes, and China has a breeding flock of 133 million, producing 1,900,000 tonnes of meat and 234,000 tonnes of raw fleece per annum.

The Rise And Fall Of The Artisan Shepherd

In my mind, the early 1980s represented a transitional period in the recent history of sheep farming in the UK. Innovative ideas were widely publicised, and a lot of effort went into improving flock performance. New breeds and breeding methods were readily adopted throughout the industry. Meanwhile, the inherent knowledge of the old school shepherds was still available to those who cared to seek it. A happy marriage of traditional husbandry methods and new technology was created. Into this arena stepped a fresh figure – the artisan shepherd. The phenomenon was not an entirely original one. The desire to tend flocks and herds seems to be something very close to the surface of the human psyche, even amongst individuals who are many generations removed from the land. The naturalists W. H. Hudson (b.1841) and Barclay Wills (b.1877) were both intrinsically attracted to the shepherding life, and another naturalist, Richard Perry, abandoned his former existence in favour of life as a shepherd in the Western Highlands and Islands of Scotland. He recounts the experience in his book *"I went a' shepherding"*, published in 1944. Thomas Firbank exchanged work in a Canadian factory for the harsh environment of a Snowdonian sheep farm (*"I Bought a Mountain"*, first published in 1940), and subsequent writers have followed suit, on an individual basis. The situation in the 1980s, however, almost constituted mass migration! Fuelled, no doubt, by the rising interest in smallholding and self-sufficiency, and inspired by characters such as John Seymour, many professional people took to shepherding. In some cases they bought or rented land, and established their own small flocks, but others simply "went a' shepherding" at the weekends. The most significant quality of the artisan shepherd was that he embraced all aspects of the sheep industry. He was equally at home tending a thousand ewes on an arable farm as he was looking after a smallholder's grassland flock of Jacob sheep. He would be present at both the local rare breed fairs and the big commercial sales. He learnt to shear, and recognized the requirements of the modern wool trade, though was quite likely to be seen wearing a textured jumper he'd created for himself using multi-coloured fleeces from primitive breeds. He sought out the knowledge of the previous generation, and also ensured he kept himself bang up-to-date with the latest industry developments. He became an all round sheep expert, and, above all, he was proud to call himself a shepherd. His services were in demand, particularly as sheep began to make a comeback in the arable areas, where their lengthy absence had resulted in a loss of the necessary herding skills.

Quite when or why the artisan shepherd began to fade out I don't really know. Maybe increased expectations of living standards meant that the younger generation weren't inspired to follow suit. Perhaps it was rising property prices that led to smallholdings being sought by a different category of buyer – many present-day smallholders do not see themselves as being part of the wider agricultural industry, and are more inclined to look at pensions or investments to provide an income stream, than find part time employment on local farms. Also, the late 1980s saw the beginning of a period of increasing public disillusionment in modern farming methods, albeit often misguided. It became the fashion in some areas for smallholders to denigrate the wider agricultural industry, and to distance themselves from their larger farming neighbours. This served only to drive a wedge through rural communities, creating a counterproductive "us and them" environment, and made it much harder for any newcomer to get a foot in the farming door. In consequence, many smallholders have found themselves isolated on the fringes of the pastoral scene.

The artisan shepherd, by virtue of his versatility, bridged the divide. He was both a traditionalist and a modernist, and wove old fashioned skills into a contemporary fabric. Today's sheep keeping smallholder would do well to take a leaf out of his book.

RECENT DEVELOPMENTS

Here I'd really like to write about things such as improved welfare standards, advancements in breeding techniques, increased use of genetic technology, new export markets for UK livestock, and so on. But I can't. It's not that these things haven't happened – they have, and progress is continually being made. Britain maintains its position as a world leader in sheep breeding and husbandry, but unfortunately these positive evolutions are permanently overshadowed by the most significant recent development of all – excessive bureaucracy.

25 years ago it wasn't too bad. There were a couple of simple single page forms to fill in each year, one being a declaration that your sheep had been dipped within the designated period, and the other (applicable only to those with more than a dozen ewes) enabled you to claim the Annual Ewe Premium. The Variable Premium, paid on fat lambs of an

acceptable standard, was dealt with by market or abattoir staff. The system of subsidising lambs at slaughter potentially penalised upland farmers, so they applied for Hill Livestock Compensatory Allowance, but the paperwork wasn't onerous.

Now the situation has escalated beyond all proportion. Sheep quotas came and went, leaving behind them a messy paper-trail of compulsory record keeping, and a series of countryside management initiatives have resulted in yet more forms to fill, with strictly prescribed stocking densities that must be adhered to. And, if you read the small print you'll find that you're pretty well obliged to abide by the guidelines, even if you're not participating in any specific project or claiming payment! The change to the current system of a Single Payment Scheme became the bureaucratic nightmare of all time, with administrative cock-ups piling upon one another like falling snowflakes. Extensive mapping exercises were carried out in order that the authorities could determine who actually farmed where, and with what. Meanwhile, farmers continued to manage their land, tenancies changed hands and fields were bought and sold, so the situation continually altered. Mapping has now become an annual headache, together with a detailed census. In depth farm inspections may be carried out at a moment's notice, and woe betide you if you put a foot wrong!

The 2001 Foot & Mouth outbreak raised the whole issue of livestock traceability, and the immediate response to this was to pile on yet more paperwork, in the form of Animal Movement Licences, which need to be completed each time sheep move between farms, and, as from next year (2010) it appears that we'll have to individually record every sheep using a Defra allocated number in an electronic identification device. The ban on the burial of fallen stock means that there's yet another piece of paper to file away each time a dead animal is collected for disposal. And so the list goes on. Where will it all end?

Unsurprisingly, many of the older generation of farmers feel that enough is enough. They resent the time they're having to devote to office work, and any suggestion that they ease the burden by computerising their records is generally met with a sneer of contempt. The logical solution is for them to give up active farming, yet continue to claim the agricultural and environmental support payments that they currently enjoy. In order for this to be possible the land must be actively managed, and maintained in good heart, and what better management tool is there than sheep? The cloud of bureaucracy has a silver lining in the shape of share farming arrangements, short term tenancies and conservation grazing projects – just the sort of opportunities that can give newcomers the chance they've been looking for. There's great potential for the establishment of a mutually beneficial symbiosis between dedicated downshifters and hereditary agriculturalists.

I think the time is right for a second rising of the artisan shepherd.

Sheep Management Systems

The Sheep Book For Smallholders

"Almost wherever I have lived, if I could only have had one domestic animal, I would have chosen sheep. The sheep provides us with incomparably the best clothing material in the world. He provides us with good meat. He provides us with the best of milk."

John Seymour

It is hardly surprising that sheep are, more often than not, the first "full size" livestock the smallholder decides to introduce to his little farm – they seem a sensible progression after poultry and rabbits. Sadly, sheep are also often the first animals on the holding to be neglected when things begin to go wrong – their reputation for being able to take care of themselves is easily abused when the going gets tough. But what could possibly go wrong? Sheep are easy to keep, aren't they? Tup them in the autumn, lamb them in the spring, shear them in the summer, right? Wrong! Given that smallholders are, on the whole, a rather unconventional set of characters, it's a shame that most of the recommendations directed at the would-be shepherd tend to be of a conventional nature, following a set format it seems. The initiate is almost overwhelmed by a surfeit of information on numerous breeds (and their fleeces), with the result that important decisions are frequently based on what they look like, not how to keep them. This is daft, really, when you consider that each breed has evolved to suit differing conditions or farming systems, so let's throw convention to the winds and first take a look at some alternative ways of keeping sheep – we can come back to talking about breed choices later!

A Non-conventional Approach For Non-conventional People

We must concede that, in this day and age, very few smallholders (except perhaps retired "hobbyists") can justify working full time on their farms, which throws the whole business of keeping breeding livestock into doubt – you simply can't look after them if you're away all day. Certainly a conventional spring lambing flock may well be out of the question. It's possible, of course, to attempt to naturally arrange lambing to coincide with pre-booked time off, but it is not likely that all ewes will give birth during that time – even in small flocks lambing can be spread over six weeks or more. This may make supervision impossible at either end of the lambing period, raising a whole lot of potential welfare issues. But don't despair! Later in this book I'll be discussing manipulation of the breeding cycle using "sponges", citing the example of a teacher who, by careful planning of the lambing period, was able to accurately time all the principal shepherding tasks of the year to coincide with school holidays.

This is just one of many "alternative" sheep farming systems. It makes perfect sense to plan a flock strategy around prior commitments, and other constraints such as available land, buildings and level of expertise. Breeds can be considered afterwards, and chosen to suit the system. Let's consider some of the options…

Flying Flocks (Ewes)

A "flying flock" is one which does not remain on the holding all year round. There is no continuity of breeding, as all sheep are sold off at the end of the cycle, and a new flock purchased each year. Bio-security is maintained by operating an "all in / all out" policy (i.e., don't overlap batches). There are a number of permutations of the basic idea, some of which rely on summer grazing (ideal if you haven't the time for feeding and lambing, or if your land is too wet in winter), and some that rely on winter keep, thus freeing up land for other purposes in the summer, such as taking a crop of hay for sale, or grazing different types of stock. Some suggestions:

• Buy a batch of ewes early in July. Sponge them to lamb at Christmas. Sell ewe & lamb couples during January. Time on holding – 6½ months.

• As above, but keep for longer. Sell fat lambs at, or before, Easter. Fatten off the ewes and sell them, too. Time on holding – 9 months.

• Buy ewes late summer / early autumn to flush and tup on aftermaths. Scan at 90 days of pregnancy and sell as "confirmed in lamb" at any time up to 15 days before lambs are due. Time on holding – 4-5 months.

• Buy ewe & lamb couples in January. This is a high input / high output system – they'll be expensive to buy, and both ewes and lambs will require supplementary feeding. Sell fat lambs for peak prices at Easter (or before), giving a relatively quick turnover. Time on holding – 10-12 weeks. Either fatten off the ewes, or sell as breeding sheep in the autumn. Or you could run them on to use as per option above – this combination would give all the pleasure of year round shepherding, with a nice little "holiday" whilst everyone else is running themselves ragged over lambing!

• Buy ewe & lamb couples in April or May. A useful option for anyone who has plenty of extensive or rough grazing, but very little

improved or "in-bye" land. Much cheaper to purchase than early lambers, and can be grass reared in the "conventional" fashion. Try to get all lambs sold off by late September, either fat or as stores, and sell the ewes in the autumn sales. Time on holding – up to 6 months.

Flying Flocks (Lambs)

The term "flying flock" is generally only applied to breeding ewes, but there are a few situations where the expression could equally well be used to describe batches of lambs – basically either store lamb finishing or ewe lamb rearing, although taking "tack" sheep could be loosely described as a flying flock. I suppose, at a pinch, we could also consider purchased "pet" lambs to be in this category – buy some orphan lambs in the spring, rear them on the bottle, then pop them in the deep-freeze in the autumn. Don't be afraid to give them all names, it doesn't affect the flavour in the slightest! On the whole though, bottle fed lambs are an uneconomical time-consuming nuisance, and best avoided.

Ewe Lamb Rearing

Ewe lamb rearing consists of purchasing female lambs in the autumn to be sold on as shearling ewes (for breeding) the following year. Always buy in a matching group, and shear them early in the year (April) to be sure to have them looking their best by late summer. Profit margins can be very good, but a lot of money will be tied up in the project for a whole year. Buy the best ewe lambs you can afford – it'll pay off in the long run. Plenty of markets hold early autumn sales of mule and halfbred lambs, though a nice even pen of well grown Suffolk X store lambs will do just as well (but make doubly sure they're all females before you start bidding!). Any poor doers in the group are the ones to put aside for home consumption. The old motto of the Women's Institute markets is as true for livestock as it is for cakes and jam – "Sell the best, eat the rest!" – don't offer sub-standard stock for sale.

Store Lamb Finishing – Short Keep

This system, as its name suggests, relies on finishing weaned store lambs in as short a period as possible. An ideal option for anyone who makes a lot of hay or silage, so has plenty of aftermath grazing available from late summer, and into the autumn. Purchase good quality forward store lambs as soon as they become available (late July / early August), or as soon as there is sufficient re-growth on the mown fields. To increase the number of "sheep grazing days" per acre, forage rape can be direct drilled into grass stubble, or incorporated into a late summer reseed, and may be grazed from about 10 weeks after sowing. Aim to market lambs within 6-10 weeks.

Store Lamb Finishing – Long Keep

Whereas short keep lamb finishing requires a quick turnover to be profitable, a long keep system relies on purchasing lambs cheaply when there's a glut (October / November) and growing them steadily through to the New Year before a final period of fattening to get them sold just as "old season" lamb prices peak in early March (shortly before the first of the "new season" lambs come on the market at Easter). Initially they'll be OK on grass, but will require some supplementary feeding and / or root crops (swedes or stubble turnips) for finishing. Or house them and shear them at Christmas, and fatten indoors for 6 weeks or so.

"Tack" Sheep

If you feel that there's a place on your holding for sheep, but aren't quite ready to commit yourself to flock ownership yet awhile, then (if your location is suitable) you could consider taking in "tack" or "wintering" lambs. Sending lambs away on tack has been an integral part of the upland farming system in the UK for many, many years. Basically, hill shepherds select the female lambs that they wish to retain in the flock as breeding ewes, then send them down to the kinder conditions of lowland farms for their first winter. Nowadays it is mostly dairy farms that take tack lambs – the sheep do an excellent job of tidying up the pasture whilst the cattle are housed for the winter – but I do know of a number of horsey smallholdings that take in 20 or 30 mountain lambs every year, which benefits their grassland no end. Payment is on a "per head" basis, and the rate is pretty good, but be aware: you won't get paid for any that die! The grazing period is usually October to April, but varies between regions.

Out Of Season Breeding

Systems that involve lambing out of season (i.e., other than during the conventional spring lambing period) are an attractive proposition for the small flock owner. Anyone with other commitments can time their lambing accordingly; pedigree breeders can lamb earlier to gain a bit of extra growth in their lambs before shows and sales later in the year; best use can be made of available buildings on the holding; direct marketing of lamb can be more structured, and so on – there are many advantages (both practical and financial), in adopting a non-standard approach, particularly on smallholdings, where specialisation and niche marketing may be key considerations. The options are:

Early Lambing

By early I mean December and the beginning of January. The usual reason for lambing at this time is to catch the early fat lamb trade at Easter, but as I've already mentioned, there are a whole lot of practical reasons why early lambing may be appropriate. A few British sheep breeds will reproduce naturally out of season, but other breeds can be encouraged to do so by the use of progesterone impregnated intra-vaginal sponges. Mating can also be brought forward slightly by the use of vasectomised "teaser" rams. The use of sponges will give a very condensed, accurately timed lambing period, so is the best choice where lambing needs to be arranged to fit in with other commitments. Details on both sponges and teasers are given on pages 67 and 68.

Late Lambing

Late lambing (mid May) is usually associated with so called "easy care" systems. The principal attraction of late lambing is that it is a low cost, grass based system, requiring little input in terms of supplementary feeding, and no housing. One notable drawback is that mating ewes late in the season has a negative impact on prolificacy. Late lambing systems are particularly attractive on extensive organic holdings, using traditional, rare or primitive sheep breeds.

Frequent Lambing

Lambing more often than once a year? We must be real gluttons for punishment, us shepherds!

A frequent lambing system is ideally suited to anyone who wishes to direct market their lamb crop, or set up a farm shop selling produce from their holding, as it enables a consistent supply of prime lamb throughout the year. The traditional approach is to lamb one flock 3 times in 2 years, but by splitting the flock into smaller sub-flocks, and by using a combination of methods, lambing can be arranged 3 times per year (in January, May and September). The techniques involved will vary according to the chosen breed, but the timing of events remains as follows:

Sub Flock 1	
Mating	August
Pregnancy Scanning	November. Transfer empty ewes to sub-flock 2
Lambing	January
Sub Flock 2	
Mating	December
Pregnancy Scanning	March. Transfer any empty ewes to sub-flock 3
Lambing	May
Sub Flock 3	
Mating	April
Pregnancy Scanning	July. Transfer any empty ewes to sub-flock 1
Lambing	September

Winter Housing

It seems appropriate to briefly mention winter housing whilst we're talking about out of season lambing – supervising outdoor lambing in May is a very pleasant occupation, but clearly not an option for a December / January lambing flock. Besides the obvious requirement of providing shelter from the elements for both sheep and shepherd, there are other reasons why housing the flock may be appropriate. On smaller acreages the principal benefit in winter housing is that it gives the land a rest – if you've only got a few small fields then you can't really afford to have them poached in the winter, or there'll be nothing to turn ewes and lambs onto except mud! Housing also has attractions for the part-time shepherd – returning home from the office on a winter's evening he knows where to find his sheep. Let's face it, if checking the sheep after work involves trudging round a muddy field with a torch in the dark and the rain, then it's quite likely to be left until the morning!

If you've got existing buildings on the holding, but not really enough space to fit all your sheep in, then consider winter shearing. After shearing stocking density in buildings can be increased by 20% (they take up a lot less space without all that wool) and trough space / ewe can be reduced. Feed intake increases, and birth weights are higher. Furthermore, lambs can quickly and easily find the ewe's teats, and get a bellyful of essential colostrum. And, of course, it's one less job to do in the summer when you're busy with hay making and other things.

Dairy Production

The real question here is whether you have the total commitment (and a fair amount of capital) that would be necessary to make a financial success of a sheep dairying enterprise, or whether you're just looking to produce milk for domestic consumption. Commercial sheep dairying was one of the things that we seriously considered when we moved to Ty'n-y-Mynydd, as our buildings and farm layout would have been particularly suitable. However this would have had to have been to the exclusion of all other enterprises, and we rather fancied a more mixed farming approach. Neither of us liked

the idea of being stuck in a cheese and ice-cream "factory" all day, and besides, the amount of capital expenditure required was rather off putting.

In most respects, health care and management of sheep kept for their milk is very similar to that of sheep kept for other purposes, although only a limited number of medicinal products are licensed for use in sheep producing milk for human consumption.

The time of year that lambing takes place will be determined by the level of production required, which in turn will be dictated by the market into which your products will be sold. Spring lambing gives a lower cost option, with the "natural factor" of milk from grass being a useful marketing tool, but lactation length will be shorter, so overall yield will be lower. Winter lambing gives a long lactation and higher yield, but costs will be much higher, both in terms of feeding and housing. All year round milk production can be achieved by having two groups of sheep lambing at different times. This makes for better use of facilities and more regular marketing.

In order for the ewes to be milked they need to produce lambs. These can be artificially reared from day one, or left on the ewe for 30 days, then abruptly weaned. Artificial rearing of lambs is costly, so sale of fat lambs is unlikely to make much of a contribution to overall income in flocks where the principal source of revenue is milk. In many cases it makes more economical sense to humanely dispose of surplus lambs at birth, although some females will need to be reared and retained as flock replacements.

The best known use of sheep's milk is in the making of the French Roquefort cheese, but there are also many other delicious Mediterranean sheep's milk cheeses. Historically, in the UK, both Wensleydale and

Cheddar cheeses were made from ewe's milk, with most of the large mediaeval sheep flocks being kept primarily for milk and wool.

Personally, I think that sheep's milk is a wonderful drink in its own right, and can also vouch for the fact that it makes fantastic ice-cream and yoghurt.

Easy Care Sheep Systems

Apart from the Easy Care breed, developed by Iolo Owen, MBE, of Glantraeth, there are two other meanings to the term "easy care" – good management practice, and damned poor shepherding!

Most of the principles of easy care selection can, and should, be applied in any flock of any breed. The basic ethos is that by breeding out certain undesirable and time consuming traits, far less intervention is required by the flockowner, and regular shepherding becomes a more relaxed affair for both sheep and shepherd. So, for example, a huge ram lamb, born with difficulty, would on no account be retained for breeding despite being a perfect example of the breed. Castrate him then and there. Any ewes giving trouble at lambing – even minor malpresentations – are marked for culling, and none of their offspring are retained. When lambs aren't doing well suspect the mother – low milk yield and poor maternal instinct are not desirable characteristics. If female lambs get dirty, and need dagging, don't keep them as flock replacements – "loose bowel syndrome" appears to be hereditary in sheep. These may seem rather draconian measures but, in time, you'll have a cleaner, healthier and happier flock, with far less dependence on the syringe and the dosing gun, lower mortality rates, and reduced costs.

And now for the other extreme. Some flockowners seem to be under the impression that "easy care shepherding" means "no shepherding". Well that's just bad husbandry, and shouldn't be tolerated. Anyone who claims they only have to check their sheep once a day (or less) at lambing time gets short shrift from me, I'm afraid. The real beauty of easy care systems is that you can spend more time **looking** after your sheep, and less time **running** after them!

And Last, But By No Means Least…

Conventional spring lambing flocks:

Despite what I've said about smallholders being an un-conventional lot, a conventional spring lambing flock remains the first choice for most. Through the forthcoming chapters of this book I'll be covering, in some detail, the management of a spring lambing flock, with plenty of practical notes for guidance. The principal husbandry tasks, such as foot trimming, dagging, dosing, lambing and shearing, are more or less the same whichever sheep keeping system you choose. It's just the time of year that varies.

THE ROLE OF SHEEP – CHOOSING THE RIGHT BREED

With over 50 British native breeds to choose from, and with more recent introductions and recognised crossbreds taking the total number of breeds found in the UK to over eighty, getting started with sheep may seem a complicated business. Large sections in sheep keeping hand books are often occupied by breed information, indeed there are whole books solely devoted to breed descriptions; breed societies have stands at all the major shows, where enthusiastic members will do their utmost to convince you that their chosen ones are the only breed worth having; the shows themselves may simply add to the confusion, with several breeds very similar in appearance, and what makes this sheep better than that one, anyway? – the owners are often too busy washing and brushing to give away much information, and the tyro may keep quiet in fear of being ridiculed for asking silly questions. And as for the big autumn sales of breeding stock, well! It would be enough to make the wannabe shepherd turn tail and run, I should think! But once you're in, you're in – one of the great beauties of shepherding is that there is no snobbery. All sheep keepers speak the same language, on the same level, regardless of the size (or colour) of their flocks, so the owner of six sheep will feel quite at home discussing problems of feet, flies or fleeces with the owner of 600. In fact, were you to join the conversation, you'd be hard pushed to tell who was who! This "brotherhood" doesn't exist to quite the same extent between large and small scale producers of other types of livestock. But I'm digressing rather. Let's get back to talking about breeds!

Here I'll be looking at making a choice based on the aims and objectives for your holding as a whole, the type of sheep keeping system you've decided to adopt, and the social interaction between your sheep and yourself (on second thoughts, perhaps I'd better leave that one out…!).

The Role Of Sheep On The Smallholding

The most fundamental aspect in deciding which breed (or breeds) would be appropriate to your situation is to identify the role of sheep on your holding, i.e., where will sheep fit into the general scheme of things? And it is as well to think ahead a bit here – you may

be intending to start with a small flock, but make a shift towards larger numbers as your experience develops, in which case start off with a small number of sheep of a breed (or type) suitable for your intended larger flock.

Let's consider some of the potential roles occupied by sheep on the smallholding, working our way through from one extreme to the other, and some of the breeds most suited to each role:

Woolly Lawnmowers

The keeping of sheep as pets doesn't really fall within the scope of this book I'm afraid, but I must just give them a brief mention, as many people do initially get involved with sheep in this way, perhaps by buying an orphan lamb as a companion for another animal or in the mistaken belief that it's a trouble free way of keeping the grass short. Choice of breed is immaterial. My reason for mentioning them here is simply to remind anyone thinking of keeping a pet sheep that they'll need to carry out many of the same routine husbandry tasks as larger flock owners – four legged lawnmowers are prone to breakdowns too, and will suffer from maggots, worms, footrot etc, just like their commercially kept brethren. I've seen some horrible sights involving "much loved" pets, including grossly overgrown hooves, sheep unshorn for several seasons, and, on one occasion, a huge (and very uncomfortable) wether still wearing a collar that had been put on him as a lamb! Hidden under a thick fleece, the collar was deeply embedded in the flesh of his neck – no wonder he looked sick!

Hobby Flocks

I'm always rather envious of anyone who keeps sheep purely as a hobby – it must be very pleasant to be able to concentrate on caring for a flock without constantly worrying about profit margins. Perhaps, when I retire…

The hobby flock owner can more or less choose any breed that takes his fancy. There are limitations of course – whilst Hebridean sheep would undoubtedly thrive in the South of England, you might be pushing your luck a bit to try keeping Southdown sheep in the Outer Hebrides! Also be aware that a heavy workload could detract from the pleasure of your hobby, so it may be better to avoid the really big long woolled breeds, though it's largely a matter of personal preference. Bear in mind what I said earlier about forward thinking – it is not impossible that your interest will develop into something more significant. If you are in a position to keep sheep as a hobby, maybe you'd consider doing your bit for some of the UK's rarer breeds.

Sheep For Showing

You may argue that sheep kept for showing are hobby flocks, but this need not be the case – one often hears of showing being described as a "shop window", and many commercial flock owners and pedigree breeders run a small string of show sheep alongside their main flock, by way of an advertisement. Despite what anyone may tell you about "oh, we only do it for a bit of fun", the principal reason for taking sheep to shows is to try to win prizes, so aim to choose a breed where, as a beginner, you will at least have a slim chance of success. I suggest that you avoid the really popular breeds, or you'll be competing in huge classes at every event – it may be years before the judges even begin to

Llinos showing one of her badgerface ewes at the RWAS Smallholder and Garden Festival

notice your presence, let alone award you any prizes! At the 2006 Royal Welsh show there were around 340 entries in the Texel section, with almost 80 sheep in one class – that's pretty stiff competition! Also avoid the more obscure types or you'll always end up having to compete in the "any other breeds" category. Many of the traditional native breeds are sufficiently abundant to have their own scheduled classes, without being so popular that the numbers are daunting. In order to compete at the larger shows your sheep will need to be registered with the relevant breed society, of which you will have to be a member. Being part of an enthusiastic group of like minded individuals can be a great source of encouragement, but beware! If you've chosen a breed with specific markings you may tire of the endless debate about *exactly* how much white/black/grey/brown should be allowed, and where, and wish you'd opted for self coloured animals!

For Home Consumption

Where sheep are kept on the holding solely in order to supply a few lambs per year for the domestic deep-freeze, then go for the trouble free options. Consider breeds such as the Wiltshire Horn, or its derivative the Easy Care. These easy lambing breeds will give you good sized carcasses off grass alone, without the hassle of annual shearing and all the other problems associated with wool, such as cast ewes and blow-fly strike. Or perhaps it's not necessary to keep breeding sheep at all, just buy in some crossbred store lambs each year for fattening. A lot of breed societies make claims about the superior flavour and eating quality of their sheep, but personally I feel that these factors are more influenced by the way in which an animal is reared and killed, than by choice of breed.

Wool

Be realistic here: how many fleeces do you actually need? There is little point in keeping a whole flock of multi-coloured animals just to provide a few fleeces for your own hand spinning needs. You may find yourself with a lot of unwanted coloured wool to dispose of each year, which the British Wool Marketing Board will only pay you pence for! (e.g., Jacob wool, 2006 BWMB price, 16p / kg, making the whole fleece worth around 40p!). On the other hand you may turn your skill into a small business, processing fleeces into a range of yarns or finished knitwear for sale, or perhaps you'd like to try felting, or rug making, or dyeing, or weaving, in which case a larger flock would be quite in order. The Shetland breed is an obvious choice as it comes in a whole range of natural colours, and the fleece is noted for its fineness. Most of the primitive breeds are coloured, but their wool is sometimes rather coarse and hairy. Coloured crossbred sheep crop up in all commercial flocks from time to time, and are often the best bet, producing softer, heavier

fleeces than their primitive ancestors. One traditional native breed well worth considering is the Ryeland, which produces a really soft springy fleece with practically no kemp. There are both white and coloured strains, and an enthusiastic breed society.

Milk

I think it would be fantastic to see more smallholders opting to keep dairy sheep for their household milk supply, but, apart from a brief flurry of interest in the 1980s, the idea doesn't really seem to have caught on in the UK. It's true that there are a number of fairly large scale sheep milk producers in this country (about 200 flocks, totalling some 12,000 ewes), but on a smaller scale people always seem to choose goats. Goodness knows why, as management of sheep (or cattle) is much simpler! So, if you'd like to produce your own dairy products, yet haven't the facilities for a house cow, then why not consider milking sheep? Any ewe that's successfully reared triplets in the past would be worth milking, but for higher yields choose specialist breeds such as the Friesland, the British Milksheep or even the Zwartbles! Other types of sheep can be bred up to dairy status by the use of rams of these breeds over several generations. The Poll Dorset and the Lleyn would both be eminently suitable ewes for upgrading in this way.

The little Icelandic sheep were, until relatively recently (1940s), the principal source of dairy products in their native Iceland, with individual ewes producing as much as 3 litres / day at peak yield. Icelandic ewes are now finding popularity as domestic milk sheep with American smallholders ("homesteaders"). They are a true "multi-purpose" breed, producing fleeces in a range of natural colours, and a small gourmet carcass, in addition to their milk.

THE PRINCIPAL ENTERPRISE

On any mixed holding, sooner or later, one activity begins to dominate and becomes the principal enterprise – the one that (hopefully) helps to pay the bills! More often than not, on grassland smallholdings, it's the sheep flock that fills this role. For some smallholders it may be a natural progression from having kept sheep as a hobby for a number of years, perhaps coinciding with taking on more land, or taking early retirement from employment, in which case you'll probably stick with the breed you already have. (Remember that forward thinking?). But, for others, the establishment of a sheep flock may represent a new venture, so, starting with a blank sheet, you've got the ideal opportunity to devise a flock strategy suited specifically to your holding, tailored to accommodate your off farm commitments, and carefully planned to ensure compatibility with other smallholding activities. And having settled on a suitable system, choose the breed (or cross) most suitable for it. This is far more likely to result in a successful enterprise than trying to devise a system to suit a particular breed.

Niche Marketing

The basic concept of niche marketing is to be selling a product that's a bit different from the normal run-of the-mill stuff on offer, so it is here that the primitive and rarer sheep really come into their own, particularly if there's a bit of a story to tell about the origins of the breed. Do carry out thorough market research before committing yourself – the simple fact that something has become a niche product implies that there aren't many customers out there! Avoid over production at all costs, for if you end up having to sell surplus stock through

Rarer breeds are suitable for niche marketing projects

conventional channels the loss sustained on those animals will probably wipe out a large chunk of your overall profit – Soay steaks may well fetch high prices amongst a small, but discerning, band of loyal consumers, but take Soay lambs to your local livestock market and you'll be lucky if you get a bid!

Another buzzword concept in livestock marketing these days is "local provenance". Hotels, restaurants and other food outlets currently seem very keen to stock meats from the local breeds, so consider keeping the traditional native types indigenous to your area, breeds such as the Shropshire, the Dorset Down, the Exmoor Horn, etc, all of which have strong geographical links.

Direct Marketing

The most important aspect of direct marketing is yourself. The whole concept revolves around the fact that it is you, selling what you have produced on your farm. That's the principal selling point, not what breed is involved, or whether it's organic, or free range, or whatever. This is where the smallholder may have an advantage over the larger scale farmer – there will probably be a whole bevy of (non-farming) friends, family and former colleagues all keen to sample the fruits of your labour. They are your first customers. From there the business should grow through word of mouth, but you may wish to consider farmers' markets, country markets (formerly W.I.) and food fairs, or even open a farm shop! Internet sales are also popular, but here you risk losing the "direct contact" aspect. Once you've got your customers it's important to keep them, which means being able to supply what they want, when they want it. Consider frequent lambing systems using breeds such as the Poll Dorset, Dorset Horn or Portland.

Conventional Production

On larger smallholdings, in predominantly livestock farming areas, a conventional approach is probably most appropriate as there will be an existing marketing infrastructure in place, into which you can tap. Prices received for your stock will not be as high as when sold as a specialist product, but then neither will you have to carry the cost of advertising, packaging, labelling, distribution, etc. You could, of course, still market a portion of your crop direct to the customer. Your local livestock market will also hold seasonal sales for various categories of breeding and store sheep, should you decide to opt for a "flying flock". As for breeds, well, the most sensible thing to do would be to pay a visit to the nearest large scale sheep farmers, whose families will have made their living from breeding sheep in your area for several generations, and ask their advice – if they haven't worked out by now what breed (or cross) best suits the current farming situation in your locality, then nobody ever will! And what if you aren't in a sheep keeping district? Then you'll be a pioneer, won't you!

What Breed Would I Recommend?

Given that there are so many breeds to choose from, and that smallholdings (and smallholders) are so diverse, it is not possible to make specific recommendations here. At the end of the day, a large part of anyone's choice of breed will be based on personal preference. However, I do often get asked to recommend a breed so, at risk of seriously upsetting lots of enthusiasts (by not listing their favourites), I'll put my neck on the line and make a few suggestions:

The very first sheep that I bought, nearly 25 years ago, when I was 13 years old, were Beulah Speckleface, and I'd have no hesitation in recommending the breed to a beginner. They're not too big to handle, hardy, fairly prolific and are good mothers. They cross well with a terminal sire for fat lamb production, with a bluefaced Leicester to produce popular mule ewes, or can be bred pure. My principal reason for recommending the Beulah is its temperament – they have a very "forgiving" nature, and will put up with the blundering antics of a novice shepherd. (If you could have seen my early efforts all those years ago...)

For early lamb production my favourite is the Hampshire Down, surely the most attractive of the down breeds. The Hampshire evolved to suit a very restricted grazing regime, that of being "folded" over arable land, so copes well with the rather confined conditions sometimes found on smallholdings – very small paddocks, often sub-divided using electric fencing. They are amenable to winter housing, but can manage equally well outdoors. One of the principal attributes of the breed is fast growth rates. When bred to a continental terminal sire, such as the Charollais, the Hampshire produces outstanding butchers' lambs suitable for the Easter market.

Hampshire Down rams are also popular as terminal sires in their own right.

At the other end of the scale, for late (May) lambing or "easy care" systems, I have a hunch that the Norfolk Horn would be ideal, though I've never tried it. Hardy, thrifty and a good forager, the breed evolved in the exposed breckland heath area of Norfolk. Numbers declined as the heathlands were ploughed and improved, until by 1969 the breed was almost extinct. Given the current interest in extensive and organic farming systems, together with restoration of traditional habitats, I wouldn't be surprised to see the Norfolk Horn returning to popularity.

For conventional crossbred lamb production I rather like the Welsh Mule. The ewes have nice clean bellies, which makes shearing so much easier.

If you intend to exhibit your stock, the Kerry Hill is a real showman's sheep. One of the UK's most distinctive looking breeds, they look particularly smart when prepared for the show ring and are frequently placed in inter-breed championships.

For a good all round, general purpose smallholder's breed you'd do well to choose the Lleyn. The Lleyn ("Hard to pronounce, impossible to beat!") is renowned for its prolificacy, ease of lambing, maternal instinct and longevity. It is hardly surprising that, from being considered a rare breed 30 years ago, the Lleyn has risen from the ashes to become one of the most popular purebred ewes in the UK.

And My Own Choice?

The Welsh Mountain, of course!

BREEDS

Clearly it's not possible to provide detailed information on all the breeds here, but hopefully the following short selection will give a general impression of the diversity of types now found in Britain, and some of the interesting links between them.

A comprehensive list of breed societies and associated organisations can be found in the appendices.

Norfolk Horn ewe. (Norfolk show champion 2009)

Norfolk Horn

The Norfolk Horn, which evolved to survive on exposed breckland heaths and the poorest of farmed land, has been recognised for nearly 400 years. During the 18th Century the breed rapidly fell out of favour in its native East Anglia, being replaced by the improved Leicesters and Southdowns. Attempts to improve the conformation of the Norfolk Horn by crossing with the Southdown led to the development of the Suffolk, and thereafter all the enthusiasm of the breeders was directed at the more modern types.

The original Norfolk Horn never had the opportunity to demonstrate its full potential, and by the mid 1800s was becoming rare. In 1948 only one flock of ewes and two rams remained, and by 1965 there were only a dozen animals left (half of which were male). These animals were so inbred that expansion of the flock was almost impossible, so a careful breeding program was developed that involved putting Norfolk Horn rams to Suffolk ewes, then back crossing each generation to pure Norfolk rams. By 1973, when the last pure Norfolk Horn ram died, lambs that were 15/16 Norfolk Horn were being produced.

The plight of the Norfolk Horn was the catalyst that led to the foundation of the RBST. Since its inception, in 1973, no breed of British farm animal has become extinct.

Southdown ewes & lambs

Southdown

The development of the Southdown began around 1780, when John Ellman of Glynde, near Lewes, began to improve the native Sussex sheep. The existing breed already possessed good hindquarters, but it was very narrow and light in the shoulder, with a poor quality fleece. Ellman increased the size of the breed, and considerably improved the conformation, by selective breeding only (i.e., no introduction of outside blood) resulting in the compact "leg at each corner" Southdown that we know today, which ultimately became influential in the

development of so many other down breeds. Popularity of the breed peaked between the mid 19th Century and World War I, with many famous flocks running 1,000 ewes or more. The Southdown has been exported all over the world – in New Zealand it was for many years a leading terminal sire, producing the famous "Canterbury lamb".

Suffolk

The first flock of Suffolk type sheep was established in 1810, and by 1859 the breed had been locally recognised – the Suffolk Show of that year included specific classes for Norfolk Horn X Southdown sheep, and called the new breed "Suffolk sheep". In 1886 there were classes for Suffolks at the Royal Show, and a breed society was formed. The first flock book was published in 1887. Thereafter, the popularity of the breed expanded rapidly, and by 1901 it had spread throughout all four countries of the United Kingdom, and was exported to Austria, France, Germany, Switzerland, Russia, North and South America, and to the colonies. The Suffolk continued to develop in line with consumer demand, and for a time this quintessentially English breed was probably the most widely used terminal sire in the world – a position it held until relatively recently.

This next group - Shetland, Hebridean, Manx Loaghtan, Icelandic and Jacob - are multi horned breeds:

Shetland

Although predominantly a two horned breed, the fact that some males have four horns reveals the Viking ancestry of the Shetland sheep. Archaeological investigations in the

Shetland Islands have also indicated the presence of a primitive breed similar to the Soay which pre-dates the modern Shetland sheep. The mouflon type markings still occur in some Shetlands, so it is reasonable to suppose that the breed came about as a result of crossing the imported Scandinavian breeds (such as the Norwegian Spaelsau) with the existing Celtic sheep in the 8th Century or thereabouts.

It is clear that the Shetland islanders began selecting for fleece quality at a very early stage, resulting in a breed that has the finest wool of any native British sheep. Unlike most fibre producing breeds, where white has become the principal fleece colour, the Shetland occurs in eleven main shades, with over 30 recognised markings. The exposed location of their island home may have made the cultivation of dye plants well nigh impossible, so perhaps the crofters relied upon the sheep themselves to provide a varied colour scheme.

Hebridean

The Hebridean is another descendant of the Viking sheep, and a similar type was once found throughout the north-western fringes of Europe, including the Western Islands and Highlands of Scotland. This primitive breed began to be displaced by the Blackface (which undoubtedly descended, in part, from them) from about the 16th Century, and, by the 19th Century the breed was restricted to a few offshore Islands. Some were brought to the mainland in the 1870s as ornamental sheep, and, by the 1970s, the breed was found only in a few parkland flocks. The original island stock had died out completely. The work of the RBST has ensured that the breed is once again on the increase, and it now seems to have established a niche as a conservation grazing animal. Around 6-8 %

Four-horned Hebridean rams

of the current population are multi- horned, and, interestingly, a few individuals carry a dominant black gene – a characteristic generally found only in Black Welsh Mountain and Jacob sheep.

> NB. The same gene that causes the horn buds to divide, giving rise to four horned animals, can also result in the occurrence of a defect known as "split eyelid", where the upper eyelid has a notch in it. This varies in severity from a small break, to a full split which may prevent the eye from closing properly. Animals showing signs of this deformity should not be retained for breeding.

Manx Loaghtan

The Manx Loaghtan is closely related to the Hebridean, although four horned sheep are preferred. Originally the breed was white, with a few other colours occurring. The loaghtan ("mouse brown") was a much sought after shade for clothing, so, as sheep numbers declined, the loaghtan colouring was increasingly selected for, resulting in the fixed breed type we see today. The loaghtan colouring, being recessive, always breeds true.

Icelandic

The Icelandic is another member of the Northern Short Tailed group of breeds, and remains genetically unaltered since the Vikings first took them to Iceland around 900AD. It was originally a multi-purpose breed, being kept for meat, hides, fleeces and milk, and, whereas other breeds have become more specialised, the Icelandic retains its truly versatile character. Indeed, it remained the principal source of dairy products in Iceland

Icelandic ram "Utlagi"

until the 1940s, and has now found favour as a milk sheep on many American smallholdings. As a result of this continued interest in dairying, and the breed's natural prolificacy, the Icelandic is a productive sheep, although its modest size means that as a carcass animal it is better suited to the production of gourmet mutton at around 3 years of age, rather than conventional spring lamb.

There is no strict breed standard for Icelandics in Britain, so there are many variations in size, shape and coat colour – the double layered fleece occurs in a wide range of natural hues, and may be patterned or spotted. Also, in addition to the more usual 2 horned sheep, you'll sometimes see Icelandics sporting four – a characteristic they share with other breeds of Viking origin.

Jacob

The piebald, multi horned, Jacob sheep shares so many genetic characteristics with these northern breeds that it is reasonable to assume that it too is of Scandinavian origin. However, the generally accepted story is that the Jacob sheep is of Middle Eastern ancestry, having descended from the spotted sheep of the Old Testament which Jacob claimed from Laban's flock, having first ensured that all the best sheep produced piebald lambs by placing part-peeled withies before their eyes at the time of conception (Genesis 30, 31 – 70). From Egypt, the breed is said to have travelled along the coast of North Africa and Morocco and eventually to Spain. This extraordinary narrative concludes with the idea that the breed appeared in Britain when animals swam ashore from the wrecked Spanish Armada in the 16th Century! (There are also stories claiming that the Herdwick, the Cheviot and the Portland sheep all swam ashore from the Armada too – were the Spaniards intent on fighting or farming, one wonders? Personally, I have not been able to find any reference to sheep in the inventory of the cargo ships that accompanied the fleet). Despite the popularity of this colourful tale there is no genetic or historical evidence linking the Jacob sheep with eastern types, beyond the fact that all domestic sheep originate from a primitive breed of southwest

Asia. It is more likely that the breed developed and became established in the UK during the 17th and 18th Centuries, when the English gentry were looking for novel livestock to grace their enclosed parklands.

Portland

The Portland breed draws its name from Portland Bill, a rocky outcrop off the Dorset coast. A similar kind of sheep would at one time have been found throughout the south west of England, but the type became restricted to Portland Bill as the mainland sheep were improved and developed by crossing with a related breed from Somerset. The Portland is horned in both sexes, and still exhibits many

Portland ewe & lamb

of the early characteristics of the tan faced pre-Roman breeds, although it is unique amongst the more primitive breeds in that it is able to breed at any time of year. The last of the original population were removed from the island in 1920, to be sold in Dorchester market, but were in such poor condition that the auctioneer struggled to get a bid! By this time the breed had become more or less extinct. In 1974 the RBST stepped in to trace the last remaining animals, and a breeding program

was established to safeguard the future of the breed. Portland sheep were re-introduced to the island in 1977. There is now a thriving breed society, with members throughout the UK, and the population of breeding sheep has expanded considerably.

Dorset Horn / Poll Dorset

The Dorset Horn stems from the same original stock as the Portland, and shares the ability to lamb out of season – one of the characteristics that has enabled forward thinking flockmasters to ensure that this old breed remains competitive in a modern world. Whereas the Portland island sheep remained unchanged due to their isolation, the mainland flocks developed rapidly, and by the 1830s the improved type had become common throughout Dorset and Somerset. A breed society was established in 1891, following some heated exchanges over which of the rival counties the breed should officially be attributed to!

The versatility of the breed was soon acknowledged worldwide. The Australians, in particular, recognised the potential of the breed to improve their existing stock. They are also responsible for creating a polled version of the breed, by crossing with the Ryeland

Dorset Horn & Poll Dorset rams

then repeatedly breeding back to the Dorset, meticulously discarding any offspring showing traces of horn. Word of this new hornless strain reached UK ears in 1956, and a society member was sent to investigate. He returned to England with some polled Dorsets, and the rest, as they say, is history! The Polled Dorset now far outnumbers its illustrious horned ancestor, but, as the Secretary of the society remarked in 1971 "Behind every good Poll Dorset is a very good Dorset Horn!"

Ryeland

The Ryeland is one of the oldest of the established British breeds. Its exact origins are not known, but it is believed to have been derived from the Spanish Merino, with the addition of Cotswold and New Leicester bloodlines. Mediaeval records show that abbeys in the Herefordshire area kept Ryeland flocks numbering several thousand ewes, with wool being exported to Flanders and Italy, where it commanded the highest price in Europe. It became the measure against which the quality of other wools was assessed. The breed further increased in popularity during the 16th Century, helped by the fact that Queen Elizabeth I was a staunch supporter of Ryeland wool – a pair of Ryeland wool stockings pleased so much that she swore she'd only wear clothing of Ryeland wool thereafter. It is believed that the Lord Chancellor's seat in the House of Lords, the "Wool Sack" was originally stuffed with Ryeland fleeces (known as "Lemster Ore"), in tribute to the breed's significant contribution to the wealth of the nation.

By the early years of the 20th Century the breed had fallen from favour, but now an enthusiastic breed society is ensuring that numbers are rising once again, and the Ryeland is an extremely popular smallholder's sheep.

Ryeland ram "Dolwen Crown Derby"

There is also a recognised coloured version of the breed, the result of the expression of a recessive gene. Coloured Ryelands can be registered in their own section of the society's flock book.

Cheviot

The Cheviot has been recognised as a hardy breed since 1372, and the Cheviot Sheep Society, established in 1890, is one of the oldest breed societies in existence.

Sir John Sinclair, the famous agricultural improver, brought 500 of these "long sheep" from the Cheviot Hills to his Langwell Estate in Caithness over 200 years ago, as part of a programme to improve the sheep stock in the area. The Cheviots were an immediate success, and displaced the existing Scotish Blackface. They became the predominant breed of the region, developing into what is now known as the North Country Cheviot. Whether or not the Cheviot may have played a part in the creation of the Border Leicester is a matter of some debate!

Welsh Mountain yearling ram

Welsh Mountain

As diverse a breed as the Welsh landscape, with many local and regional variations, each promoted by its own breeder's association. Types range from the wiry tan faced ewes of North Wales, to the longer bodied, white faced sheep of Mid Wales (such as the Tregaron strain), and to the massive Nelson and Talybont-on-Usk types of South Wales. There's also a variety known as the "pedigree" Welsh, a smaller, less hardy type found in small flocks throughout the principality, and over the border.

Brecknock Hill Cheviot

In 1885, the McTurk family moved from south west Scotland (with their sheep) to take over the running of the Cnewr Estate at Sennybridge, driving their flock of Cheviots by road from the nearest railway station – at Crewe! The ewes were fitted with special leather shoes for their long journey. Crossing these sheep with the local strain of Welsh Mountain led to the formation of a new breed, the Brecknock Hill Cheviot, now found throughout the Brecon Beacons National Park, and beyond.

Wiltshire Horn

The Wiltshire Horn is a very old breed – archaeological remains of similar sheep have been found dating back to Roman times – and is clearly descended from the same early ancestors as other white face, short wool breeds found in the south of England. Until the end of the 18th Century it was the predominant breed of the Wiltshire Downs, and played a part in the creation of the Hampshire Down sheep. The Wiltshire Horn has a fleece of short wool and hair which it sheds in the spring. Shepherding costs are kept to a minimum, with no shearing or dagging required, and a much reduced risk of flystrike.

Easycare

The Easycare is a "composite" breed, created from a cross between the Wiltshire Horn and the South Wales Mountain. Breed development began in 1965, after Iolo Owen of Glantraeth, Anglesey, identified the need for a low input, yet productive ewe, without all the encumbrances brought about by wool – a costly by-product in today's farming environment. The new breed was based upon his father's prize winning flock of Wiltshires, founded in 1911. After a slow but thorough breeding program, sales of Easycare sheep commenced in 1985, and the breed has proved very popular both at home and abroad. The attributes of the breed are many, but possibly the key attraction is that shepherding costs are reduced by over 80%, when compared to other breeds.

Hampshire Down

This breed is said to have emerged between the 1830s and the 1850s as a result of crossing

Hampshire Down rams

Texel ram lambs

the Wiltshire Horn, the Berkshire Knot and the Southdown. However, it is perhaps more likely that a sub-type had previously been developed from the Berkshire / Wiltshire cross, as the Berkshire Knot was already extinct by the mid 1830s. This sub-type, when bred to the Southdown, probably resulted in the Hampshire. The breed has developed as a terminal sire, producing early maturing, high quality butchers' lambs. It is also capable of breeding out of season, making Hampshire ewes an ideal choice for early lambing flocks.

Hamps have been exported to more than 40 countries, thriving in both the heat of South Africa and the cold of the Andes!

Texel

Originating from Texel, one of the north-western islands of Holland, this breed has been known since Roman times. Through the centuries the breed's potential has been improved by careful selection and the infusion of British blood, principally from the Lincoln, but also from the Border Leicester, Southdown, Hampshire Down and Wensleydale. The Texel is now the most popular terminal sire in the UK, and hit the headlines recently when a ram lamb sold at the Scottish National Texel sale for a world record price of 220,000 gns (£231,000). This exceeded the previous record (also held by a Texel) by 100,000gns.

Charollais

The Charollais originated near the town of Charolles in the Saore Loire region of France. It was developed in the 19th Century from a number of local breeds which were crossed with the British Dishley Leicesters. It is now 30 years since the Charollais was imported into the UK, and in that time it has risen to be one of the 3 most popular terminal sires in the country. However, to my mind, one of the principal attributes of the Charollais is ease of lambing. Even on small native breeds and primitive sheep the Charollais gives no problems at lambing time, yet produces a fast growing offspring with excellent conformation.

Charollais ram

The Shepherd's Year
Autumn

You could be forgiven for thinking that lambing time is the most important time in the shepherd's calendar, but you would be wrong. Certainly it's the time when resources and skills in stockmanship are pushed to the limit, but the really critical time, in management terms, is tupping time. It is well worth making the effort to get it right – the effects of poor autumn management may be felt for a full 12 months, or even longer!

Although it is often said that the shepherding year begins at tupping time (which, for a conventional spring (mid-March) lambing flock, would be mid-October) in reality the year begins at least six weeks prior to turning in the tups, when the flock is "made up" at the beginning of September.

I once asked an extremely successful and well respected ram breeder what he considered to be the most significant factor in producing top quality stock for sale: "The ewes" he replied. "You've got to get your ewes right, or you're wasting your time. You can spend as much as you like on a fancy tup, but he'll get you nowhere if your ewes aren't right!" So, armed with that piece of sound advice, we'll start the shepherding year by "getting the ewes right!"

SHEEP IN SEPTEMBER

Making Up The Flock

Basically this means having a jolly good sort-out, to ensure you've got the optimum number of ewes of appropriate ages, and in good health, to take your flock forward into the forthcoming farming year. For newcomers to sheep keeping, starting from nothing, making up the flock will involve a trip to the sales – more about that in a minute – but in established flocks there'll be plenty of head scratching and soul searching, as each ewe is assessed on her relative faults and

merits, and a sometimes tough decision made as to whether or not she remains in the flock for another season. It is as well to have a clear idea of the long term objectives for your flock, and a simple written down health & management plan (it needn't be a complicated document – brief notes in an old diary or year planner will do) to aid you in making consistent decisions. For the purpose of this September sort-out you'll need to have made up your mind as to your culling criteria, and have devised an appropriate replacement policy. Include these strategies in your written flock management plan, and adhere to them as closely as you are able. It may take several years for the consequence of a particular course of action to be noticeable within the flock – hopefully the effect of having a sound policy will be seen in the form of an overall improvement in the standard of your stock, but it is a good idea to re-assess the plan every 3–5 years, and make adjustments where necessary.

CULLING POLICIES

Culling By Age

In many hill flocks it is the usual practise to sell all ewes at four years old, for further breeding on lowland farms (often for crossing to produce mules and halfbreds). These middle aged sheep are not technically described as culls but are called "draft" ewes, and make up a large proportion of the total number of breeding sheep sold each autumn in the UK. Drafting out of 4 year olds is not generally appropriate in smallholders' flocks, but may be relevant where rapid genetic improvement is required, for example in the early years of establishing a purebred breeding flock or when grading up commercial stock to pedigree status.

Culling By Teeth

"Broken mouth" is probably the commonest reason for culling from commercial flocks, but again may not be appropriate on smallholdings. Given kinder conditions and a bit of feed, a "broker" may continue to produce and rear lambs for several years to come. Almost all the ewes I've ever bought have been broken mouthed culls, and most have produced at least three more crops of lambs before either dying of old age or being sold for killing at about the same price I paid for them (£5). A sheep that has lost all of its teeth will generally "do" better than one with a couple of wobbly teeth remaining. Of course what I'm talking about here are the front teeth. The back teeth – the molars – are a different matter altogether, and are often overlooked. Ewes that have problems with their back teeth are not easy to spot initially, and aren't generally noticed until the animal looks pretty poorly. Even then diagnosis is often confused by the fact that the symptoms are easily mistaken for those of chronic liver fluke infection. (Bottle jaw, mealy nose, dry coat, progressive weight loss, etc). Before reaching for the dosing gun and pouring an expensive flukicide down her throat, check out her cheeks! By running your finger tips along the side of the ewe's face you'll be able to feel a prominent uneven ridge if the molars are abnormal. In advanced cases the overgrown teeth will have lacerated the inside of the cheeks, and ewes will (not surprisingly) show signs of discomfort as you make your examination, so do be gentle. Affected ewes may spill half-chewed cud from the side of their mouths, staining the wool on their necks, and food often collects in the cheeks, giving a hamster-like appearance. (Cud spilling and impacted cheeks are also symptoms of one form of listeriosis. Confusing, eh?). These sheep must

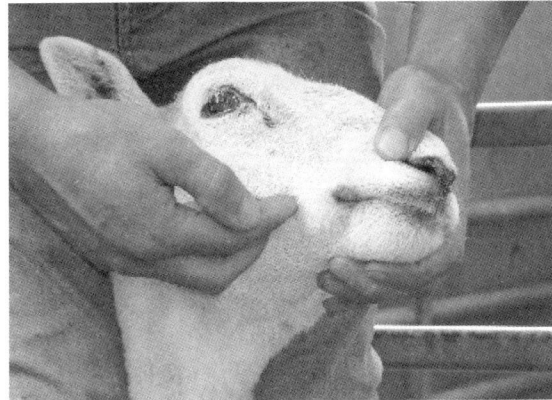

Checking a ewe's back teeth. Note the staining at the corner of her mouth where she has been spilling cud

be culled, and I wouldn't recommend retaining their offspring in the flock either – better to breed the problem out and be done with it, as far as is possible.

Whilst you're going through the teeth of your flock (particularly when selecting replacements) keep an eye out for under- or overshot jaws. Minor deviations can be tolerated where all lambs will be destined for slaughter, but if you intend to either sell or retain breeding stock you'll need to be more strict. Once you allow the problem into your flock it can lie hidden for several generations, only to reappear at a later stage. If you sell a ram you've bred, and he subsequently produces poor mouthed offspring, your reputation as a breeder will plummet. Rapidly.

Culling By Udders

If in doubt, chuck it out! Retaining ewes with suspect udders will cost you in terms of both time and money. Lambs will be compromised from birth due to insufficient colostrum, and there'll be a lot of tedious bottle feeding to be done, not to mention veterinary expenses. And if full-blown mastitis is allowed to develop it'll be most unpleasant for the ewe, with potential complications such as maggots, gangrene and sloughing away of large areas (or even the whole udder). I don't know if a predisposition towards mastitis is hereditary, but physical characteristics, including udder strength and attachment, probably are, so breeding for better shaped udders could result in fewer clinical cases of mastitis in the flock. We are very strict about this in our own selection, and to date have not seen acute mastitis in any of our homebred ewes.

Culling By Feet

Lame sheep will never perform well. You can keep on top of lameness in the flock by routine hoof trimming, spraying, footbathing etc., and the worst affected can be treated with antibiotics, but there always seems to be a few persistent offenders whose feet never really come sound despite regular attention. These animals serve as a reservoir of infection for the remainder of the flock, so should be culled. If you are in any doubt about the wisdom of expelling a ewe for seemingly so small a crime, consider also that one of the bacteria responsible for footrot (*Fusobacterium necrophorum*) can cause navel-ill and liver damage in young lambs.

Culling Barreners

Many flock owners routinely sell off any barren ewes at scanning or after lambing, but I believe this to be a false economy – in my experience it's cheaper to keep a barrener through the summer than it is to buy a replacement ewe in the autumn, even after taking into account her slaughter value. Only if a ewe is barren for two consecutive years should she be marked for culling, but this will only rarely be the case. A ewe that hasn't reared

a lamb will go to the ram in tip-top condition, and in all likelihood conceive twins. Probably she just needed a year off!

Culling By Production

Whether or not to cull by production will depend upon the specific aims and objectives for your flock and, in some cases, may depend upon the aims and objectives of the relevant breed society. Do you feel that you need to take steps to increase the prolificacy of your flock? If so, consider selling off ewes that only produced one lamb, and only retain ewe lambs that were one of twins. On the contrary, if your grazing consists largely of rough mountain land, you may consider too many twins to be a bit of a nuisance, and prefer good strong singles, in which case you'll be selling off the twin bearing ewes and not keeping their daughters. And while we're talking about lambing I'll also mention that you should be culling ewes that have had difficult births, and not breeding from their offspring, either.

Culling By Performance

Estimated Breeding Values (EBVs) are largely of interest to pedigree producers of terminal sire breeds, but are becoming increasingly popular as a selection tool in all sectors of the sheep industry, so the concept should be taken seriously by anyone who intends to sell or retain stock for breeding purposes. I've dealt with the subject of performance recording in some detail on page 70, but, in the meantime, here's a simple performance related check that anyone can apply to their flock as an aid to identifying their better ewes:

It has been shown that there is a strong correlation between a ewe's maternal ability and milkiness, and the weight of her lambs at

8 weeks of age. Weigh all lambs at eight weeks, then calculate averages for both twins and singles. Reject ewes whose lamb weights fall well below the average, and select replacement females from amongst lambs of above average weight (for their litter size). Maternal instinct and milk yield are both highly hereditary, so considerable improvements in flock performance can be achieved by this strategy. (Some allowance must be made, however, for young ewes in their first lactation).

Ill Thrift

This covers a multitude of sins! Once you've ruled out feet and teeth, you'll have to accept the fact that some sheep are just "poor doers", and will need to be culled. If the number of cases of ill thrift within the flock seems disproportionately high then there may be some underlying problem that should be investigated, such as a trace element deficiency.

There are times when you have to accept the fact that some sheep are just poor doers!

Culling By Type

Selection for type is a contentious issue, often leading to warring factions within breed societies, where some members will insist that commercial characteristics are more important than breed characteristics, and vice versa, and that conformation is more important than colour, and as for the size of their ears or the length of their socks, well! Who'd have thought that grown men (and women) would get so up-tight about it all? But the fact is that they do, and if you want to get involved in all that then that's up to you! At the end of the day it's going to be a personal decision, but the bottom line, I feel, is that you must like the look of the sheep you're working with, so if there's a couple of odd looking ewes in your flock that really stand out and spoil the appearance of the bunch then sell them on – they may be just the type your neighbour dotes upon! (Beauty is, after all, in the eye of the beholder!). If your sheep are pleasing to your eye then you'll spend more time looking at them. More time spent looking at your sheep means better shepherding. Better shepherding means better sheep, and so on.

FLOCK REPLACEMENTS

The question is simple: do you buy them in or breed them yourself? The answer, I'm afraid, is not so straightforward! At a glance the solution may seem obvious: you simply keep female lambs from your existing ewes to replace any members of the flock that die or need to be culled, and on many farms that is exactly how it's done. In fact, in hill flocks where ewes are drafted at four years old the majority of the ewe lambs will be retained, the principal outputs being sales of wether lambs and draft ewes. However, on other farms (particularly smaller farms and smallholdings) it is often more

INBREEDING

If you retain home bred ewe lambs then, in small flocks, you'll sooner or later find yourself asking the question "is it safe to use a ram on his own daughters?" Well, you might be OK to use a ram on his daughters, or you might not! Congenital defects do not appear from nowhere, so if there is nothing "nasty" in the genetic makeup of the parents, then nothing will materialize in the offspring, no matter how closely related they are. The trouble is you don't know 'til you try! Problems (e.g., undershot jaws, entropian, etc.) may remain hidden for several generations, then pop up all over the place following a certain combination of sire and dam. The chances may be greater where inbreeding takes place, but, having said that, some of the finest stock in the world has resulted from close matings, followed by a very strict culling policy. In a pure (or pedigree) breeding flock I would only mate a ram to his own daughters if both the sire and the offspring were of exceptional quality, or in order to safeguard a threatened bloodline. I certainly wouldn't do it very often, and I'd cull out anything not 110%. A more usual policy would be to put a ram back on his own mother. This can help to consolidate type, without running the risk of dispersing problems too widely.

If you can be absolutely certain that all offspring will end up in the freezer, even if they look OK, then I see no reason why a ram shouldn't be used on his own daughters under normal flock circumstances (i.e., not within a specialised breeding program), but, if you've only got a few, the best thing would be to run the daughters with someone else's ram, in return for running some of theirs with yours. In this way, both of you can extend the useful life of a ram in your flock. Or change your ram every 2 years.

Inbreeding is a perfectly natural occurrence in wild flocks and herds, where the harem of the dominant male will comprise of all the best females, which probably includes his mother, his sisters and some of his daughters. It is only when we apply human values to animal behaviour that we find it distasteful.

appropriate, in both practical and economical terms, to buy in the required number of replacements annually.

Situations Where Home Bred Replacements May Not Be Appropriate

• Ewe lambs kept as replacements must be at least as good (if not better) than the sheep that they are replacing. If you are disappointed with your ewe lambs, perhaps because you made a poor choice when purchasing a ram the previous autumn, then don't keep any. It would be better to hang on to those old ewes for another season, than to take a backward step. (And get rid of the ram!).

• Where you are using two pure breeds to produce crossbred offspring for sale. For example, if your enterprise consists of using a Bluefaced Leicester ram on Swaledale ewes to produce North Country Mules, then your flock replacements will be bought-in draft Swaledale ewes. (Unless, of course, you also keep a purebred flock of Swales, but that's starting to get a bit complicated for a small scale venture!)

• Where a terminal sire is used on crossbred ewes to produce fat lambs for slaughter (eg, Suffolk ram on Mule ewes).

• On very small holdings: Rearing your own replacements is not cheap or simple, and they'll take up valuable grazing land that may be better utilised by keeping more adult breeding ewes.

Situations Where Home Bred Replacements Would Be Most Appropriate

• Where a holding has common, moorland, or open mountain grazing, ewes will have to have been bred on the farm. Without a hereditary instinct hefting them to that particular heath, bought in ewes would just wander off. (Mind you, my sheep do that anyway!)

• Where a "closed" flock is being maintained to attain high health status accreditation.

• On organic holdings – it can be difficult to source organic breeding stock.

• If you're a sentimental old fool like me, and like being able to trace back the lineage of your ewes to that one old favourite that you were given as a wedding present, to get your flock started, all those years ago!

Situations Where Purchased Replacements Would Be Appropriate

• When insufficient numbers of home bred females of suitable quality are available, for example when a rapid flock expansion is taking place to coincide with additional grazing land becoming available.

• To safeguard vulnerable bloodlines, in rare breeds for example, by dispersing them through

TIP! A livestock dealer is a very useful contact for the smallholder to have, as you will often find yourself with small numbers of stock to sell, rather than the large batches of evenly matched animals favoured by the major buyers. A dealer working in conjunction with a number of smallholders can, in effect, co-ordinate the marketing of stock from those holdings. Many livestock dealers are themselves smallholders who have diversified into the "buying and selling" side of things to generate a bit of cash flow and help make ends meet.

a number of flocks.

• To introduce new bloodlines into a pedigree flock. This gives more flexibility and control than simply going out and buying a different ram.

• When replacements can be bought in at a lower price than the cost of rearing your own. (Believe it or not, this is usually the case).

A Trip To The Sales

Whether you decide to buy in replacements, or breed your own, there will be some point in your sheep keeping career when you'll need to purchase females. This may be just a "one off" exercise to get you started, or could form an essential part of the annual routine. When establishing a pedigree flock, many people (particularly if they've got sheep to sell) will advise you to buy directly from an established breeder and, although this method of purchasing does have a lot to commend it, there are some potential drawbacks. It rather limits your choice for a start, and, if you source all your foundation stock from one breeder,

TIP!

When purchasing breeding ewes at auction, don't bid on the best, no matter how smart they look. They've obviously been well fed, and may "go back" (i.e., lose condition) when you get them home. Try to buy sheep that are just a little bit poorer than your ground will support, so they'll immediately begin to improve, and go to the ram in good rising condition.

then your flock will simply be a representation of his own. Any error of breeding policy in that flock will be replicated in yours, and, whilst an experienced stockman would soon spot the problem and breed it out, a novice shepherd may be blissfully unaware of anything amiss. Better to buy from several sources to ensure genetic diversity or, better still, attend the annual show & sale organised by the relevant breed society. The societies rely on these events to raise the profile of the breed and to generate some funds, so members who insist on always selling from home are not being very supportive!

If you've opted for a crossbred flock (or non-pedigree purebreds) then getting started may simply involve a visit to your local livestock mart. All markets will hold weekly sales of store and breeding sheep throughout late summer and autumn, with occasional "special" sales which will attract sellers and buyers from a wider area. These are the ones to go to, but with so many sheep on offer – often several thousand – it can be a daunting experience for the beginner! Try to arrive at the sale yard early enough to discuss your requirements with the auctioneers. They should be able to point you in the direction of some suitable lots, but bear in mind it's a busy day for them, so don't pester

them with unnecessary questions. Once you've spoken to the auctioneer he'll recognise your face in the crowd, which will be a help when it comes to bidding, but you may like to consider commissioning a local livestock dealer to bid on your behalf – you needn't run the risk then of being identified as a greenhorn by the vendor's friends, and run up for sport! At the bigger sales sheep are presented in batches, often twenty or more at a time, which may be more than you wish to purchase. This is another situation where a friendly livestock dealer comes in handy – he'll undoubtedly be able to find another buyer with whom you can split a group, or perhaps he'll buy the remainder himself, for re-sale at a later date.

TIP!

Be wary of anything that seems cheap, and generally avoid animals that are "free to a good home" - they always seem to end up costing more in the long run! If an animal is any good then the person selling knows it is good, and will ask a fair price. You will know it is good, and should expect to pay a fair price.

When making up a breeding flock, either in the first instance or when doing the annual sort out, ensure that there is a good cross section of ages represented, or you may be faced with the problem of having all your ewes growing old at the same time. However, avoid groups of sheep offered for sale in "flock ages" (i.e., mixed) as it is probably a group of odds and sods that have been batched up for convenience – all the escapologists and trouble makers that someone is desperate to get rid of!

The Age Of A Sheep By Its Teeth

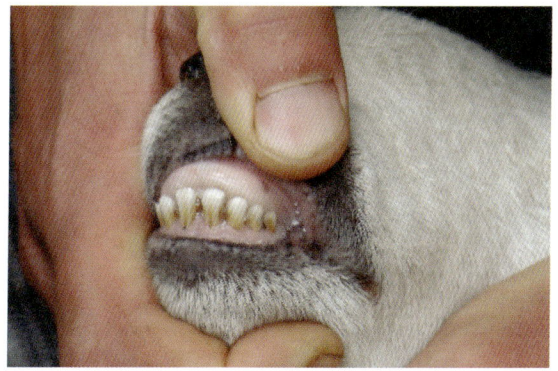

Lamb's teeth.

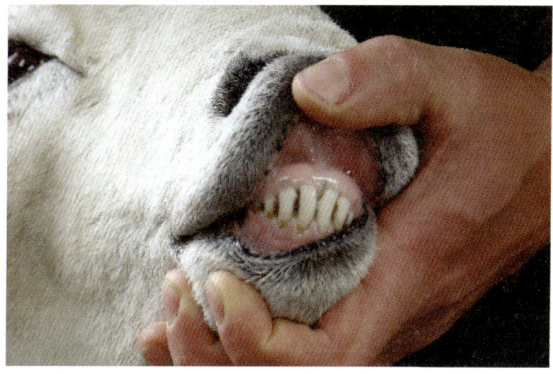

6 Broad teeth = 3 years old.

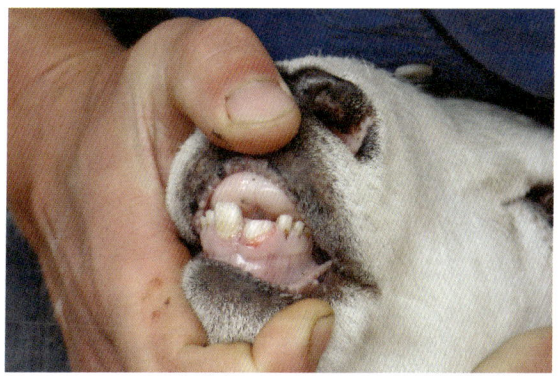

Broad teeth erupt at 15 – 18 months of age. These sheep will generally be termed "shearlings" (i.e., they've been shorn once), but terminology varies according to region. These will be the most expensive to purchase.

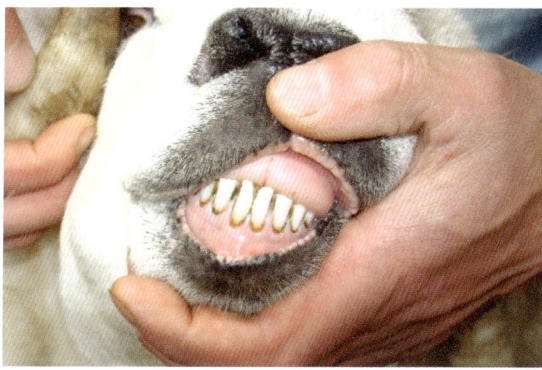

*Full mouthed (8 broad teeth) = 4 years old and over. Genuine 4 year olds are a good buy, but **caveat emptor** - many full mouthed ewes sold as 4 year olds may in fact be much older.*

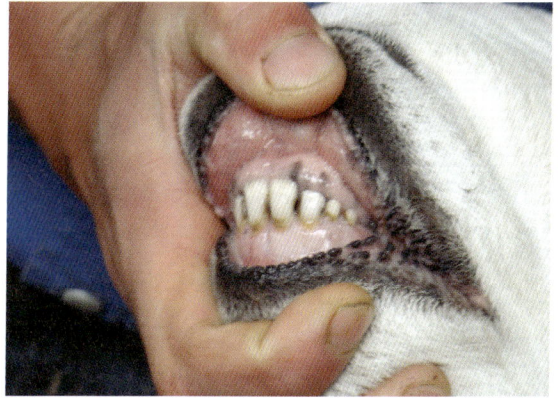

4 Broad teeth = 2 years old.

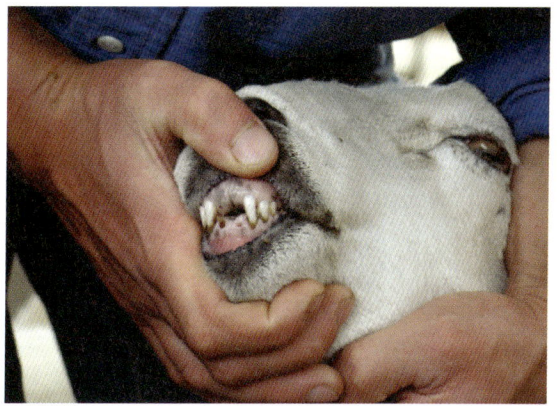

Broken mouth. She'd be culled from a commercial flock, but is good for a few years yet on the the smallholding!

EWE M.O.T

In the previous section I looked at making up the flock in terms of age structure and overall health, but the annual autumn appraisal doesn't really end there. At some point in the run up to tupping time (at least 2 weeks prior to turning in the rams, but preferably much sooner) you'll need to go through the whole flock, one by one, giving each ewe the individual attention she requires. Although lambing time is still a long way off (the ewes aren't even pregnant yet!), the success of the flock at lambing is influenced considerably by pre-tupping management.

Feet

Each ewe must be turned over, her feet thoroughly checked, and trimmed where necessary. Give every hoof of every ewe a good squirt of an appropriate foot spray. Any really lame sheep may require an antibiotic injection. In my experience, this individual approach is far more successful than using a footbath where, invariably, any sheep that really needs attention will hold up an infected hoof, keeping it well clear of the chemical solution in the bath! There is a vaccine available to control footrot, but it should never be used as an alternative to good routine husbandry.

It is vital that hoof trimming is carried out properly before tupping, as turning over in-lamb ewes to attend to lameness may do more harm than good. What's more, conception rates will be lower in flocks where lameness is prevalent, leading to a higher number of barren ewes, and less lambs.

Hoof Trimming

Hoof trimming is one of the jobs I really enjoy doing. Now you probably think I'm mad – how could anyone possibly enjoy trimming sheep's feet? Well I do, principally because it's one of those jobs which, when carried out correctly, can considerably improve the health and wellbeing of any sheep flock – on a number of occasions I've gone to work on a farm where the sheep's feet have been sorely neglected, and with a bit of dedication I've been able to transform not only the health of the flock, but its productivity too.

If you take the time to attend to your sheep's feet properly, twice a year (pre-tupping & post-lambing), you should only rarely see lame animals in your flock, if at all.

TURNING SHEEP OVER

The first challenge to overcome is that of turning the sheep over into a sitting position. This is something that a lot of people seem to struggle with, but once you've got the knack you'll wonder what all the fuss was about! Best likened to judo for sheep, we use the animal's own weight and point of balance to achieve our aim, with very little effort on our part at all. In fact, with a bit of practise (and by starting with your right leg slightly further back), it can easily be accomplished using only one hand – very useful if you've just scrubbed up to attend to a difficult birth, or if you don't want to put down your cup of tea!

It's difficult to explain the correct technique on paper – I'd rather get you all "hands on" and show you how it's done, with a good big flock of sheep for everyone to practise with – but I'll do my best with annotated illustrations:

It never ceases to amaze me, the number of people who are in the habit of lifting sheep bodily off the floor in order to turn them over! This involves a lot of wasted effort, and seems to me to be simply asking for a slipped disc or hernia. I don't suppose it's very comfortable for the ewe either. Don't do it.

Here I am correctly positioned to turn this ewe over. With the animal pressed against the side of the pen (not essential, but much easier) I've got my left hand under her jaw, and you can just see my right hand holding her dock (the base of the tail). My legs are snug up against the sheep, my left at her shoulder and my right at, or slightly forward of, her hip. It is critical that there is no gap between me and the sheep – trying to turn an animal over at arms length always leads to a tussle. My upper body leans over the sheep, and everything is well balanced. I always tell students that they should feel "as one" with the sheep when in this position.

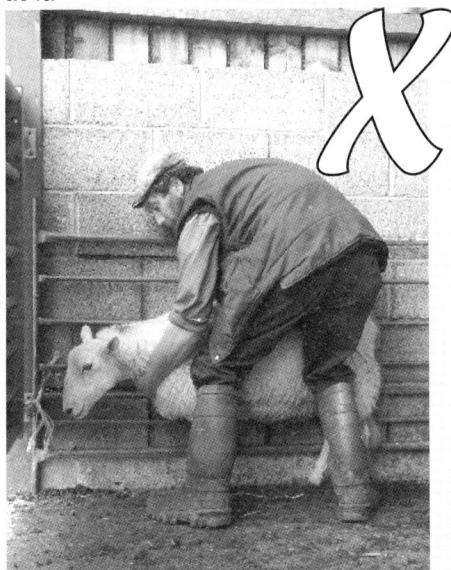

A common mistake. With my left hand under her neck, rather than her jaw, the ewe can get her head down and fight. Compare this picture with the picture above right, and you can see that here the situation is not balanced. Trying to turn her over from this position is stressful for the animal, and will give me a screaming back ache!

Not only am I holding the ewe incorrectly but, with my back against the hurdle, I've got no room to manoeuvre. The ewe, on the other hand, has got plenty of room to get away from me!

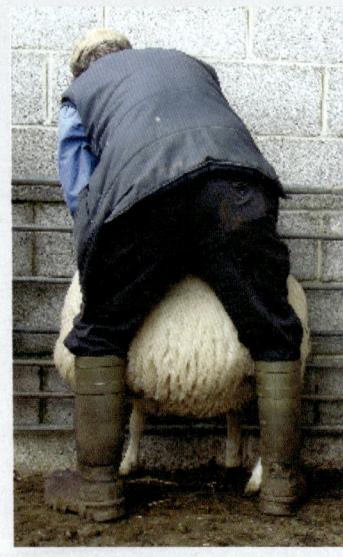

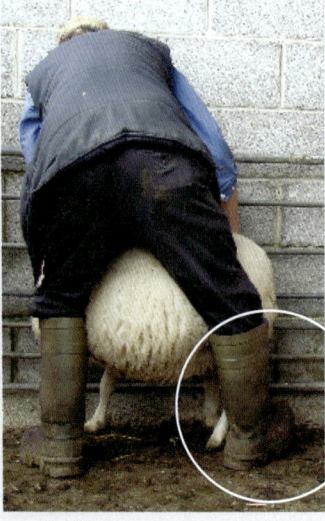

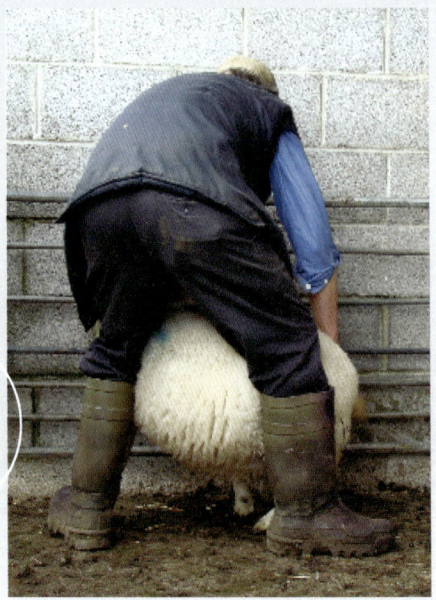

Having restrained the sheep correctly I begin the turning over procedure by bringing the head and dock together, quite literally folding her in half!

Notice how I'm beginning to turn out my right foot. My left foot remains in the same position throughout.

As the ewe begins to lose her balance I turn my right foot a little further out, and move it back slightly, allowing the ewe's back end to slide gently down my right leg. Do not push down on the ewe at all - she will try to resist that, and in the struggle that ensues she'll probably jump up and run away!

Here she is, immobile on her side on the floor, with no effort on my part whatsoever! (Note that my left foot still hasn't moved).

NB: Instructions as given here are for a right handed person. It may seem strange at first that, being right handed, I'm suggesting restraining the ewe with the left, but it will become apparent in due course (particularly when we move on to talk about lambing) that this is the most practical approach.

FROM THE FRONT ANGLE

Folding the head and dock together.

My right foot turning out and moving back as the ewe begins to lose her balance.

Down she goes!

Now all that remains is to lift the ewe up into the sitting position.

I reach across with my right hand, and grasp the wool above her brisket (or the left front leg if the sheep is newly shorn).

I straighten up and throw the weight of my body over to the right, whilst bringing my left leg in to the animal's side. This is the first time in the whole process that my left foot has moved!

Job done! With the weight of the ewe resting on her left hip, the left hind hoof is the first one to trim.

SOME COMMON FOOT DISORDERS

•   Scald: Inflammation of the skin between the cleats, looking remarkably similar to athlete's foot! Usually the precursor to footrot. Prevent scald by topping stemmy pastures (see page 230) to reduce abrasion, and address any trace element deficiencies. Treat affected animals with an antibiotic foot spray.

•   Footrot: Characterised by a foul smelling grey pus, separation of the outer horn from the soft tissue underneath, and, of course, severe lameness. In summer, infected hooves may become flyblown, with maggots often transferring to the body of the animal as it spends so much time lying down. First and foremost, affected sheep should be isolated. Treatment is by antibiotic injections and topical sprays. Trimming of the hoof should be carried out in stages.

•   Contagious Ovine Digital Dermatitis (CODD): This is quite a newly recognised condition which can easily be introduced to a flock by bought in sheep. Superficially it appears similar to a moderately advanced case of footrot, except that there is no strong smell, and the horn separation starts at the coronary band rather than between the cleats. The separated horn often splits down to the toe, and the soft tissue underneath bleeds. In extreme cases the whole shell of the hoof may drop off. Treatment is by antibiotic sprays, and Micotil injections.

•   Toe Granuloma: Toe granulomas are usually brought about by excessive use of the foot shears, particularly when treatment is carried out by a shepherd of the old school who believes it's necessary to "trim to blood" in order to do any good! A granuloma is rather like a fleshy wart, which may be as big as a strawberry and bleeds at the slightest excuse. The hoof horn will grow around the granuloma, resulting in a horribly misshapen foot. Toe granulomas can be surgically removed, and the wound cauterised, but they do tend to reoccur.

Warty growths between the cleats are also a common cause of lameness, particularly in rams. Why rams should be particularly affected I don't know. Perhaps it's because they spend most of the year lying around doing nothing, so their hooves don't wear in the fashion of more active sheep

•   Shelly hoof: Shelly hoof is a minor separation of part of the hoof horn, as is often found when carrying out routine foot treatment. The trouble is that, if left untreated, the pocket of horn will become filled with soil and small stones, exacerbating the problem and leading to other, more complicated, disorders. In the early stages it's a relatively simple matter to clean out the dirt and trim back the loose piece of horn. Trace element deficiencies may be responsible for hoof wall separation.

•   Foot abscesses: Abscesses that occur in the hoof are difficult to locate, but generally burst out at the coronary band in the end. Affected feet can be poulticed to encourage this. Abscesses should be drained and treated with topical sprays and injectable antibiotics.

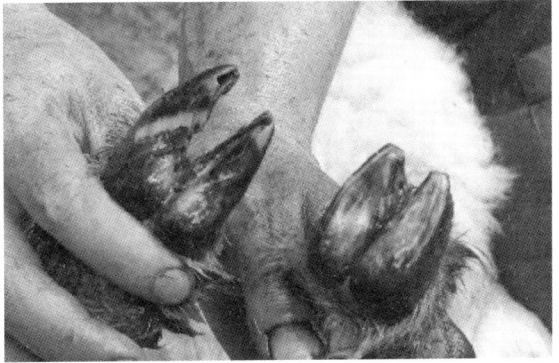

Here is a fairly typical overgrown hoof, such as you are likely to encournter during your routine twice yearly inspection. I've trimmed one back already so that you can see what it should look like. Now the other needs trimming back to match.

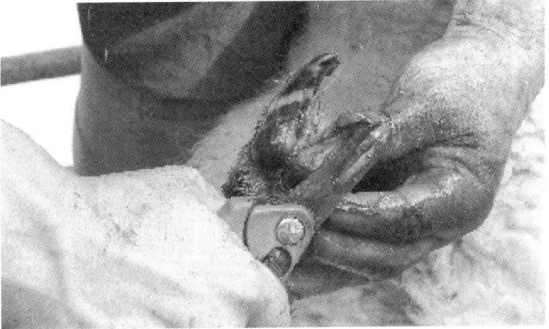

As there is no footrot, scald or any other complication here, it's a relatively simple matter to trim back the flaps of overgrown horn. The sole of a sheep's hoof is relatively soft, and the weight of the animal is largely born by the outer rim of horn, so don't trim back more than is necessary or you'll do more harm than good. (This is quite different to trimming goats' feet, where it is usually necessary to rasp back the sole). In particular, avoid over-zealous trimming at the toe – it is all too easy to inadvertently cut to the quick and cause bleeding. Often, wart-like growths called granulomas will develop following careless clipping, and any animal thus affected may never be sound again.

TIP! Assuming that you've turned the sheep over correctly, start by clipping the left hind foot - with the weight of the ewe on her left hip, she'll not be able to kick that leg (much!). Before moving on to clip the right hind foot, transfer the animal's weight to the right hip, by a simple movement of your knees.

TRIMMING FEET

Equipment

The best tool for the job is a really good pair of serrated hoof shears. I've been using the same pair for the past twenty years, but sadly broke them recently. The clippers I'm using now are of a type often sold as being suitable for lamb's feet. They're small, light, and wickedly sharp – ideal for women and children. I keep a few pairs in stock for students on our courses to use, but they're not strongly made, and won't last long. You'll also need a bucket of water and an old washing up brush (to clean any really dirty feet – sometimes it can be difficult to tell where the muck ends and the hoof begins), a topical foot spray, and possibly an injectable antibiotic. Some disinfectant solution is useful for cleaning the shears between animals, or squirt the blades with the foot spray from time to time.

Method

Again, this is something that isn't easy to explain in writing, but I'll do my best with photographs:

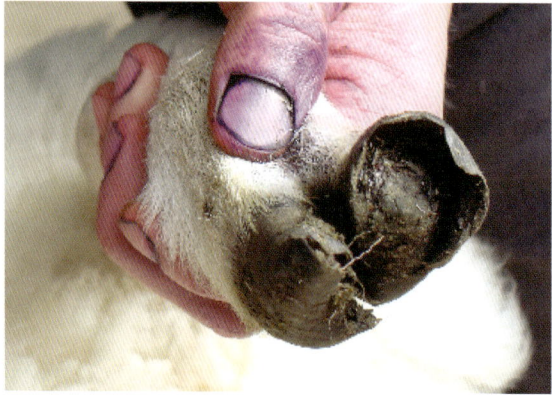

Unfortunately not all hooves are so easy to deal with! Where foot care has been neglected for some time, or when infections such as footrot are present, sheep's feet can become hideously overgrown and misshapen. It can be difficult to know exactly what to do with a foot like this.

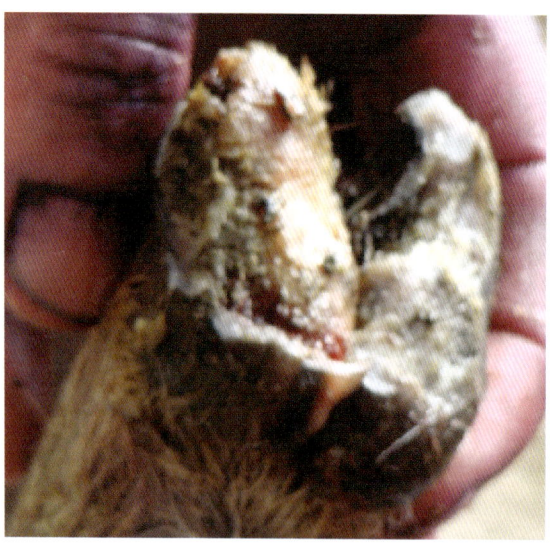

Just as when trimming healthy feet it is important not to cut back too much at once.

It may be necessary to pare back a little at a time over a period of several weeks, in order to get the hoof back to its correct shape.

I've had to cut a lot of loose horn off this hoof, but at least the tissue underneath looks healthy and I've not caused any bleeding. With appropriate treatment it'll recover fairly quickly.

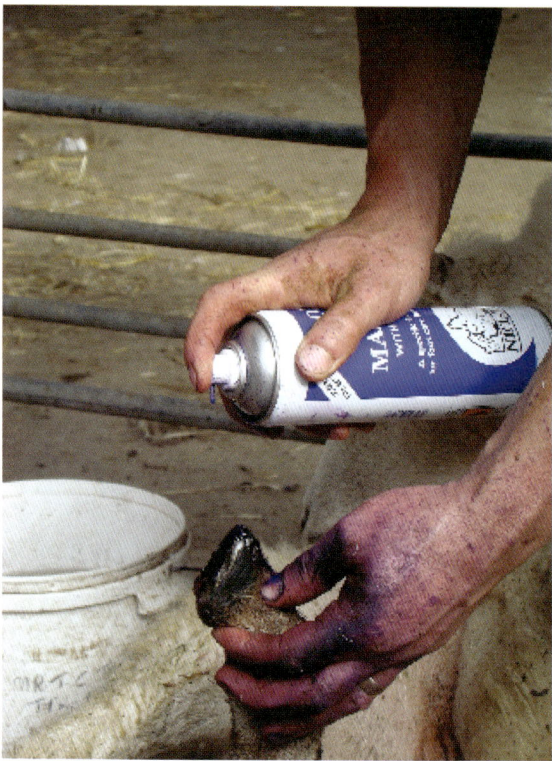

A good coating of an appropriate foot spray completes the job. Of course, I should really be wearing gloves for this. However, the irremovable purple rim around the left thumbnail has become the universal emblem by which shepherds recognise one another when pressed into uncharacteristic surroundings such as weddings, funerals or dinner parties. It makes these occasions far more bearable when you discover that there's someone else to "talk sheep" with!

CRUTCHING

Many flock owners routinely trim away the wool from around the backends of all their ewes before tupping, but I do not recommend this as it can make some rams reluctant to mate. Just clip any particularly dirty ewes. If a high proportion of the ewes have mucky bottoms, you should be asking yourself: Why?

DOSING

Worms

It shouldn't be necessary to dose all ewes in the autumn, as adult sheep generally have quite a high natural resistance to worms. Just dose any thin or dirty ewes. Unnecessary use of anthelmintics may lead to wormer resistance (see "Quarantine" over the page).

Correct Use Of The Dosing Gun

Always dose sheep with them standing up. I know it's tempting to carry out dosing while you've got a sheep in the hoof trimming position, but don't do it, or you risk drowning the poor creature. The best thing is to have a whole group stood in a race, but you can manage just as well in a small pen. Stand astride them if it helps (as shown here). Restrain each animal with your hand under its chin, lifting the head slightly, and gently enter the nozzle of the dosing gun at the corner of the mouth. It passes through the gap between the front and back teeth, and slides over the back of the tongue. (If you find this awkward then adjust the angle of the nozzle by slackening off the locknut where it attaches to the gun and twisting it around a bit until it's correctly positioned). Do be careful how you handle the dosing gun – injuries to the larynx caused by rough dosing are all too common, particularly in lambs, and may be fatal.

If the nozzle of the gun is correctly placed over the back of the animal's tongue the whole dose will be swallowed immediately on administration.

Liver Fluke

Whether or not you need to dose for fluke at this time will vary from year to year, and from farm to farm. In all likelihood you'll need to fluke dose at least once, probably twice, during the period tupping – lambing.

TICKS, LICE AND MITES

Until relatively recently all sheep would have been dipped in the autumn. Dipping remains the most effective way of dealing with ectoparasites, but amidst concerns for the

environment and human health it is no longer a legal requirement. There are a number of pour-on products available to treat and prevent ticks and lice, and certain injectables can be used against scab. Ensure that the flock is free from external parasites, as itchy sheep will lose body condition very quickly.

CONDITION SCORING

You will have condition scored all your ewes at weaning, (see page 167), and sorted them into respective groups for management purposes, but it is important to continue to monitor body condition right up to, and throughout the mating season. Aim to have lowland ewes at CS 3 – 3.5, and hill breeds at CS 2.5 – 3, by tupping time. This body condition must be maintained until the ewes are well in lamb, hence the need for regular monitoring. It is too late now to rectify problems, so any ewes which still fall far short of the target CS should be run as a separate group for tupping, with a bit of supplementary feeding. This does of course necessitate having more than one ram.

QUARANTINE

It's a good policy to quarantine purchased replacements for four weeks. In a mid-March lambing flock, this would mean buying in new stock at the very beginning of September, to give a sufficient period of isolation before incorporating them into the main flock in time for flushing at the beginning of October. The following should be attended to whilst the sheep are in quarantine:

Feet: As page 51.

Crutching: Shouldn't be necessary, as you won't have bought any dirty bummed sheep, will you?

Dosing: Dose sequentially with both a clear (macrocyclic lactone) and a yellow (levamisole) drench. This will hopefully remove any anthelmintic resistant worms. Efficacy will be increased if the sheep can be yarded for 24 – 48 hours. Anthelmintic resistance is a very real threat to the health and welfare of your flock, and for a number of reasons may be more likely to develop on smallholdings than on larger farms, so please take these guidelines seriously.

Also dose for liver fluke, but be aware that some flukicides shouldn't be used in the breeding season.

External parasites: As before. If you use an injectable product against scab mites, then you can omit the clear drench from your quarantine dosing strategy.

Vaccination: Never make the assumption that bought in sheep will already have been vaccinated against clostridial diseases. In fact, even if the vendor assures you that they've been done, do them again! A primary course of two injections will be required, with an interval of 4 – 6 weeks between them, so inject all ewes as they come onto the farm, and again as they reach the end of their quarantine period.

N.B. All these quarantine requirements apply equally to bought-in rams.

FLUSHING AND TUPPING

The Ram

There is an old saying, that you'll no doubt

have heard, that "The ram is half the flock", which is pretty nonsensical when you come to think about it! It would be more appropriate to describe him as half the crop, for that is what he is – half the genetic make up of the current year's lamb crop. If you keep him for a couple of years, and retain his daughters for breeding, then of course his influence is more long lasting, but the ewes remain the really important element of the flock. If you are unhappy with the offspring of a particular ram, then don't keep any, and sell the ram on – his undesirable characteristics then have no impact on the future quality of your stock. On the contrary, a problem in the ewes can take many years of careful selective breeding to eliminate. Having said all that, the purchase of a new ram is indeed a significant undertaking, and, with luck and judgement, the right animal will have a positive and lasting effect on your sheep, perhaps earning himself the tag of being considered "half the flock".

Buying A Ram

A big ram sale is an exciting place to be, and there's a real buzz around the ring when top quality tups are changing hands for large sums of money. Set yourself a budget though, and try not to get carried away by the atmosphere! The really high prices are usually the result of pre-arranged deals, where the same pot of money simply circulates within a tight knit group of breeders (farmer A buys from farmer B, farmer B buys from C, and C buys from farmer A). It looks good in the papers, and may help promote the breed, but in reality means very little. Just occasionally a newcomer gets carried away, and ends up paying over the odds for one of these tups, which leaves the little syndicate rubbing its hands with glee over the fresh injection of cash! Don't get drawn into this end

Welsh Mountain rams at the NSA Wales and Border Region sale, Builth Wells

of the market – at least, not yet awhile!

Always arrive at the sale early enough to carry out a full inspection of any lots that take your fancy. It amazes me how many buyers will cheerfully bid on a sheep they've never handled! This is most evident at smallholder's or rare breed sales, where potential purchasers will pour over pedigree details going back generations, make a decision based on history and cosmetics, and then buy the sheep without

TIP! A breeder once gave me the following piece of good advice: "When you're at a show or sale, and you see a tup you fancy, don't buy him! Go onto the farm where he was bred, and buy his twin brother."

even entering the pen! I recently heard of a ram being sold through the ring at a pedigree sale with only one testicle – this was in spite of having been examined by an inspector appointed by the relevant breed society, and having presumably been subject to whatever pre-purchase perusal the buyer felt necessary! Even more surprising, perhaps, is the fact that the vendor was also unaware of the fault. Makes you wonder, doesn't it?

By all means make a shortlist based on appearance and genetics, but narrow this down by getting in the pens, handling the sheep, and making a thorough and ruthless appraisal of all their physical strengths and weaknesses. It's even better if you can arrange to visit the farms of a few vendors a month or more before the sale, to view their stock without the gloss. Don't take this as an opportunity to try to buy stock privately in advance of the sale in which they're entered though – this would be a serious breach of etiquette. It is, in many respects, far better to buy a tup privately, straight off the farm where he was bred, but not one that's already entered for a specific forthcoming sale.

POINTS TO CONSIDER

Terminal Sires

A terminal sire is a ram used to produce fat lambs for slaughter, i.e., his progeny are the terminal generation. Here, the key considerations will be conformation and growth rate. Always buy a pure bred animal (as cross bred rams never seem to throw consistent offspring), though he needn't be pedigree. In fact, he needn't even be a particularly good example of his breed, provided he has the characteristics you're looking for. Despite the need for a meaty body shape, don't let your desire to produce the ideal butcher's

lamb overshadow other issues, such as ease of lambing. Until recently, the Suffolk was the most widely used terminal sire in the UK, although the Texel has now become more popular. Both are ideal on big crossbred ewes, but may lead to difficult births if used on the smaller native sheep, as lambs are born with broad heads and shoulders. The Charollais is an excellent choice for use with smaller ewes, or on ewe lambs of larger breeds, due to the "torpedo" shape of the lambs at birth – they just pop out with ease from even the tiniest of ewes, then simply get up and grow! Enthusiasts of more traditional breeds should consider the Hampshire Down as a terminal sire. Breeders have made huge progress in developing the commercial characteristics of this old breed, through the use of Signet recording and EBVs, without losing any of its native charm.

Crossing Sires

Primarily we are talking about the Bluefaced Leicester and the Border Leicester, although any ram used to produce first cross female offspring for further breeding could be termed a "crossing" sire. Here we are most interested in maternal characteristics, though it is of course impossible to judge a ram's feminine side while he's strutting his stuff in the sale ring! Performance data and Estimated Breeding Values can be used to take a bit of guesswork out of the game, e.g., the higher the maternal EBV, the better that ram's daughters can be expected to perform as mothers.

Pure Breeding

Any ram used for pure or pedigree breeding must be a cut above average, not only in terms of conformation and the potential performance of his offspring, but also in breed

"The ram you need now must be spot on in every respect, and that's the type that doesn't come cheap"

characteristics. Pedigree details may help you make your choice, and so too will EBVs, but, no matter how much research you do, on the day of the sale your head will be ruled by your heart, and you'll bid on the chap that catches your eye! Initially you'll be looking for a tup that will cancel out any weaknesses in your flock, so, for example, if your ewes are rather narrow, make a point of looking for a broad chested ram. Ewes rather fine in the bone? Too light in the shoulder? Producing effeminate ram lambs? The tup you need will be an ugly old sinner, but he'll throw you nice strong lambs! Or perhaps you feel that your sheep are rather coarse, in which case you'll be looking for a finer headed ram. Fleeces too open? Buy a ram with a good tight coat, even if it's a bit extreme for the breed. And so on.

However, there will come a point in the development of your flock when the above advice ceases to apply, and, in order to continue to move forward, you'll need to breed like with like. This is quite a defining moment. It is also the point at which you are going to have to start paying much higher prices for the right tups! You are no longer looking for a specific characteristic to correct deficiencies in your ewes; you've done that already, slowly, over a number of years. The ram you need now must be spot on in every respect, and that's the type that doesn't come cheap.

Testicles

Should be two in number, even in size, large, firm, and hanging well clear of the body. The epididymides (the "lump" at the bottom of each testicle where semen is stored) should be well defined, but there must be no unusual bumps or swellings anywhere in or on the scrotum. The scrotum should not be excessively woolly – in some breeds it may be necessary to shear the scrotum in the autumn in order to maintain correct semen storage temperature. Avoid rams with small, soft testicles, carried high up.

Feet

Often there'll be so much straw in the pens at a ram sale that you can't even see the animal's legs, let alone their feet! If at all possible, ask the vendor to get the ram you fancy out of the pen so you can see how he stands and walks on the concrete. (Much easier if you've arrived early, before the aisles are crowded). Avoid sheep that appear down-at-heel, or are in any way unsound.

Teeth

The incisor teeth must bite squarely onto the dental pad. Some deviation may be acceptable in a terminal sire, where all offspring will be going to slaughter, but for pedigree or pure breeding the teeth need to be as near to perfect as can be. In a good mouthed yearling tup the front two teeth should be nearly as wide as they are long.

Fertility

It is thought that as many as 1 in 10 rams may be sub- or infertile, so the chances of buying a dud are actually rather high. No matter how carefully you examine your potential purchase, you cannot determine whether or not he is fertile, although any ram that hasn't "coloured up" (i.e., the skin in his groin should be flushed red as the breeding season approaches) is suspect. The terms and conditions of most sales allow 10 weeks for claims to be made in the event of a ram

proving to be infertile. The ultimate test of a ram's fertility is whether or not ewes hold to service, by which time, of course, your breeding programme will be well and truly mucked up, should he prove incapable! For peace of mind, take your new ram to your local vet (well within the 10 week time window) where it will take only a matter of minutes to examine semen obtained by electro-ejaculation. This service may cost as litle as £20, and gives an on-the-spot verdict.

Health Status

When attending shows and sales you'll probably have noticed that the sheep are divided into "accredited" and "non-accredited" sections, with a definite gap between the two. On the one hand will be seen most of the popular terminal sire and crossing breeds, and on the other will be mostly hill breeds, rare breeds and primitives. The accreditation refers to their maedi-visna (MV) status – accredited flocks will have been regularly tested for, and found to be free from, maedi-visna, as part of a flock health accreditation scheme. If your chosen breed society encourages accreditation, then in all likelihood you'll have to join in, otherwise you may find yourself unable to sell at society sales or to compete at major affiliated shows. You'll then have to ensure that you only buy in accredited stock. Other health schemes include screening for the presence of caseous lymphadenitis (CLA) and enzootic abortion (EAE), and scrapie monitoring.

Insurance

The purchase of a quality ram for a small flock represents a huge outlay in terms of £ per ewe mated, so in the case of a very valuable animal, you may wish to consider taking out

MAEDI-VISNA

MV is a highly contagious viral disease that affects both the lungs and the nervous system. It has a long incubation period (up to 4 years) so by the time an outbreak is confirmed a large proportion of the flock is likely to be infected. Symptoms include pneumonia, paralysis, mastitis and wasting. Arthritis is also a recognised symptom, although this form of the disease has not yet been reported in the UK. The disease is also responsible for high mortality rates in both lambs and adult sheep, decreased conception rates and premature births. Lambs born in affected flocks are infected almost immediately after birth by suckling infected milk and by inhalation of droplets in the breath of infected ewes. Owners of accredited flocks must undertake to carry out double fencing of boundaries in order to avoid contact with non- accredited neighbouring flocks.

There is no cure for maedi-visna, but, having said that, I've never seen a sheep with MV, and I don't recollect meeting anyone who has, so clearly it's not the sort of problem that crops up in the average smallholder's flock.

A few of the bigger breed societies that have encouraged MV accreditation for many years are now considering leaving the scheme, as there are doubts over whether the high cost of accreditation is justifiable, given the restrictions that it imposes. Sheep for export would still need to be accredited, but this could be carried out on an individual basis, provided that suitable isolation facilities were available on the premises of departure.

CASEOUS LYMPHADENITIS

CLA is a bacterial infection causing abscesses at lymph nodes throughout the body. These abscesses are most commonly seen around the head and neck. Transmission occurs via skin damage, so is easily carried from sheep to sheep on contaminated shearing equipment. Carcasses of infected animals are likely to be downgraded or condemned, and any sheep showing signs of the disease will be refused entry at shows and sales.

an insurance policy. However, insuring an individual animal against untimely death or loss of use is very pricey. If you've just blown your life savings on a topping tup, then by far the best form of "insurance" is to have semen collected and stored for future use by artificial insemination.

While on the subject of artificial insemination (A.I.), it is perhaps worth mentioning that small scale pedigree flocks are one area where A.I. of sheep really can be cost effective as a standard procedure. Consider the cost of purchasing a ram, on a £ per ewe basis. Even a fairly run-of-the-mill pedigree tup may cost you the best part of £1000. (Commercial rams may cost a lot less, but we're talking about pedigree breeding now). Within a year your ram could be worth as little as ⅓ of his purchase price, or even less. If you've only half a dozen or so ewes, that's a big cost to bear for very little genetic gain. A.I, on the other hand, gives you – the novice breeder – access to the very best bloodlines available. Semen from rams valued at tens of thousands of pounds can be purchased for as little as £25 - £50 per dose! On farm laparoscopic insemination costs vary according to the number of ewes, with reduced charges for larger numbers, so the price could be kept to a minimum by teaming up with other local breeders. For batches of less than 25 ewes, a minimum fee of around £375 (based on 25 @ £15 a ewe) will be charged, together with a visit fee in the region of a hundred pounds. Therefore, if we assume that you've got only 6 ewes (everyone has to start somewhere…) the total cost of using a top quality tup by A.I. will be in the region of £625, or £104 per ewe, or less than £45 per ewe if you can join forces with three other breeders, each with a similar number of ewes, or if you take your ewes to an A.I centre to be inseminated. The poorer quality bought in tup will cost you £111 per ewe (if he lives) or £166 per ewe (if he doesn't), plus whatever you've spent on feed etc. to keep him. Considerable additional benefit will be gained from the increased value of the offspring by the better ram, together with a reduced risk of introducing disease into a high health status pedigree flock.

If you are a member of a sheep breed improvement scheme, you may be eligible for grant funding to help cover the cost of A.I.

Straws of frozen semen, stored in liquid nitrogen

Depreciation

As I've mentioned above, one of the biggest costs associated with owning a ram is his depreciation. A shearling ram costing many hundreds (or even thousands) of pounds may, at the very best, be worth only half that amount in twelve months time, and much less after two years, which is the usual period of his usefulness in a flock where females are retained for breeding. That is if he lives that long – sheep have an uncanny knack of dying at costly moments, such as just before a sale! The exception to this declining value situation is where a ram's offspring have shone out in the show ring, or been otherwise outstanding in their field, in which case a premium price may be asked.

One possible solution is to purchase an older ram in the first place, so someone else will already have carried the cost of the depreciation. We used to cross our Welsh ewes with a Suffolk, and would buy old rams for around £15. When we came to sell them on in a year or two the slaughter value would be about £10, so each ram actually only cost us a fiver! On a smallholding where he can be mollycoddled a bit, an older ram may give several years of useful service, and his offspring will be genetically every bit as good as the lambs he produced in his prime. What's more, you'll have added a couple of years to his lifespan, a leisurely semi-retirement that otherwise he'd never have had!

FLUSHING

Hopefully, your ewes will have been gaining condition steadily (or losing a bit, in the case of very fat ewes) since weaning. Now, some 2-3 weeks before tupping, ewes should be within

C.C.N.

Cerebro-cortical necrosis (CCN) seems to be the only disease of sheep that doesn't have a colloquial name, even though it is relatively common. I suspect that this is due to the fact that symptoms were usually attributed to "worm in the brain" or "gid" (tapeworm cysts) which would have occurred far more frequently in the past, when farm dogs (and others) were allowed to scavenge on carcasses, and regular worm treatments were unheard of. CCN can also be confused with listeriosis.

Despite the similarity of symptoms (circling walk with head tipped right back, apparent blindness and eventual collapse, followed by convulsions of the limbs, with the head still craned back onto the shoulder) CCN is in fact caused by an imbalance in the microflora of the rumen that are responsible for the production of certain B vitamins. Usually it occurs following a change from coarse herbage to lush grazing, for example after moving hill ewes onto better pasture for flushing. Treatment is by injections of vitamin B1 (about 6ml for a ewe) which will need to be repeated several times. I also inject oxytetracycline and, if the ewe is pregnant, calcium borogluconate. Affected animals will need careful nursing. Only if the condition was identified in its early stages is treatment likely to be successful. To prevent further cases occurring, consider returning the rest of the sheep to their original pasture.

half a condition score of the pre-tupping target – the proverbial "fit, but not fat". At this point it is necessary to significantly increase the ewes' plane of nutrition, usually by moving the flock onto better pasture. (Aftermaths are ideal for the purpose, but not red clover swards, as red clover has been found to depress fertility). This process, known as flushing, ensures that all ewes go to the ram in good rising condition, leading to increased ovulation and higher conception rates. In our own flock we move the ewes onto better grazing on the 2nd October, on the same day that we introduce the teasers. We also provide molassed mineral licks containing a high level (9.5%) of phosphorous. The combined effects of flushing, teasing, and phosphorous supplementation result in a very compact mating period and, subsequently, in the spring, most of our ewes give birth over 10 days, without resorting to artificial methods.

Teaser Rams

Teasers are vasectomised rams that are run with the ewes for 14 days before introducing fertile tups. The response is most marked where teasers are introduced slightly in advance of the normal mating season for the breed, provided that ewes have been kept well away from rams during the preceding month. Pheromones, or scent hormones, secreted by the teasers stimulate sexual activity in the ewes, resulting in a roughly synchronised oestrus some 17 days after the teasers were first introduced.

This phenomenon isn't limited to sheep – the presence of a male stripper in a girl's boarding school would initiate a similar response!

Breed Of Teaser

It may seem rather odd to be discussing

Having teasers of a different breed or colour makes identification very simple

breed of teaser as, being vasectomised, he won't be producing any offspring. Nonetheless, breed choice is important, and will have a bearing on the success of the teaser effect. What we are looking for here is a breed that will be in full sexual condition early in the year. The teaser must be fully active when introduced to the ewes, so it would be no good, for example, expecting a rough, tough Welsh Mountain ram to exert a teaser effect on lowland ewes in late August – he simply wouldn't be up for it, the normal breeding season for the Hardy Welsh Mountain being late October / November. Rams from early lambing breeds, such as the Poll Dorset, are an obvious choice, and, as the teaser is a non-breeding animal we can take advantage of hybrid vigour by using a crossbred – we want our teaser to be as active and vigorous as possible, for maximum impact. The best teasers I've ever seen in action were Dorset X Lleyn. Another valid reason for using a different breed of ram as a teaser is that

you can never make the disastrous mistake of putting the wrong rams in at the wrong time. Our stock rams are all white, and our current teasers are black, which makes identification very simple.

TUPPING

On day 14 (i.e., in our case, on the 16th October) the teasers are removed, and the breeding rams turned in. Opinions differ as to the optimum ram / ewe ratio, but 40 – 50 ewes for a mature ram, and up to 20 for a ram lamb would be OK. Where oestrus has been synchronised, either by artificial means or by efficient use of teasers, the ram : ewe ratio may need to be increased to as much as 1 : 10. It should go without saying that the rams must be fit and healthy but, sadly, the stock rams are often the most neglected sheep in the flock! It's no good just chucking them in with the ewes and expecting good results – put the rams through a full M.O.T., paying particular attention to feet, well in advance of tupping time, and aim to build up body condition to CS 4, or thereabouts.

It is best to raddle the rams, even in small flocks, so that you have a clear indication of what is going on. In the short term it enables you to see that the ram is indeed working, and in the longer term you'll be able to sort your ewes according to their expected lambing date. The type of raddle most commonly used is a wax crayon attached to the ram's chest by means of a harness, but for ram lambs (or tups of very small breeds) it is better to smear a coloured paste on the animal's brisket. Raddle paste can be purchased in powder form, for mixing with oil or grease, or make your own by mixing children's powder paints with margarine!

Rams of heavily bearded breeds, such as

Ewes in season will gather round the ram, seeking out his attention, particularly where oestrus has been synchronised by the use of teasers or sponges

the Nelson (South Wales Mountain), will need their whiskers trimmed, or the long hair becomes plastered down over the crayon, preventing it from marking the ewes.

The colour of the raddle should be changed after the ram has been running with the ewes for a while. Exactly when will vary, according to your own management system, but in our case we change the crayons in a fortnight, which takes us neatly up to the end of October.

All being well, November can be quite a quiet period in the conventional shepherding year. Most, if not all, of the ewes should have held to the ram in October, but ensure that the flock remains under close observation until you are certain that this is the case – November won't be such a peaceful month after all, if you're rushing about, desperately trying to source a replacement tup!

Changing Crayons

Each oestrus cycle in the ewe lasts for 17 days, consisting of 1 - 2 days when she is "on heat" or "in season", and will allow the ram to mate her, followed by 15 days when she will not. If she is not mated during the heat period (or if mating was unsuccessful), the ewe will continue to come on heat at 17 day intervals throughout the breeding season. This is known as repeating. If repeat matings also fail then the ewe will remain barren – a non-productive passenger in the flock. If a significant number of ewes are seen to be repeating then there is clearly a problem with the ram, and action must be taken quickly to replace him. Sensible use of different raddle colours enables the mating activity of the flock to be closely monitored, and any problems can be spotted straight away.

Initially a fairly pale coloured crayon should

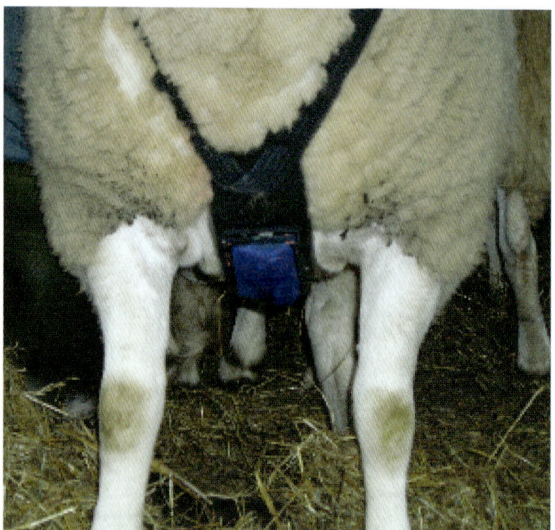

Raddled ram

Changing crayon from blue to red on a Brecknock tup

be used (we start with blue), with the crayon being changed (in our case to red) after 14 days. Invariably a few ewes will be in season on the day that the colour is changed, so may be marked twice, but it is from day 16 (two days after changing crayons) that you need to be particularly vigilant. Some ewes will have been going off heat when the ram was first turned in, so although he may have mated them it was probably a rather half-hearted affair – these ewes are likely to repeat, so don't be too alarmed if you see one or two red over blue marks. However, if more than half a dozen or so ewes in a group of 20 are seen to have repeated then something is clearly wrong. Where two or more rams are running with the ewes, an infertile ram may never be spotted, as most ewes will have been mated by more than one ram, but where single sire mating is carried out, for example in very small flocks, pedigree flocks, or performance recorded flocks, an undetected infertile ram will wreak havoc with not only your breeding plan, but your whole timetable for the coming year. Raddling the rams is a simple way of ensuring problems are easily identified, and nipped in the bud. As I

mentioned earlier, it is well worth having new rams fertility tested by your vet before tupping commences, however there may be other reasons why an apparently fertile ram fails to get ewes in lamb.

The colour of the crayon should be changed again (to black) after another 2 weeks, in order to identify any late lambers (there's bound to be a few!). These can then be run as a separate group through the latter part of the winter, as they'll not need supplementary feeding until 4 - 6 weeks after the rest of the flock start their pre-lambing rations.

In very large flocks, or where space in the lambing sheds (or fields, if lambing outdoors) will be at a premium, it is often the practice to change the crayon colour weekly, in order that the flock can be subdivided into smaller groups for lambing. In this case, look out for repeats when changing from the second to the third colour, and again when changing from the fourth to the fifth.

While changing the crayons, check the fit of the harness. A hardworking ram will lose a considerable amount of body condition during tupping time, and his harness may need to be tightened from time to time – a loose fitting harness will chafe under his "armpits" and cause brisket sores, resulting in severe discomfort and a reluctance to mate.

Where several groups of ewes are run separately with different rams, it is quite a good idea to amalgamate them all after a month or so, just to ensure everyone gets a fair crack of the whip. This does mean, however, that you'll not be able to identify the sires of your later born lambs. Using different coloured crayons on different rams as a means of identifying which ewes each one has mated does not work! Where several rams run together, most ewes will be mated by two or more. If it is necessary to accurately determine parentage, then ewes must be single sire mated. Use a paint mark, or record ear tag numbers, to identify which ewes ran with which rams.

Breeding From Ewe Lambs

Should sheep be bred from in their first year, or not? Some years we do breed from our ewe lambs, and some years we don't. In the years that we do, we wish we hadn't, and in the years that we don't, we wish that we had! On balance, I feel that putting ewe lambs to the ram in their first autumn (to lamb down at one year old) isn't a good idea, and we shan't be doing it again. Having said that, our Welsh Mountain sheep are fairly slow maturing. Ewe lambs of faster growing, larger breeds may well be fit to take the ram in their first year, provided that they have reached 60% of their expected adult weight. Ewe lambs should be put to the ram slightly later than the main flock, certainly no

earlier than the beginning of November (Guy Fawkes night is a good time – it'll give you lambs on April fool's day!), and should run with the ram for only one oestrus cycle. If you leave the ram in for longer then the teaser effect of his presence will bring the less mature lambs into season, with the result that they too will be mated, and may become pregnant, when really they'd have been better off remaining barren.

If you are thinking of breeding from ewe lambs, consider the following pros and cons, before committing yourself:

Pros

• Sheep bred from in their first year tend to have much easier lambings, compared with sheep giving birth for the first time as shearlings, which are notorious for protracted parturition.
• They'll make much better mothers later in life.
• Enables more rapid progress to be made in breeding projects, for example when upgrading to pedigree status, or when developing a higher genetic merit flock through performance recording.
• The extra lambs gained by breeding from ewe lambs being kept as replacements may be seen as a bonus crop, over and above the normal output for the flock. This could help to justify the purchase price of a ram for very small flocks, and can be offset against the cost of rearing replacement ewes, rather than buying in older animals.

Cons

• Breeding at a young age places tremendous strain on an immature animal.
• Ewe lambs which are bred from always suffer a setback in growth, and never seem

to blossom into such strong breeding ewes. Skeletal development in particular appears to be affected.

• Pregnant ewe lambs will need to be run as a separate group from other in-lamb ewes and barren ewe lambs, making winter grazing management rather complicated, particularly where acreage is limited.

• Around half of all ewe lambs run with the ram may conceive, and will, on the whole, carry singles. The only way to determine which ones are in lamb is by scanning, but as it will be necessary to scan two sheep for each lamb expected, this may be a rather expensive exercise.

• It can be difficult to encourage ewe lambs to accept supplementary feeding in late pregnancy (particularly with hill breeds), leading to low birth weight non-viable lambs, little or no colostrum, excessive drain on the metabolism of the ewe and, potentially, pregnancy toxaemia and death.

• Extended lambing season – it is difficult to give late lambing sheep the attention they deserve, when other spring tasks on the smallholding are making pressing demands on your time.

• Yearling ewes that are rearing lambs may need extra feed throughout the summer.

• The "bonus" crop of extra lambs may, in fact, prove to be more of a liability than an asset! These late born lambs from immature mothers will often be seen wandering round the farm in the autumn, pot bellied and stunted, long after the remainder of the year's lamb crop have been sold.

TECHNOLOGICAL FLOCK MANAGEMENT

Controlled Breeding

Although many sheep keepers may see controlled breeding techniques as useful tools to aid them in achieving better management of their stock, for some flock owners the thought of tinkering with the breeding cycle of their animals is just too much. Maybe they feel, like Miriam in Stella Gibbons classic *Cold Comfort Farm*, that "'Tes wickedness! 'Tes flying in the face of nature!" (though of course it was a different kind of birth control she and Flora were discussing), and perhaps it smacks just a little of intensive agriculture. Unfortunately the words "extensive" and "smallholding" do not go together. Extensive farming, often seen as a greener method, requires by its very nature, a large acreage to be successful, and large acreages are what smallholdings, by their very nature, do not have! For anyone aiming to make their smallholding their principal occupation, a fairly high level of output / acre or per animal will need to be maintained, and perhaps the production of early or out of season lamb would be a suitable high output enterprise.

Usually the manipulation of the breeding cycle of sheep is carried out by "sponging" – the insertion of a progesterone impregnated sponge into the vagina of the ewe – although there are of course breeds such as the Dorset Horn or Polled Dorset that will naturally breed out of season. Commercial farmers will weigh up the costs associated with the process, the purchase of sponges, etc, against the benefits of out of season or condensed lambing, such as more effective use of paid labour or higher price of early born lamb. Perhaps for pedigree breeders the extra period of growth gained by lambing two months before rival flocks may make all the difference to prices at the autumn shows and sales.

But these are all financial considerations. For many smaller flock owners the financial element is secondary to the pleasure of keeping sheep, and provided that they are not ridiculously out of pocket and there is a regular supply of good quality lamb for the freezer or fleeces for spinning, they will be quite content. It is my intention, then, to put money matters aside and take a look at this subject from the point of view of good shepherding, which for all of us is the most important factor, whether we intend that our flock should contribute to overall income or not.

Many smaller flock owners will be part timers and, whilst it is possible to carry out most of the routine flock tasks and general shepherding in the evenings and at weekends, during lambing time it is a different story. Supervising the lambing of a flock of ewes, no matter how small, is not something that can be carried out satisfactorily on a part time basis. Regardless of how carefully you tend your charges in your spare time, an awful lot can go wrong between 9 and 5 when you are otherwise engaged. OK, so perhaps you ask a neighbour to pop in and have a look around from time to time during the day. Well, that is a start, but to my mind there is no substitute for a 24 hour round the clock vigil. This has been our policy for several years now and the results speak for themselves – mortality rates have been reduced to a level described by one vet as "unheard of". It is not that we need to intervene very often – typically only 3 or 4 ewes out of every hundred will need any significant level of assistance to give birth – but it is the fact that we are on hand to sort out any minor muddles or mishaps

as they occur that makes all the difference. We feel we owe it to the sheep to be there. It is, after all, due to our own intervention that the ewes are being bred from in the first place. It was us that opened the gate and let the ram in five months previously, so we have a duty of care to see the job through. If given the chance I'm sure the majority of ewes would gladly drop motherhood from the annual routine!

This is all very well of course, for us who are lucky enough to work more or less full time on our holding, but not so easy for others. Even in small flocks, lambing may typically be spread over six weeks or more, and no employer, however sympathetic to the cause, is likely to allow that much time off work. How about it then, if lambing time could be so arranged and condensed as to coincide with a pre- booked period of (paid?) holiday? If that idea appeals to you then perhaps you should consider sponging.

It may be possible by careful timing of the lambing date to plan the whole shepherding year in such a way that it can be fitted in around other commitments – a teacher for example, if he were to sponge his ewes on the first day of the long summer vacation, could pregnancy scan in the autumn half-term break, lamb them at Christmas, sell fat lambs during the Easter holiday and carry out his shearing over Whitsun!

Sponging

Sponging refers to the use of small sponges impregnated with a progesterone substitute which are placed in the vagina of the ewe for 12 to 14 days, effectively preventing her from coming into oestrus until the sponge is removed. It can be seen, then, that removing sponges from all ewes in a flock at the same

Close up of sponges (with 20 pence for scale)

time causes them to come on heat more or less simultaneously, resulting in a very compact mating (and hence lambing) period. Where it is intended to breed from ewes outside the normal season, for example for the production of very early spring lamb, it may also be necessary to inject ewes at the time of sponge removal with PMSG (pregnant mare serum gonadotrophin). Your vet should advise as to the correct dosage of this product as it will vary according to the breed and time of year – an incorrect dose could result in large litters of undersized lambs being produced.

Method

The small sponges, which have strings attached to facilitate their removal in a fortnight's time, are inserted using a simple applicator consisting of a tube and rod. Make sure that the strings are not tangled or wrapped around the sponge, then place it into the wider end of the applicator tube and push almost through using the rod. Insert the loaded applicator gently into the vagina of the ewe using plenty of lubricating gel. The tube should easily slip in to about ⅔ of its length. If it does not, or if any resistance is felt, then that animal should not be sponged (NB: It is not generally advisable to try to insert sponges into

maiden ewes, although there are some sponges marketed as being suitable for this purpose). By holding the rod whilst pulling back on the tube the sponge will be ejected from the applicator and remain in the vagina close to the cervix.

Sponges, lubricating gel, 2 types of applicator and antiseptic

Withdraw the applicator and rinse it well in a dilute solution of an antiseptic such as Dettol, before using on the next ewe. The strings should be visible hanging out of the vulva.

Leave sponges in place for 12 to 14 days, then withdraw by pulling gently but firmly on the protruding strings. This is not a very pleasant job as some evil smelling fluid will have accumulated behind the sponge, which will squirt out onto your hand as the sponge comes away. You may wish to wear surgical gloves. Sometimes the strings will not be visible and it will be assumed that the sponge has fallen out. In reality, very few sponges will be lost in this way (less than 1%). If no strings are visible outside the vulva they will usually be found to have been drawn into the vagina, and can generally by hooked out with a fingertip. Very, very rarely the strings will break, or the sponge be irremovable for some other reason, in which case ask a vet to retrieve it using a speculum and forceps. I have inserted several thousand sponges over the last few years, and only 2 have needed to be removed in this way.

A period of 36 to 40 hours must elapse after the sponges are removed before introducing the rams, which should all be mature proven animals, used at a rate of at least one ram to every 10 ewes. I find it preferable not to use raddle harnesses on rams at this time, but to observe their activity closely to ensure mating is taking place. If crayons are used then the colour should be a light one. The rams remain with the flock for four days during which time all ewes will be mated, though not all will be in lamb – conception rates to 1st service after sponging are typically around 70%, so the smallholder wishing to produce early synchronised lambs from a flock of, for example, 25 ewes will need to have sponged around 35. It goes without saying that all ewes should be sound and in good health – don't waste sponges on thin, poorly, or lame animals.

The question now is how to identify the "empty" ewes, and what to do with them. There are three practical options:

1. Introduce a raddled teaser ram to the flock from 10 days after the entire rams were withdrawn. He will soon identify, and mark, those ewes which are not in lamb, which can then be removed and sold immediately, leaving a flock of around 25. This option is preferable where acreage is the limiting factor to the size of the flock.

2. Introduce raddled entire rams to the flock from 10 days after the end of the first mating period. Any ewes not having held to the first service will be mated now and hopefully become pregnant. Following scanning at 90 days this group is sold as "in- lamb" ewes, leaving a flock of around 25 to lamb over a six day period as planned.

3. Where it is intended to lamb very early, e.g., before Christmas, neither of the above options

will be appropriate, as ewes will return to their off-season state rather than continue cycling. In this case barren ewes should be picked out at scanning, which for a pre-Christmas lambing flock will be around 14th October (assuming rams to have been turned in on 14th July). This should be just in time to catch the last of the autumn breeding ewe sales, where these empty ewes can be sold on to someone with a conventional lambing flock, again leaving a flock of 25 to lamb to the sponges.

Not all flock owners will be happy with the concept of sponging. Some may object on ethical grounds, and for others it will just seem too fiddly. We do not sponge our own flock of Welsh Mountain ewes (except for a chosen few used for A.I. or embryo transfer work), yet there are still considerable benefits to animal health and welfare – such as less time for disease build up in lambing pens and sheds – to be gained by having a shorter lambing period. By paying particular attention to detail at tupping time we have succeeded in condensing our lambing into little more than a 10 day period, which considerably reduces the number of sleepless nights suffered every spring, without having to resort to artificial means. The key to achieving this level of synchronisation from a conventional mating system is that the timing of events must be strictly adhered to. If you are a day or two late removing the teasers because you didn't get time, or if the raddle crayons were not changed on the appointed day, then I'm afraid you will be disappointed with the results.

Performance Recording

What is it, and is it relevant to the smallholder?

Performance recording forms a basis for positive selective breeding of livestock, based on a whole range of economically important criteria. This may just sound like basic common sense, but it is worth considering that at one time, selective breeding of livestock was anything but positive. Quite the opposite in fact! Every winter, the biggest, fattest and healthiest animals on the farm would be slaughtered, and the poorer, thinner specimens would be spared the knife, to live and breed the following year. It's hardly surprising that there was very little progress made in livestock farming throughout a lengthy period of British history! Then came the 18th Century, the Agricultural Revolution, and characters such as Robert Bakewell of Dishley (1725-95).

Considered by many to be the "founding father" of pedigree livestock breeding in the UK, Bakewell abandoned the rather haphazard breeding practises common at that time, and carefully separated his male and female stock. Matings were only allowed to take place between animals he considered to be of superior type, and, in an extremely forward thinking way (made possible only by meticulous record keeping), he carried out careful evaluation of the progeny of each union. Using the data he collected he was able to breed for specific characteristics, even to the extent that close relatives would be bred together in order to consolidate the improvement. Principally working with Shire horses, Longhorn cattle and Leicester sheep, the latter were his greatest success – the Dishley Leicesters were used in the development of almost all the local types of longwool sheep, including the Romney, the Lincoln and the Border Leicester, and went on to achieve international fame.

After Bakewell, George Culley developed the Border Leicester, the Collings brothers used their local Teeswater cattle to create the Shorthorn, Watson & McCrombie pioneered the Aberdeen-Angus and the Tomkins family, working closely with other breeders, established the Hereford as an ideal beef animal. Some of the basic principles of what we today call "performance recording" had been born and, within a very short space of time, Britain had come to be regarded as the livestock centre of the world.

Now, fast forward to the 21st Century...

We live in fickle times. Fashions and tastes change in a twinkling, and, for livestock breeders, it's nigh on impossible to keep up with current consumer trends using traditional selection methods. And margins are tight. Farmers must constantly aim to reduce costs, whilst increasing the output per animal. There's also an historical and sentimental issue at stake here – we have a tremendous diversity of breeds and types of livestock in the British Isles, they are part of our heritage, and we don't want to lose them. So, how can we combine all of these factors – traditional breed characteristics, modern tastes, higher output and reduced costs – into one breeding program? The answer, I believe, lies in performance recording.

Purchasing a ram by eye alone is a risky business. If you buy the smartest looking shearling on offer, you may well strike lucky and end up with the animal with the best breeding potential, but, 9 times out of 10, you will actually be buying the oldest, best fed, single born ram at the sale. None of these factors will have a positive influence on his offspring's performance. In fact quite the opposite may be the case – it has been shown that rams from multiple births sire more prolific female offspring, so by purchasing a single born ram you may in fact be selecting for lower lambing percentages!

When you make a visual assessment of potential breeding stock, much of what you see is a factor of management, such as feeding – it is often said that "half the breeding lies in the feeding", meaning that the so-called "well bred" animals, that win all the prizes in the show ring, may better be described as "well fed!" When exposed to a more rigorous regime, such as earning their keep as breeding stock, these individuals often disappoint.

Well, help is at hand. Using the technology and software now available, combined with better understanding of genetics and heritability, it is possible to "see through" the effects of management, and accurately estimate the true breeding potential of an animal, leading to relatively rapid improvements in flock performance.

Signet Sheepbreeder

Performance recording in the UK is carried out by Signet Breeding Services, a bureau based department of the AHDB (Agriculture and Horticulture Development board, formerly the Meat & Livestock Commission (MLC)). In its early days, performance recording was largely concerned with muscle depth, and it was only possible to make within flock assessments. However, things have moved on, and evaluations are now carried out on a range of economically important characteristics, and combined into a single EBV (Estimated Breeding Value) for each recorded animal. Furthermore, the use of BLUP (Best Linear Unbiased Predictor – the statistical procedure

that disentangles the effects of management from the influence of genetics) enables comparisons to be made between different flocks of the same breed. Societies or local breeding groups can now work together, with common goals, for the overall improvement of their chosen breed, ensuring that that breed has a place in the future of British agriculture.

What Characteristics Are We Looking At?

Seven Estimated Breeding Values are calculated for each animal, with some additional EBVs introduced for assessment of breeds where these traits are relevant. In order to calculate EBVs for each animal, BLUP uses all the available performance data on that animal, together with data from any identified relatives. It also takes account of the degree of correlation between traits, and their heritability.

Growth And Carcass Traits

• 8 week weight, in kg.
• Scan weight – measured in kg at 21 weeks of age.
• Muscle depth – measured in mm by ultrasonic scanning at 21 weeks of age.
• Fat depth – measured in mm by ultrasonic scanning at 21 weeks of age.

Female traits

• Mature size – pre-tupping gimmer weight, in kg.
• Litter size – number of lambs born.
• Maternal ability – total weight of lambs reared, in kg.

Additional EBVs

• Carcass Lean Weight, Carcass Fat Weight and

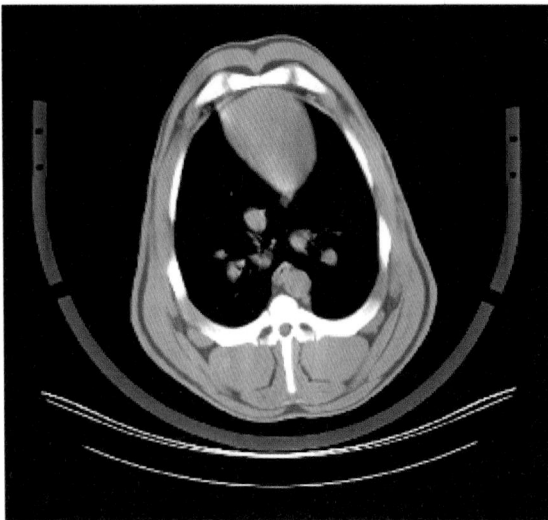

CT scan image

Muscularity – assessed using CT image analysis of live animal.

• Faecal Egg Count EBV.

This is a relatively new selection criteria, but potentially the most significant of the lot. The rise in the presence of anthelmintic resistance in parasite populations is cause for considerable concern. Equally concerning, perhaps, is the fact that we have to turn to ever more potent synthetic products to control these parasites. Well, the solution to the problem may actually lie in the genetics of the sheep themselves. Calculated by counting the number of nematode eggs present in dung samples taken from lambs challenged by parasite infection, the FEC EBV will identify sheep whose genetic make up confers a degree of natural resistance to *Nematodirus battus*. The ability to selectively breed sheep with this natural immunity is great news, not only for sheep keepers who have experienced anthelmintic resistance in their flocks, but also for organic farmers, and anyone else wishing to cut down on the use of expensive chemical drenches.

It is acknowledged that not all of these traits are of equal importance in different breeds,

or under different management regimes, so Breeding Indexes have been designed to help flock owners improve several traits at once, in a way that is most relevant to their own breeding objectives:

Terminal Sire Index

(e.g., Suffolk, Hampshire Down)

For the selection of superior terminal sires. This index is designed to enable breeders to select for an increase in lean meat in the carcass, without any associated rise in fatness. The index is calculated using weight data, and ultrasonic measurements of muscle and fat depth. In recent years Computed Tomography (CT) has also been used, to accurately measure the potential yield of lean meat and fat in the carcass. Being able to carry out these assessments on live animals (rather than on carcasses at the abattoir) enables the best individuals to be retained for breeding.

Maternal Index

(e.g., Lleyn, Poll Dorset)

Designed to increase lamb survival rates and early growth, by improving maternal ability, this index is calculated using the individual EBVs for litter size, 8 week weight, mature size and maternal ability. The maternal index is particularly relevant in flocks where home-bred females are kept as replacements.

Longwool Index

(e.g., Bluefaced Leicester, Border Leicester)

The longwool index acknowledges the very specific role of the longwool breeds as crossing sires, in the production of popular commercial ewes such as the Mule. Calculated using EBVs for muscle depth, fat depth, scan weight and litter size, high index longwool rams will sire ewes of an appropriate mature size, that produce lambs with good growth rates and superior carcass conformation. Prolificacy is maintained at current levels.

Hill-2 Index

(e.g., Cheviot, Blackface, Swaledale)
Use of this index when selecting female replacements in hill breeds will lead to an increase in ewe mature weight, maternal ability, longevity and the number of lambs reared to weaning. Lamb growth rates and survivability are the most important factors here. Some breeders have expressed concern that increasing the mature size of the ewes may lead to higher costs associated with their maintenance on sparse mountain grazing, however they are able to select animals with low mature size EBVs to counteract this problem.

Welsh Index

(e.g., Beulah, Welsh Mountain)
The Welsh Index identifies hill sheep with superior breeding potential for maternal ability, lamb growth and carcass quality.

The index encompasses two breeding targets, which are weighted equally:
• Maternal Ability – assessed through the performance of a ewe's lambs up to eight weeks of age, and represented by the Maternal EBV.
• The lamb's own potential for growth and carcass composition – this is determined by scan measurements of fat and muscle depth, together with weight at scanning. These are similar criteria to those used in calculating the Terminal Sire Index.

HOW DOES THE SYSTEM WORK?

Any flock owner who keeps purebred sheep, and keeps accurate records of their breeding (even if not technically "pedigree"), will already have a lot of the information needed to set up a "within flock" recording project. Even better would be to set up a breeding group or sire reference scheme in conjunction with the relevant breed society, or with other like-minded local breeders. It may well be that there is already a scheme for your chosen breed – contact Signet to find out, and to request a membership application pack. If you are hoping to set up a new project, discuss your ideas with Signet and with other breeders who may be interested in forming a group. Also check out what funding is available. As with everything, there is a cost involved, but you may find that grants are available. Here in Wales, performance recording is financially supported by HCC (Hybu Cig Cymru / Meat Promotion Wales).

Together with a membership application form, Signet will send you a number of fairly simple forms to fill in with all your flock details. If you keep accurate records, it should be a relatively simple matter of transferring data from your own private flock book. The more information that you are able to provide, the more accurate the resulting figures will be. This is particularly pertinent where several flockowners are setting up a scheme together – ideally you need a degree of "connectedness" between the flocks. Full pedigree details will enable the links to be identified. Connectedness can be increased by the use of a "reference sire" – a ram that is used on a number of ewes in each flock of the group. The presence of his offspring in each flock provides a baseline against which the performance of other rams in the group can be assessed.

Throughout the year it is down to you, the flockmaster, to collect the relevant data relating to lambing, birth weights, 8 week weights, etc, and complete the necessary paperwork. It is not onerous, and what's more, Signet will provide you with a dinky little pocket-sized notebook for all lambing records and so on, so you don't have to take a load of paperwork out in the wind and rain!

A date will be arranged for a technician to come out to your farm to do the necessary back fat scanning, when the lambs are about 21 weeks old, and, following that, all the data collected during that season is submitted to Signet for analysis – EBVs for each trait (together with an overall index) will be calculated for every animal in the flock or group.

The resulting report makes fascinating reading. Your best sheep will still be your best sheep, but you'll probably be startled by the range of the results! In our first year of performance recording, our flock average index was 98, with the best animal having an index of 204, and the poorest a measly minus 12! Initially the degree of accuracy of the results will be relatively low, but as more information is fed into the system each season, accuracy will improve. Having said that, even in the first year, the figures provided can be used to help plan out your breeding schedule, choose replacement ewe lambs, identify high index ram lambs as potential stock sires, and decide which ewes should be culled.

Ultrasound scanning of Hampshire Down lambs, to measure backfat and muscle depth

Is Performance Recording Relevant To Smallholders?

In a word, yes.

Despite the fact that it all sounds a bit complex, there are a number of very good reasons why this technology should be embraced by smallholders who keep purebred sheep:

• Improving performance will lead to better returns from a limited acreage.

• The UK sheep industry depends, to a large extent, on quality purebred rams being used on large flocks of crossbred or hill ewes. Smallholders with small flocks of pedigree sheep are often the ones at the top of the tree, producing the purebred tups for crossing.

A Performance Recording Success Story

In 1998, the Hampshire Down Sheep Breeders' Association was at something of a cross roads in the history of the breed – they had either to fight to regain their place as one of the best terminal sire breeds, or surrender their commercial dreams and head to the hobbyist, and perhaps the "rare breed" tag. The society's Publicity & Marketing Committee, and its Breed Development Committee, met to determine the best way forward. They knew already that the Hampshire Down had improved a great deal from the rather "dumpy" version of the '60s and '70s, and it was decided that the best way to demonstrate this to the commercial ram buyer was to use Signet's independent analysis of the breed's leading flocks. A Sire Reference project was soon established, with the criteria being the standard Signet terminal sire parameters of growth rate, muscle depth and fat cover. This was combined with the basic selection criteria of breed type and minimum standards. A recent development has been the use of CT scanning of the leading ram lambs, giving an even greater degree of accuracy to the prediction of breeding potentials. None of this has been detrimental to the breed's continuing success in the show ring, as breeders have taken great care over their selection of breeding stock.

The overall improvement in the breed's performance as a direct result of this project has been quite outstanding – average index has risen from around 100 in 1998, to over 180 today, with Hampshire lambs weighing, at 19 weeks of age, an average 5.27 kg above the average at the start of the scheme, with 2.13 mm more muscle depth.

The Hampshire Down has firmly re-established itself as a market leading terminal sire breed, without losing any of its traditional native charm

• Marketing of traditional breeds – the traditional sheep breeds tend to be very popular with smallholders, and claims are often made relating to the "superior" qualities of these breeds. However, these claims are largely anecdotal and unsubstantiated. Performance recording could provide the facts and the figures to support these claims.

• Preservation and development of rare and minority breeds – Most of the interesting breeds now classified as rare owe their scarcity to the fact that they have been uncompetitive in the modern market place. This problem has been further exacerbated, particularly in the case of coloured or marked breeds, by over selection for specific characteristics (such as facial markings) that are relevant only to the show ring. However, one of the beauties of performance recording is that the selection for economically significant traits is based on "invisible" factors, and need not be detrimental to the characteristic appearance of the breed. Personally, I feel that it would be a wonderful thing if the breed societies of some of the rarer British sheep could encourage their members to set up performance recording projects. Who knows? They may be sitting on a hidden asset! We might all be pleasantly surprised by the results…

The Shepherd's Year
Winter

DECEMBER / JANUARY

By the beginning of December tupping will be complete. If flushing and teasing were carried out correctly the majority of the ewes will be at about six weeks gestation. The rams have worked hard, and will be looking rather ragged and footsore by now, so should be separated from the ewes for a well earned rest. Make a note of the date (always December 2nd in our case) so that you can calculate when the lambing period will end. This is very handy to know, as it enables you to plan ahead for other spring tasks on the smallholding secure in the knowledge that you won't be called away to attend to a lambing ewe! Examine the rams carefully for sore patches when removing the raddle harnesses, and treat any chafed areas immediately with an antiseptic spray or ointment. Clean up the harnesses or they'll be hard and crusty when you come to use them next year. Part used crayons can be kept for future use, but make sure you store them in a rodent-proof place, or you'll be finding multicoloured mouse droppings all over the place. (My sheepdogs once ate a pile of green raddle crayons, with quite alarming consequences...!) Check the rams' feet, trim where necessary, and either spray each hoof or stand them in a footbath. Fluke dose if required. Any very thin rams will need some supplementary concentrate feeding from now until the spring, but if they've held their condition then a bit of hay every day may well be sufficient to see them through the winter.

EWES IN DECEMBER

On the day that the rams are removed it's a good idea to go through the ewes and make more permanent marks than those left by the raddle crayons – the crayon mark alone may become almost undetectable by the time you need to sort the flock into groups for lambing. A neat spot of marking fluid (commonly known as "pitch", as that is what shepherds used in years gone by) of a corresponding colour should be applied to the animal's rump with a small stick about the thickness of a pencil. Push the stick into the fleece, twist, and withdraw to leave a tidy mark that will last 'til lambing time. Similarly make a small mark on the ewe's shoulder according to which ram she ran with, if it is necessary to record the parentage of lambs born. We also put a mark on the back of the neck of all our shearling ewes at this stage, making them very easy to spot at lambing time. This is extremely useful when we're running our lambing course, as I'm able to point out to students the different behavioural characteristics of the first time lambers, and teach appropriate handling techniques for dealing with these young mums.

The ewes will probably need to be fluke dosed at the beginning of December – there is no need to use a combination product, a fluke only drench will be fine, but do check that it's OK for use in early pregnancy.

Once the rams are out, and you've done all the marking and dosing that's required, the ewes can pretty much be left to their own devices through December. The developing foetus is minute at this stage, and is placing no significant demand on the ewe's metabolism, so supplementary feeding of concentrates or hay shouldn't be required yet. The flock can be used to "scavenge" around the holding, eating down untidy headlands, clearing unkempt corners, and levelling out pastures that have held cattle or horses during the summer. Remember to concentrate grazing onto the fields furthest from the homestead, as you'll need the paddocks closest to the farm buildings at lambing time. Only if grazing is in really short supply, or if there are significant early snowfalls, will it be necessary to introduce extra rations in December to ewes expected to lamb from mid-March. If feed is introduced, then keep it to a minimum – overfeeding now may lead to problems later on. (Clearly any very thin ewes will need to be treated differently – use your judgement here).

THE FLOCK IN JANUARY

Early January can be another fairly quiet period in terms of the amount of attention the flock requires, but by the middle of the month various factors are beginning to influence management and, even if no immediate action is required, we need to be very aware of the fact that from now on, the growing lambs will be placing a rapidly increasing demand on the metabolism of the ewe.

Body Condition And Nutrition

It is well worth regularly monitoring body condition score throughout pregnancy. There are times when some loss of condition is acceptable, and times when it is not. There are times when it is inevitable, unavoidable, and even useful!

During early pregnancy, i.e., throughout the whole of the period that the rams were running with the ewes, body condition should have been maintained to avoid embryo losses. Through December and early January there is very little taking place in terms of foetal growth (by mid-January, developing lambs will have reached only about 15% of their eventual birth weight), however this is the critical period for placental development – poor placental development may result in non-viable lambs, regardless of the feeding strategy later in

pregnancy. Although in most cases it will not be necessary to provide additional feed during the period mid-December to mid-January for ewes due to lamb from mid-March, nutritional stress (i.e. sudden or prolonged reductions to the plane of nutrition) must be avoided. On the whole, ewes can safely lose ½ to ¾ of a condition score (depending on breed) during this period, provided that it is a steady loss, and that they were in good condition to begin with. If a significant portion of the flock is seen to be losing condition, or if the loss is rapid, then provide a low level of supplementation in order to stabilise the situation, such as high nutrient density feed blocks, especially formulated to optimise forage utilisation. Hay or silage should only be required if there are lengthy periods of snow cover, or if pasture is completely bare, or waterlogged. Don't be in too much of a hurry to introduce extra feed – once you start, you'll have to carry on right through to lambing time. From mid-January, however, the situation begins to change. At this stage, 90 days of gestation, placental development is complete, and the growth rate of the foetus begins to accelerate rapidly. This is when we scan the ewes, in order to determine the number of lambs that they are carrying, and adjust the management accordingly.

Pregnancy Scanning

Ultrasound scanning of in-lamb ewes is well worth carrying out, even in small flocks. It's not expensive (43p per ewe in January 2009), and the cost will be almost immediately covered by savings on feed. However, it is at lambing time that the benefit will really be seen. In flocks where scanning is not carried out, feeding has to be based on guesswork, and is usually a compromise. This means that single bearing ewes may be overfed, leading to big lambs and

Pregnancy scanning trailer. The technician is seated inside on the right (out of sight). Sheep are held in the race on the left for scanning

A ewe looking flukey, in desperate need of dosing

difficult births, and twin bearing ewes don't get enough to eat, so their lambs may be small and weak, and they won't produce much colostrum. Pregnancy scanning solves this problem, as all ewes can be fed according to the number of lambs they are carrying, and there's no need to waste feed on barreners, either! Scanning is most accurate (98 – 99%) if carried out 90 days after the date that the rams were turned in. It can be done earlier, but any later than this and accuracy is significantly reduced.

Ewes must be handled with care when presented for scanning. The technician will have some sort of race, crate or yoke to ensure that each animal stands correctly whilst he applies the transducer and interprets the image, but you must have adequate handling facilities to ensure that sheep move smoothly through the system. The contractor who carries out our scanning has designed and built an ideal set up on a trailer, complete with an adjustable width race, and gates operated by pedals, allowing him to hold and release the sheep without assistance. He sits in a little cubby-hole, with just his hand protruding holding the transducer, and he can apply an appropriate paint mark to the ewe's backs via a small slot at

the right height! With this arrangement it takes less than an hour to scan 200 ewes, with no stress.

Directly after scanning, separate the flock into groups according to the number of lambs they are carrying. Singles can continue on the same grazing / feeding regime that they've been on, for the time being. Move twins to a better place, and perhaps up their level of nutrition slightly, by introducing high energy licks (for hill breeds), and / or ¼ of a pound per head per day of concentrate feed (for heavier breeds). Get racks out ready to start feeding hay or silage (if you haven't done this already). Triplets can be run with the twins for now. Continue to monitor body condition score – this is easy to do if your sheep are tame, as you'll be able to put your hand on their backs in the field. There's no need, yet, to further subdivide the flock according to when they are due to lamb, unless you've got any particularly early or late lambs expected. If you didn't fluke dose earlier in the winter (or if ewes are looking flukey – see photo) then do it now, while you're sorting the flock. Barren ewes can be dealt with as you see fit. Many people sell them off at this stage, but we don't. I like to give them another chance. They don't need any extra feeding, so can be run with the ewe lambs, or any later lambing ewes.

THE SPRING LAMBING FLOCK IN FEBRUARY.

Feeding

By the beginning of February the pace is hotting up! The rapidly approaching lambing season is foremost in our minds, and the requirements of the flock in late pregnancy dictate all management decisions. It's a critical time – 70% of foetal growth occurs during the final 6 weeks of gestation, placing considerable strain on the metabolism of the ewe. When managing the flock through this important period there are, to my mind, 3 principal factors, of equal importance, that must be taken into account, (together with other lesser concerns):
1. The demands of the rapidly developing foetus.
2. The health of the ewe.
3. Udder development.

Sorting The Flock

• Separate off ewes carrying triplets, for preferential treatment and closer shepherding.
• Any very thin twin bearing ewes (brokers etc.) can run with the triplets, for a bit of extra TLC.
• Ewes that took the ram on the final raddle colour can be put aside, as they are still some way off lambing, and don't require such close attention. If you're short of fields for all the separate groups, then run them with the barreners and / or ewe lambs. (We turn ours back to the mountain for a couple of weeks).

Sorting ewes according to scan results and raddle colour

This leaves the bulk of the flock to be sorted thus:
• Twin bearing ewes on the first raddle colour. This is the principal group at this stage, in management terms. In a prolific lowland flock it will also be the largest group.
• Single bearing ewes on the first raddle colour. This will be the largest group in hill and primitive breed flocks.
• Twin bearing ewes on the second raddle colour. Sort these according to condition score. Leaner ewes can be put with the earlier lambing twins. Fatter ewes can run with the earlier lambing singles, for now.
• Single bearing ewes on the second raddle colour. It can be a bit of a job to know what to do with this group. If flushing was effective, and if the teasers did their job properly, there should only be a few, in which case run them with the first group of singles until housing.

Now, clearly it won't be practical on all

A Brief Note On Lambing Percentage

It is not possible to calculate lambing percentage from the scanning results. To attempt to do so is mere speculation, and far worse than counting chickens before they hatch! The pregnancy scanning will tell you how many lambs your flock is carrying on that day only. This figure can be used to calculate prolificacy. Prolificacy is the ratio between the number of ewes in-lamb and the total number of lambs carried. It makes no allowance for barren ewes or subsequent losses, so gives an idea of the maximum potential output of the flock, in a good year.

Lambing percentage is the ratio between the number of ewes put to the ram and the number of lambs reared, so takes full account of barren ewes, abortions, still births and losses of lambs after birth. We calculate this at the beginning of May, on the day we turn the flock to the mountain, though some people take things one stage further and count lambs at weaning, or even at the time they are sold.

The aim of every shepherd must be to ensure that the two figures – prolificacy and lambing percentage – are as close as can be.

holdings to sort the flock as I've described here, particularly in small flocks or where the number of available fields is limited. However my suggestions will at least give you a good idea as to where your priorities (in terms of the level of shepherding required) should lie. Split the flock up as best you can, in order to make optimum use of available feed, forage and time.

Concentrate Feeding – Stepped Or Flat Rate?

Due to the rapidly increasing weight (and therefore increasing nutritional requirement) of the developing lambs, it has long been the practice to gradually increase the amount of concentrates fed to ewes in late pregnancy, in a series of "steps", in line with foetal growth. However, this method of feeding makes no allowance for the fact that the lambs take up considerable space inside the ewe, gradually reducing rumen capacity and suppressing appetite. It may become physically impossible

for the ewe to consume the "correct" quantity of a carefully calculated ration during the final week or so of gestation, and, as the level of feed has been stepped in line with requirements, she has had no opportunity to accumulate body reserves early on that she can draw on now. The alternative is to feed at a "flat rate" throughout the final six weeks of pregnancy, so initially, whilst the lambs are still small and the ewe's appetite is good, you are providing a higher level of nutrition than is actually required. This enables the ewe to maintain good body condition, which in turn gives her something to utilise when nutritional requirements rise above the level of consumption. The total amount fed per ewe, over the six week period, remains roughly the same.

Either way, there is no getting away from the fact that an energy deficit in late pregnancy is a very serious thing. It is the pre-disposing factor in the onset of pregnancy toxaemia (twin lamb disease), which usually results in death. Even the fittest, healthiest, well cared for ewes may be on a knife edge, metabolically speaking. The

Feed should always be placed in troughs outdoors

key to efficient mobilisation of body reserves to satisfy the energy deficit lies in the level of protein in the diet, so, rather than trying to increase the total volume of concentrates consumed each day, consider switching to a higher protein feed during the final week or two, or provide high DM silage instead of hay. A rise in the level of protein in the diet at this stage will also satisfy the requirements of udder development. Any change should be gradual, to avoid a check in intake. The energy deficit can be further offset by housing ewes for lambing, as the ewe's own energy requirements will be considerably reduced.

There appears to be considerable variation between breeds in terms of their ability to safely utilise their own body reserves in order to satisfy the needs of the developing lambs. It is possible, in some cases, with careful monitoring of condition score, to actually incorporate a predicted level of utilisation into the overall strategy. The ultimate performer in this field must surely be the Welsh Mountain. Bred over many generations to lamb unsupervised on sparse upland grazings, the Welsh ewe can lose a massive ⅓ of her own body weight during late pregnancy, with seemingly no ill effects, and still produce a lamb that's well grown at birth. Don't try putting that much pressure on any other breed of ewe – she'll die!

How Much Concentrate Should We Feed?

In practice, it is probably most appropriate to devise a feeding strategy somewhere between the two extremes of flat rate and stepped, comprising of a couple of long, shallow steps. It is as well to be aware that initially, particularly in hill and primitive breeds, ewes may be shy about coming to the trough. It's no good taking enough feed to the field for all the ewes, and having only half of them eat it! When the flighty ones do eventually join in, the bolder characters (who've been enjoying more than their fair share of the grub) will suffer a sudden drop in nutrition, which is just what mustn't happen. In order to avoid this scenario we generally begin taking small quantities of feed (not more than ¼ lb per head) out to the ewes quite early on, and counting how many come for it. We adjust the amount accordingly, and gradually, over a period of a week or so, by the use of this "tit bit" level of feeding, we encourage all the ewes to come up to the troughs when called. Feeding proper begins, for us, during the first week of February, some six weeks before lambing. As we have sorted the flock according to their due dates we are able to start feeding each group at the most appropriate time, and at the appropriate level.

Feed should always be placed in troughs

outdoors, not sprinkled in a line on the field. This is quite the opposite from feeding indoors, where it is advantageous to broadcast compound pellets onto the straw bedding (though this is clearly not possible when using home mixed rations or coarse blends). As soon as the sheep have finished eating, the troughs should be turned upside down, to prevent contamination by birds and to avoid the risk of a ewe getting cast on her back in a trough. Roll them over the same way each day, so that the line of troughs gradually moves down the field, minimising the damage caused by poaching. Don't place the troughs at the entrance to the field, or the ewes will always be hanging around the gateway, making access very difficult, particularly when carrying a heavy bag of feed. The same applies to hay racks and ring feeders – place them away from the gate, and either move regularly, or have a designated "sacrifice" area.

TIP! Choosing the best compound feed for your ewes is confusing, with so many brands available, each with a slightly different analysis. It can be difficult for the uninitiated to decipher the information given on feed bag labels. However, the better quality feeds can be identified by their lower levels of fibre and ash – if these two figures added together come to more than 20 then it's likely to be a pretty poor feed for in-lamb ewes. Having said that, when you find a feed that your sheep like, stick with it, even if the analysis isn't the best – there's nothing worse than splashing out on a more expensive "better" feed, only to have the ewes turn their noses up!

Feed Requirements – A Rough Guide

It is not possible to give precise instructions here, as there are so many variables that must be taken into consideration, such as forage quality, type of feed, breed of sheep and so on. However, the following information should help, based on my own "semi- stepped" regime. NB. These figures are based on my own experience, and are for guidance purposes only.

	Compound feed requirements for ewes in late gestation. Kg per head per day					
	Light breeds <50 kg (Shetland, Portland, Hebridean, etc.)		Medium 50 - 70 kg (Lleyn, Shropshire, Dorset Horn etc.)		Heavy breeds >70 kg (Oxford Down, Suffolk, Wensleydale, etc.)	
	SINGLE	TWIN	SINGLE	TWIN	SINGLE	TWIN
6 weeks before lambing	0.25	0.30	0.35	0.45	0.50	0.75
2 weeks before lambing	0.35	0.45	0.45	0.60	0.75	1.00
Feeding	When increasing the ration, split the total amount into two equal feeds per day					
Forage	Good quality hay or high dry matter silage fed ad-lib throughout					

Hay and silage should be fed in racks or ring feeders, not spread on the ground

Hay And Silage

Exactly when supplementary forage is introduced will vary considerably from year to year, and between breeds and flocks. Certainly no later than the beginning of February the twin bearing ewes on the first raddle colour should have access to good quality hay or silage (fed in racks or ring feeders, NOT spread on the ground), with the same being introduced to the remainder of the flock shortly afterwards. This should be available more or less ad-lib, but not to the extent that there is wastage – the ewes should be expected to clean up all that they've been given, but not allowed to go hungry. Don't fluff up hay in racks, always feed in wads or even whole bales if the racks are big enough (simply cut and remove the strings). This prevents selective feeding and reduces wastage. Sheep will generally consume 1–1.5kg of hay per day (or roughly the equivalent amount of dry matter, if silage is fed), though this is very variable. The weight of individual bales also varies considerably, as does the length of time over which supplementary feeding will be required – it may be as long as 10 weeks (or more) if weather conditions are severe or

lambing is drawn out. As a rough guide, 30 ewes will eat more than one, but less than two small bales of hay per day. When buying in hay, allow 3 – 4 bales per ewe per year. That should be ample. When feeding silage, we allow 2 big bales per 100 ewes each week. Under certain management regimes it is possible to feed good quality, leafy barley straw in place of hay or silage, with the uneaten portion going under the sheep as bedding.

I've discussed, in some detail, the relative merits of both hay and silage on page 244 (and following), so there's no need to repeat it all here. Suffice to say that high dry matter silage is a far superior feedstuff (see table on page 250), and in many cases is more convenient both in storage and use. Small bale hay, on the other hand, is very practical when feeding smaller numbers of sheep, particularly when housed or in individual pens after lambing. Both hay and silage have potential disease risks associated with their use (Listeriosis in the case of silage contaminated with soil, and Toxoplasmosis in the case of hay bales contaminated with cat faeces), with neither one nor the other being any better or worse in this respect.

Clostridial diseases

These diseases are caused by a group of bacteria called Clostridia which are commonly found in soil. Healthy animals also harbour Clostridia, with disease outbreaks being triggered by various causes. In the event of an outbreak the bacteria multiply very rapidly, producing toxins which are invariably fatal. In years gone by these diseases were responsible for huge levels of loss, both in young lambs and in adult sheep. I should think that the flockmasters of old lived in dread of braxy, blackleg, tetanus, black disease, struck, lamb dysentery and pulpy kidney. We are extremely fortunate nowadays in having inexpensive vaccines that provide effective protection against all the major clostridial diseases, with the added benefit that vaccinated ewes pass on a temporary immunity to their offspring, until such time that the lambs themselves are old enough to be given a course of injections. As a result, clostridial diseases are something that the conscientious modern shepherd never encounters. Occasionally one may hear stories about outbreaks attributed to "product failure", but closer inspection usually reveals improper vaccination procedure to be the likely cause.

Vaccinations

At some point during February it will be necessary to give the ewes their annual booster against the clostridial diseases. A simple subcutaeneous injection of 2ml, given 4 – 6 weeks prior to lambing, ensures that sufficient antibodies are present in the colostrum of the ewe, providing protection for her lambs early in life. Any previously unvaccinated sheep will require two injections, 4 – 6 weeks apart, with the second injection being given 4 – 6 weeks before lambing. Having said that, unless you've recently purchased in-lamb ewes, there shouldn't be any previously unvaccinated sheep in the flock at this stage.

Remember to give a booster to the rams, and any non-breeding sheep on the holding too.

Indoor Or Outdoor Lambing?

For early lambing flocks (December - January) there is no question that they'll need housing at lambing time. Similarly, there is no questioning the fact that late (May) lambing "easy-care" flocks will not. However, in spring lambing flocks, the situation is not so clear cut. Outdoor lambing can be fantastic when the weather's good – there's nothing to beat being amongst the sheep in the lambing field on a calm and dewy spring morning, watching the sun rise, and instinctively knowing that all is well with the flock. At times like that you wonder why anyone brings their sheep in! There's less opportunity for disease build up outdoors (you'll rarely see scours or joint ill in lambs born outside), and less chance of mis-mothering, too. But that's when things are going well. Unfortunately it isn't always like that. I've lambed ewes outdoors in torrential rain, and driving snow, and sworn I'll not do it again. I've seen lambs, only partially born, with tongues pecked out by gulls and crows. I've seen a newly lambed ewe distraught at the loss of her only offspring, carried off by a fox under cover of darkness, before he'd even taken his first tottering steps in the direction of her udder. And I've seen lambs frozen to the ground, or simply washed away in a downpour.

We need to be realistic here. Balance the potential losses outdoors due to adverse conditions, against the costs of housing. Consider the potential disease risks indoors, yet also consider the ease and speed with which you can tackle a problem, if the animal concerned is already penned up. The list of pros and cons goes on and on, and there's no one answer to the conundrum, no simple solution to suit every situation. However, factor "shepherd comfort" into the equation and housing begins to look much more attractive, regardless of the cost!

Despite the fact that there is currently a move towards low cost outdoor sheep systems, I feel that, on balance, the smallholder would be well advised to house the flock for lambing. It gives the land a rest, for a start – an important consideration on a limited acreage. It'll simplify husbandry for any one trying to juggle shepherding with other commitments – no more trudging round the fields with a torch

Hypocalcaemia (Calcium deficiency)

The last few weeks of pregnancy place a considerable strain on the metabolism of the ewe. The demands of the rapidly growing foetus, together with the production of colostrum, require significant quantities of calcium. Usually the requirement is satisfied by calcium obtained from the diet and from mobilisation of skeletal reserves, but, during a brief period of stress such as bad weather, nutritional inconsistency, or gathering the flock, the ewe's body system is unable to cope. This leads to a rapid fall in blood calcium levels, and clinical cases of calcium deficiency will occur. An affected ewe will generally be found lying with her forelegs tucked under her in the normal fashion, but with her back legs partially extended behind, as if she's toppled over forwards. Her nose will be close to, or touching, the ground, and she'll probably be dribbling a clear fluid. Breathing may be scarcely perceptible, and, as the ewe's condition deteriorates further, no response will be obtained if you wave your hand in front of her eyes. Her ears will feel very cold. The good news is that even ewes that are pretty far gone will usually respond rapidly to treatment (see page 88). Typically calcium deficiency is seen in the last couple of weeks of gestation, but occasionally occurs much earlier, in which case affected ewes seem to become "calcium junkies", requiring regular repeat treatments right up until the time that their lambs are born.

If the diet in mid-pregnancy contains particularly high levels of calcium (i.e., when it's not really needed) this appears to muck up the ewe's ability to regulate the release of skeletal reserves at a later stage. The key to avoiding calcium deficiency therefore lies in stress-free handling, with any supplementary feeding being restricted to late pregnancy.

after work. It gives greater scope for subdividing the flock according to their nutritional and management requirements. It makes it possible to supervise the flock through the night, and gives a real opportunity to reduce mortality rates, which, on average, are much too high. For the novice shepherd it can take away a lot of the anxiety about dealing with complications – no need to pursue the poor animal around the field, just quietly catch her up in a corner of the shed, with no stress to man or beast. Above all, it'll make it a pleasure to be with the sheep, whatever the weather, and at any hour of the day or night.

Housing

By early March many of the ewes are looking decidedly heavy, and some will be showing reasonable udders – they are beginning to "bag up" as we say. Shepherding needs to be stepped up, for we're fast approaching the critical point at which feed intake alone may be insufficient for the requirements of the growing lambs, and ewes will begin to draw on their own body reserves. Increase the level of feed, and divide the total amount into at least two small feeds per day in order to stimulate the ewes' declining appetite. Be on the alert for any ewe that's not looking her usual perky self, and act swiftly

Pregnancy Toxaemia (Twin lamb disease)

Pregnancy toxaemia is caused by an energy deficit in late pregnancy, and, as the name suggests, is more common in ewes carrying multiple lambs. The symptoms seen are a result of low blood sugar levels and ketone toxicity. In an attempt to correct the energy deficit ewes will begin to utilise body reserves, and ketones are a by-product of excessive tissue breakdown. Affected ewes will initially appear to be quite excitable, and, if disturbed, will run off rather erratically in a curious high-stepping gait, with their heads held up. As the condition progresses they'll go down, and are usually found with forelegs stretched out in front as if they've slumped backwards. The presence of ketones produces a characteristic smell of pear-drops on the breath (although not everyone is able to detect this). The energy deficit is further exacerbated by the fact that affected ewes stop eating. Once these symptoms become apparent the outlook is pretty poor. You can drench with proprietary high energy drinks, but better results may be obtained by using a mixture of milk, molasses and egg yolk. Ewes that abort their lambs have a better chance of recovery, as the foetal load is removed. Digestion may have to be kick-started using rumen contents drawn from a healthy ewe. A ewe with a jaded appetite can often be persuaded to eat by offering sprigs of tamarisk or ivy.

Prevention is undoubtedly better than cure! If cases of twin lamb disease occur this should be seen as a "whole flock" problem, and management adjusted accordingly. Pregnancy scanning is a particularly useful aid to ensuring that all ewes in the flock receive the correct level of nutrition. Although primarily seen in flocks that are undernourished, pregnancy toxaemia can also affect ewes that are over fed, due to decreasing rumen capacity – and hence appetite / intake – in late pregnancy.

Contented ewes on a clean straw bed

it's settled, crisp and dry, we leave them out as long as we can, but prolonged periods of wet weather will have the ewes looking hunched up and miserable and losing condition rapidly. They really are better off indoors, I think, and the collective sigh of pleasure from the newly housed ewes as they gratefully subside into a deep, dry, straw bed, tells me that they agree!

The worst of both worlds: Cobwebs and condensation in a traditional building re-roofed with corrugated iron

to forestall the onset of metabolic disorders. Don't wait "to see if she looks a bit better in the morning" – administer calcium borogluconate by subcutaneous injection (80ml split between 4 sites) as a matter of course to any ewe that looks off colour at this stage, *then* worry about whether your diagnosis is correct! Ewes suffering from calcium deficiency often respond with startling rapidity to this treatment, and may be up and away within 20 minutes, but pregnancy toxaemia (twin lamb disease) is more serious, and affected ewes will require careful nursing and further treatment.

Given that there is a need to have the ewes under much closer observation at this stage, we feel that the beginning of March is an appropriate time to house the flock, allowing them at least a week to settle down before the first lambs are due (although in practice there are always a few early arrivals that put in an appearance almost immediately after housing, and catch us on the hop!). Some years we house the twin bearing ewes slightly earlier, depending on how they are looking – occasionally we've brought them in as much as four weeks before lambing, and sometimes they've stayed out right up until four days before the first lambs were due!

We base our judgement on the body condition of the ewes, and on the weather – if

The Sheep Shed

Unfortunately, traditional outbuildings are notoriously bad places for housing livestock. Many were designed and built at a time when it was believed that animals should be kept warm and "cwtchy", with the result that the poor creatures dwelled in a sort of fuggy miasma of moist, germ laden, stale air. A lot depends on the local style – you might be lucky, and have open fronted cattle courts or cart sheds which would be ideal for lambing in. Step inside and look up at the rafters. If they're festooned with cobwebs, then forget it – there simply isn't enough air flow for sheep.

More recently constructed corrugated iron sheds have their own problem. Condensation. On cold nights, the warm damp air rising from the sheep condenses on the underside of the roof, only to come dripping down onto the backs of the animals as soon as the mid-March morning sunshine begins to warm the tin. In poorly designed buildings, the fleeces of the sheep may become perpetually damp, which, coupled with fluctuating temperatures and poor air flow, will soon lead to respiratory problems.

Regardless of whether you're building a new shed or adapting an old one, ventilation is the key consideration. Anyone who's shivered their way through the nights on one of our lambing courses may say I've taken this too far, by building a shed that resembles a wind tunnel, but I no longer hear my sheep coughing as they used to in the old tin shed!

The inside of the sheep shed should not feel any warmer than the outdoor temperature. Sheltered, yes, and dry, but not warm. And there must be plenty of air movement above sheep height. The basic principle of ventilation in livestock housing works like this: Warm, stale, damp air rising from the animals and their bedding exits through the ridge (or at the higher, open front in the case of mono-pitch buildings), to be replaced by fresh air drawn in through the slatted sides. You can test it by setting light to some oily rags in a wheelbarrow – try placing this at different locations within the building, and watching where the smoke goes. Don't blame me though, if your sheep all get lung cancer!

Factors Affecting Ventilation

• *Roof pitch* - Warm air in the building rises, carrying with it a lot of moisture. If its progress is impeded by a flattish roof, condensation will form. Slope should be 15° for an apex roof, and no less than 10° for an open fronted mono-pitch.

• *Roof height* - In buildings that are too high

Building a simple mono-pitch pole barn using a mix of new and re-claimed materials

(more than 5m to the apex) warm air cools and falls before reaching the outlets at the ridge.

• *Exit at ridge* - If ventilation is poor, remove alternate capping pieces from apex roofs, or even strip off the lot, leaving the ridge completely open. On mono-pitch buildings (or apex roofs with insufficient pitch), consider re-laying the roofing sheets, leaving a side-to-side gap of a couple of inches between each vertical run. It sounds daft, I know, but it really does work.

• *Inlet at sides* - The walls only need to be solid up to a height of 4 feet, to prevent draughts at sheep level. Above that they can be completely open, if facing away from the prevailing weather, or semi-closed using Yorkshire boarding (7:1 ratio) or a wind break mesh. So called "ventilated" cladding sheets (like roofing sheets with perforations) aren't much good.

• *External factors* - Look at external factors that may impede air flow to the building – I can remember making a huge improvement to the internal environment of one lambing shed, simply by cutting back an overgrown hedge that grew alongside! Consider the effects caused by other buildings close by – this is particularly pertinent if constructing a lean-to on an existing building, as the new shed may seriously reduce air flow to the old one.

In spite of these important considerations, housing for sheep need not be elaborate. For very small flocks it will be quite acceptable to utilise existing structures, as stocking densities will be low. For larger flocks, a simple mono-pitch pole barn (built from reclaimed materials, where possible) will provide excellent accommodation. However, where the sheep flock is the principal occupation on the holding, or where it is intended to make maximum alternative use of the lambing shed at other times of the year, I would heartily recommend a purpose designed agricultural building. If you already have a suitable site, and plenty of family labour to help keep construction costs to a minimum, the total expenditure can be kept to a reasonable level – an "off-the-peg" steel framed sheep shed, 60′ x 30′, delivered in kit form, can be purchased for a little over £2,000.

HOUSING THE FLOCK

Try to get the building prepared in advance, so that you can take advantage of the first suitable weather window to bring the sheep in. If, like us, you put the building to other uses when it's not full of sheep, then there'll probably be a fair amount of sorting out to do – stored machinery may need to be dragged out, to be temporarily parked on the yard under tarpaulins, accumulated odds and ends will need to be re-homed, and, in our case, we need to sell the cattle that fill half the building, dismantle the cattle yards, and thoroughly muck out.

Setting up the accommodation for the sheep shouldn't take long, provided you've worked it all out carefully beforehand – there's nothing worse than a higgledy-piggledy arrangement of hurdles, old pallets and string, cobbled together on the spur of the moment. You'll be spending a lot of time in the lambing shed over the next few weeks, and extreme tiredness will lead to a shortened temper, so do ensure that the working environment is as good as you can make it. My recommendation is that you fix gate hinges to every available upright in your shed (install intermediate uprights if required, preferably removable posts that fit into sockets in the floor), and construct yards using lightweight galvanised gates of various lengths. This is a very versatile arrangement that can be added to and adapted year on year as your flock develops. It looks tidy and professional too. At the end of lambing it is a simple matter to lift out the whole lot, giving one clear area for easy mucking out.

Hook-over style hay racks can be hung on the gates that form the sides or front of each yard, where they won't take up valuable floor space.

Even in small flocks you'll need at least four yards, enabling you to separate singles and twins, and early and late lambers. The maximum number of sheep per yard should ideally not exceed 30. You'll also need a fifth area specifically for individual "bonding" pens. If a significant number of triplets are expected you may need yet another area for them, but in many cases they can be housed with the twins.

Housing Guidelines

• Never House Wet Sheep
Even in a well ventilated modern building they'll take days to dry, and in traditional buildings it may take a week or more. In the moist environment created by all those damp, steaming fleeces, respiratory problems are inevitable. Try to get the sheep in at mid afternoon on a bright and breezy day.

Our new sheep shed at Ty'n-y-Mynydd. It was fabricated to my design and delivered to the farm in kit form. Inset: The old tin shed, which was a health hazard to livestock

• *Once They Are In, Keep Them In Until They've Lambed*

There's nothing to be gained by letting housed sheep out during the day. They don't need the exercise (even in a crowded shed ewes may walk as much as 2 miles per day), they don't need the grass (which is very poor quality at this time of year, and will only serve to fill their bellies, leaving no room for more nutritious feeds), and they may get soaked in a shower, meaning that you can't bring them in at the end of the day.

• *Bedding*

Straw is most popular. Make sure you choose nice crumbly spreadable stuff. Long, clean, bright yellow straw isn't much good at all, despite the fact it looks so lovely, as it's not very absorbent. Tell your straw merchant that it is for bedding sheep, and if he's worth his salt he'll know exactly what's required. Alternative bedding materials include:

Rushes. These are not very absorbent, difficult to muck out and take a long time to break down. Having said that, rushes are readily available at no cost. When used in conjunction with straw, they are an excellent way of eking out the more expensive material.

Bracken. Dried bracken is a popular (and traditional) livestock bedding material on the small hill farms of Wales. It is very absorbent, and when mixed with dung it breaks down to a beautiful peaty consistency, making excellent compost. However, bracken is poisonous to stock (if consumed in quantity, or over a long period), and its spores are carcinogenic.

Woodchips. These are potentially useful, but probably not cost effective.

Slats. These shouldn't be considered for lambing ewes, but may be suitable for housing store lambs.

Whichever material you choose to use, be generous with it. Remember, this is the bed on which the sheep will be giving birth, and it should be sufficiently clean that you wouldn't mind lying down on it yourself!

- **Access**

It's easy enough for a supple live animal to turn sharp corners, pass through standard width doorways, and walk down narrow passageways, but could you get a dead animal out? This is more likely to be an issue if you're adapting existing traditional buildings, and is particularly important if you'll be using the same accommodation at different times of year to house other (larger) animals. The body of a sheep is awkward enough, but it's not much fun if you have to cut up the carcass of a much loved family pony in order to get it out of the shed in which it died! If you divide up your building using gates, as I've suggested previously, you can simply remove any obstacles between the casualty and the door, and drag it straight out.

- **Water**

I once met some smallholders who spent a total of more than four hours every day carrying buckets of water to their animals. When I finally persuaded them to fit automatic drinkers, it nearly changed their life! Installing self-filling water troughs will save you a lot of time, time that is better spent watching the sheep. On no account though, should they be seen as a "fit & forget" solution – they must be checked regularly. The troughs needn't be a permanent fixture in the building, and at the end of lambing time they can be disconnected, thoroughly scrubbed out, and stored ready for next year.

- **Hay Racks**

The number of racks required (i.e., the total length of feeding space) will depend upon how you intend to feed your hay or silage. If it will be available on an ad-lib basis, each ewe will feed little and often, so fewer racks will be needed for a given number of animals. If, however, you'll be rationing the forage part of the diet then there must be room for all the

Walkthrough feed troughs are nothing but a waste of space, in my opinion!

Floor feeding prevents bullying, and keeps the ewes occupied for ages

ewes to feed at once, so allow 15″ – 18″ per head, depending on breed.

- **Floor Space**

The minimum space requirement (1.2m² per ewe) that you'll see quoted in many books on sheep keeping is largely academic, unless you intend to house the flock for an extended period either before or after lambing. In practice, you run the sheep in until the stocking density looks about right then add a few more to each yard, for luck! The number of sheep in each yard won't remain constant anyway – ewes that have lambed will be moved out, later lambing ewes will be drafted in, and from time to time you'll shift sheep around to make better use of space. The most significant limiting factor to the number of ewes that a given area will comfortably house is trough space (see facing page).

- **Trough Space**

I am firmly of the opinion that feed troughs (in particular the walkthrough variety) are nothing but a waste of space in the lambing shed. In addition to limiting the number of sheep that a yard will hold (in terms of inches of feeding space required per animal) they take up too much room, are always in the way, and generate a stressful (and potentially damaging) environment for the ewes. My preferred option is to broadcast compound feed onto the straw bedding. In this way I can utilise the entire available area for feeding. The ewes spread out nicely and there is no bullying, no pushing and shoving, the shy feeders are able to compete successfully, the sheep don't eat too quickly, it keeps them occupied for ages, and there is no wastage as there is with fixed or walkthrough troughs, where spilt feed tends to accumulate underneath, attracting vermin. By adopting this feeding method, and by providing forage ad-lib, we have been able to comfortably stock ewes using less than two thirds of the "recommended" floor space. NB. Floor feeding is not appropriate if coarse rations or home mixed feeds are used.

- **Winter Shearing**

If you're really pushed for space, or are having to make do with poorly ventilated traditional buildings, then consider winter shearing of housed ewes. Shorn ewes take up less space, suffer less heat stress, eat more, produce more colostrum and give birth to heavier lambs. Observation of ewes in labour is made simpler, and new born lambs can find the ewe's teats more easily. Don't worry about them being cold – they adapt very quickly by producing a layer of subcutaneous fat, and can be turned out to pasture after lambing, in the normal fashion. The principal downside is that shearing heavily pregnant ewes is most definitely not a job for an amateur!

- **Lighting**

I must confess that our use of electricity at lambing time can at best be described as extravagant, though in defence I'd like to add that we're very frugal at other times of year! The inside of the lambing shed is fully illuminated from dusk 'til dawn by a couple of big sodium floodlights – one at each end of the building – angled in such a way that there are no shadows, anywhere! This may sound like a minor detail, but it makes such a difference to be able to see all the sheep clearly, from any location within the building, and not have to worry about what disasters may be occurring unseen in darkened corners!

- **Tidiness**

Keep a broom in the lambing shed, and use it! There really is no excuse for allowing walkways to become cluttered up with straw, bits of string, and empty feed bags. Make it part of your routine to sweep up twice a day, and also immediately after doing jobs like bedding down or filling hay racks. Fussy, perhaps, but worth it.

SOME PROBLEMS ENCOUNTERED IN HOUSED EWES, PRE-LAMBING

- **Pneumonia**

Principally a problem in poorly ventilated buildings, or if sheep were damp when housed. There are some extremely effective antibiotic treatments available for pneumonia nowadays, such as Micotil (which must be administered by a vet), but for less severe cases you can inject a general purpose oxytetracycline based antibiotic, such as Tetroxy LA, which you should have in your medicine cupboard.

- **Vaginal Prolapse**

This may appear quite startling to the

uninitiated, though it's not a complicated condition to attend to. It generally occurs in ewes that are overweight, and Texels and their crosses appear to be particularly prone. I am also convinced that competition for space at the feed troughs can be a contributory factor – another sound reason for floor feeding! The everted vagina (which may be about the size of a grapefruit) should be gently returned to its proper place, using plenty of lubrication, with the ewe standing up. Be warned: as it slips back into place the ewe will urinate copiously, most of which will go straight up your sleeve or into your wellingtons! Fit a retainer spoon or truss to prevent the ewe pushing it out again, and provide long acting antibiotic cover. It is also worth injecting calcium borogluconate, as a low level of calcium deficiency may cause poor muscle tone.

• *Abortion*

Ewes that abort when the flock is outdoors will often pass unnoticed, but when it occurs indoors you should spot it easily. Not all abortions are caused by an infection (again, this is a condition where I believe that physical damage caused by bullying at the feed trough may at times be responsible), but you must treat each one on the assumption that it could well be transmissible. Isolate affected ewes, clear up dead lambs and placenta and dispose of properly, spread hydrated lime onto the contaminated bedding, and thoroughly cover with fresh straw. If more than one or two ewes are seen to have aborted then consult your vet, and arrange for tests to be carried out to determine the cause. Do not cull affected ewes, as infection imparts subsequent immunity.

Above all, be aware that some forms of abortion are zoonotic, and pose a very serious risk to pregnant women.

If test results do show the presence of either toxoplasmosis or chlamydial (enzootic)

abortion, then I'm afraid that you are in for a pretty miserable lambing season. Attempts can be made to reduce the level of impending loss by injecting all ewes still carrying lambs with a long acting oxytetracycline (in the case of chlamydial) or by feeding decoquinate (a coccidiostat that can be incorporated under prescription as an additive to compound feeds or blocks) where toxoplasma is the causative agent. Abortions can be prevented in future years by vaccinating ewes 1 – 3 months before tupping. This provides lifelong immunity, so, after the first year (in which it will be necessary to inject the entire flock), only flock replacements will need to be vaccinated.

• *Lameness*

It is best to turn a blind eye to lameness at this late stage, and leave dealing with the ewe's feet until after they've lambed.

• *Not Drinking*

This is a most infuriating situation. Some sheep simply refuse to drink still water from a bucket or trough, and if they're not drinking then they'll more or less stop eating dry feed. If you aren't aware of the problem they can actually starve before your eyes! Look out for ewes licking the outside of the water trough, or gates and hurdles, or even the walls in damp weather. You may well have to turn these individuals out again. Fitting small self-fill bowls, rather than large troughs, can help, as the contents are replenished regularly, and the sound of the water running in may tempt the more fastidious drinkers.

It is a particular problem amongst the hill breeds. One hill farmer I met had diverted water from a nearby mountain stream, to run through a length of guttering attached to the inside of the back wall of his sheep shed. This passed through each pen of ewes, providing fresh running water for all. Until he did this, he had been unable to house any of his sheep.

INDIVIDUAL LAMBING PENS

Sooner or later you will need to start thinking about lambing pens – individual pens, often referred to as "mothering" or "bonding" pens, used to contain a ewe with her newborn lambs for the first few hours after birth. If your sheep are of a hardy type, or an "easy-care" breed, and you intend to lamb them outdoors, then simple shelters constructed of straw bales or windbreak netting situated in the least exposed part of the lambing field will probably suffice. Even so, it is as well to be prepared and have some pens under cover – bad weather may set in, resulting in the need to keep new families penned up together for longer.

Indoor lambing provides the solution to many problems, but also introduces a few new ones. It is recommended that you erect one individual pen for every 10 housed ewes, but I warn you this is never enough! 1:10 is an OK ratio for outdoor lambing, as some of the more sensible mothers with single lambs probably won't need penning at all, but, for housed sheep, a ratio of 1:8 or even 1:6 is more appropriate. Where the flock has been synchronised, for example by sponging (see page 67), as many as one pen to every three ewes may be needed.

When I started my own sheep flock (nearly 25 years ago now, but it still seems like yesterday!) the preparation for lambing time consisted of a trip to a local depot in search of old pallets. Augmented by bales of straw, wire netting and some other bits and pieces, these were cobbled together to form my lambing pens. It worked, but was far from ideal and is not what I would recommend! Lambing time is about looking after sheep, not about bodgery, and time spent lashing things up with string is time wasted.

I continued to employ these unsatisfactory arrangements for a further 12 years or so (in defence I will state that at least pens made of pallets and straw can be burned after use, so no chance of disease carry over!), but every spring I also "contract lambed" other flocks, each with it's own solution to the penning problem. Some were good, some were not so good, and some were bloody awful! Based on this experience I was gradually forming a mental image of the ideal set up. One day, perhaps, when I got settled with a farm of my own, I'd put it all into practice, giving my sheep the accommodation I felt they deserved…

Within a couple of years of moving to Ty'n y Mynydd we had graduated from pallets-and-string to galvanised metal hurdles, purchased second hand at a farm dispersal sale. It felt like progress (despite the fact that we still needed to lash everything up with baler twine), but still fell far short of the ideal I longed for. Our fundamental problem at this stage, however, was not so much the individual pens, but the whole lambing shed! It really was on its last legs, and there seemed little point in putting up posh pens in a building that looked like it might collapse at any moment!

All this changed in 2003, when, as a result of a successful application to the Welsh Assembly for a farm improvement grant, we were at last able to design for ourselves a building that would suit our purposes. We lost no time in demolishing the old tin shed which, by this stage, had been condemned by our vet as being "totally unsuitable for housing livestock".

One of the most marked observations I had made whilst working with other flocks at lambing time was the fact that, on farms that had a designated separate area (or in some cases a different building) devoted solely to individual penning, the whole operation ran more smoothly, with far fewer welfare lapses. On the contrary, where pens were scattered about here and there (presumably with the intention of making the best use of every available space) things often seemed to degenerate into a bit of a muddle. I incorporated this observation into my design by drawing a lean-to on the main shed. At lambing time the 35′ x 15′ lean-to would be given over entirely to individual pens. It took a couple of years of use of the new shed to develop an appropriate and cost effective penning arrangement.

Standard 6′ sheep hurdles were ruled out quite early on. Set up in squares, they make pens which are too big to be practical. It simply isn't true that large pens are required to avoid lambs being squashed by their mothers. Good motherly ewes will not lie on their lambs. Poor mothers may well do so, regardless of pen size – I've seen a Border Leicester ewe squash her lamb flat in the middle of an eight acre field! (Motherliness is hereditary – cull the poorest, and don't keep their offspring as flock replacements). Flighty ewes in large pens may dash about at the slightest excuse, often trampling their offspring underfoot. In narrower pens they tend to stand quietly across the back.

The problem of pen size is easily overcome – metal sheep hurdles are available in a range of lengths, so a combination of, say, 6′ hurdles for pen sides and 3′ hurdles for front and back could be used, but there is also the considerable issue of cost. 6′ galvanised hurdles will set you back by about £20 apiece. Three footers cost in the region of £15 (all + VAT, of course). Four foot "lambing hurdles" (which are half meshed so lambs can't slip from pen to pen) sound perfect, but at £22 each it would cost well over £1000 to set up the twenty or so pens we require.

A double row of pens, with a wide enough alleyway to work in

Clearly I would have to make my own, designed to fit the space available.

Solid sided pens would have been nice – I've seen some lovely ones made by welding together a frame of 1″ box section, to which was riveted galvanised sheeting, but I'm no metal worker. I did manage to create a steel door for our chemical store but it took me weeks! With lambing time rapidly approaching I needed to manufacture nearly 70 hurdles in less than a month, during which time I was also working 7 days a week lambing someone else's sheep! Our homemade hurdles would have to be constructed of timber, a material I feel comfortable working with.

Still considering the advantages of solid sided pens, I remembered seeing some that had been made using 10′ x 5′ sheets of marine ply, each being cut into four equal pieces. The resultant panels 5′ long by 2′ 6″ wide had hooks attached to the ends enabling them to be joined together in such a way as to form a very neat set of pens. The attraction of this method was the speed I'd be able to make them – a piece of ply could be sliced up in minutes! However the cost was prohibitive – 10′ x 5′ marine ply is over £50 a sheet! Also, although 2′ 6″ high pens would be fine for heavy lowland ewes they'd scarcely be likely to contain our lively Welsh sheep. Our hurdles would need to be at least

3′ high, so couldn't be cut economically from sheet material. I decided to fall back on the traditional "five bar gate" style railed hurdle.

We needed at least 20 individual pens, so planned to place a row of eleven down each side of the lean-to, leaving a passageway through the middle wide enough to work in.

Each pen was to be 5′ deep by 3′ wide. This is somewhat smaller than the recommended minimum of 1.5m² but, as I mentioned earlier, I don't really think it matters a great deal. Besides, by removing an intermediate hurdle we could always create a double size pen for any "long stay clients".

Now a few calculations. To create our 22 pens we'd need to make 24 five foot hurdles and 44 of the three footers. Each hurdle would be 3′ high, have five horizontal rails and be braced diagonally…so, total length of timber required would be 1,976 feet! It looks a heck of a lot written down like that! I managed to source 3″ x 1″ larch battens from a local forestry sawmill in 10′ lengths. They cost £1.20 per length. In fact they are cut oversize then trimmed off to 10′ to square the ends, so, by agreeing to take them untrimmed, I got battens about 12′ long for the same price. Every little helps. I also took home all the outer pieces with bark still on, free of charge, so actually only paid for 150 lengths. So far, so good – a saving of £900 over purchased hurdles. Just the small matter of making them to deal with now…

As I had barely enough timber it was important to plan the cutting list carefully to minimise wastage. Where possible I wished to avoid creating off-cuts, although any leftovers of more than 18″ would undoubtedly come in useful for something else.

Each five-foot hurdle would require 5 x 5′ rails, 2 x 3′ uprights and 1 x 5′ 6″ diagonal brace. The three-foot hurdles would need 7 x 3′ lengths (five rails & two uprights) and a 4′

diagonal.

Therefore the timber needed to construct a "stand alone" pen of four hurdles (2 x 5′ + 2 x 3′) would be cut from twelve of my 12′ battens.

Seven battens cut as follows made the two five footers:-
1. 2 x 5′ lengths + 24″ off-cut
2. 2 x 5′ lengths + 24″ off-cut
3. 1 x 5′ length + 1 x 5′ 6″ length + 18″ off-cut
4. 4 x 3′ lengths
5. 2 x 5′ lengths + 24″ off-cut
6. 2 x 5′ lengths + 24″ off-cut
7. 1 x 5′ length + 1 x 5′ 6″ length + 18″ off-cut.

The two three-foot hurdles were cut out of 5 battens like this:-
1. 4 x 3′ lengths
2. 4 x 3′ lengths
3. 4 x 3′ lengths
4. 1 x 3′ length + 2 x 4′ lengths + 12″ off-cut
5. 1 x 3′ length + 9′ carried over to next set of hurdles.

(Of course it was only the first pen of each run that required four hurdles. Subsequent pens were added on using one long and two short hurdles until the row of eleven was complete).

The first hurdle of each size was assembled with the utmost care and attention, particularly to ensure all angles were accurate, and that the spacing between rails was correct (closer together at the bottom to prevent lambs hopping from pen to pen). These prototype hurdles were then used as patterns to set up a jig, speeding up the construction of the remainder no end. Initially the jig was made by screwing blocks of wood to a sheet of shuttering ply I had kicking around. It was a simple matter, then, to drop the parts for each hurdle into place on the plywood sheet where

I was able to set up a jig on the workbench to construct the smaller hurdles

they were held securely without the need to re-measure or check angles. I also carefully measured and cut one each of the four different timber lengths required. These pieces served as templates for cutting the components of each hurdle, and were finally incorporated into the last one! Part way through the project I used the sheet of plywood for something else (more of that anon), but as I had finished all the longer hurdles by this stage I was able to make a jig for the remainder by screwing blocks to my workbench.

As to the method of construction, well I simply nailed them together! Basic I know, but really there wasn't time to be experimenting with more sophisticated jointing techniques, and besides, I had about a hundredweight of 2½″ galvanised nails left over from building the shed. (There must have been a miscalculation somewhere along the line I think!). Whilst assembling the first few hurdles I noticed that, when fixing close to the ends of the battens, the act of knocking in the nails was causing splits to appear in the timber – thereafter I drilled a pilot hole for each nail and experienced no more trouble. The nails were knocked in to about ¾ of their length whilst the component parts were held securely in the jig, and driven fully home after lifting the hurdle off. The

Hurdles joined using eyebolts and a length of bamboo

protruding ends were then hammered over. This may sound like rather crude carpentry, but in fact "clench nailing" is a perfectly acceptable method of joining two bits of wood together, as any traditional boat enthusiast will tell you! It's quick, simple and strong, and if you take a bit of care to orient all nails the same way, and thoroughly bury the points, it'll look neat and tidy too!

It didn't take long to settle into a sort of "production line" routine, with each hurdle taking less than 10 minutes to complete. Even so, it took me a few weeks to do the lot, making a small batch each evening when I returned home from lambing work on another farm. The next question was how to join them all together. Our own sheep had begun to lamb before all the hurdles were finished, so in order to begin using the pens I just tied up the first few with string initially (oh, will I ever be able to shake off this dependence on baler twine?). A number of suggestions were considered and, as is so often the case, the option first thought of turned out to be the most suitable – two eyebolts at each end with a pin to drop through. That meant 272 eyebolts. I quickly abandoned the idea of making them by welding washers to lengths of studding – as I said before, metalwork just isn't my thing! "Screwfix" had some that looked ideal but when they arrived they weren't anything like the catalogue description, so I sent them back. The problem was to find some where the shank was threaded almost up to the eye – this was very important as at one end of each hurdle I needed to be able to place a nut on either side of the timber in order to set the eyes out a bit, or the hurdles would foul one another and it wouldn't be possible to fully open the gate of each pen. Eventually a local builders' merchant ordered some for us – they cost nearly 30p each, but were exactly what we'd been looking for, so I didn't grumble. Following a bit of trial and error with some off-cuts, the first four hurdles were soon fitted with rings at each end. I hadn't yet acquired any mild steel rod to make the pins, but a rummage in the workshop unearthed three short lengths that would do for a trial run. I dropped a piece of bamboo through the fourth set of eyes. Bamboo! Why hadn't I thought of it before? Not only was it a better fit and looked tidier than the metal pins but, in the event of a pen gate being forced back too far, for example by a determined ewe trying to push past the open front of the pen in which I wished to contain her, then the bamboo cane would give way in the nature of a shear-pin, thus avoiding bent eyebolts or damaged hurdles. A quick sortie to the vegetable garden and a raid on last year's runner bean sticks supplied all the rods required. Steadily I worked through the night shift whilst watching the ewes, replacing strings with rings & pins, until our set of pens

Pannier style hay rack, made from off-cuts

was complete. From now on, when making the annual preparations for lambing, there'll be **no** scrounging pallets, **no** hunting the hedgerows for old hurdles used to block holes and, above all, **no** baler twine! For most of the year the pens stack neatly down one side of the lean-to, out of the way of other activities, but I know, as lambing time approaches, that it's a simple matter of half-an-hour or so to get the whole lot set up and ready for occupation. We are then free to turn our attention to more important things in life – the sheep!

Accessories

As the project progressed we started to think about the "accessories" we wanted, to finish off the whole set up, namely hay racks and bucket holders for each pen.

Hay Racks

Pannier style hay racks were what we needed. They sit on the partition between two pens, allowing sheep to feed from both sides. For our 22 pens we'd need 10. We already had a couple made of galvanised mesh that we use for the calves, so we tried them on the new hurdles. Perfect! Only another 8 required. Well, it seemed that the price had gone up somewhat since we bought those two, as we were quoted over £20 apiece, putting them well beyond our reach. The solution? Make our own, of course! Remember that sheet of ply? And all those off-cuts? It didn't take long to design pannier hayracks with plywood ends and larch slats. By employing the same "production line" type construction methods as I was using to manufacture the hurdles I could turn them out very quickly – there was soon a neat pile of 10 matching hayracks outside the workshop door. I used every single off-cut of 18″ or longer, so overall wastage from my 150 purchased battens was minimal.

Bucket Holders

We knew exactly what we wanted; bolt-on, drop-down galvanised rings for 1¼ gallon buckets. You would think that every farmers' co-op and smallholders' suppliers would stock such a thing, wouldn't you? Not so, it seems. We could have bought any number of different designs of rack, bracket or hook for 2 or 3 gallon buckets (too big for lambing pens), or hook over / fixed holders for smaller buckets, but no bolt-on, drop-down rings of the size we required. Eventually, after many fruitless enquiries, we contacted a fabricator in East Yorkshire, and received excellent service. "No problem" they said, and made and delivered exactly what we wanted in less than a week, without waiting for an up-front payment. The bill, when it came, was a mere fraction of what other bucket holders would have cost!

Bolt on, drop down, galvanised bucket rings

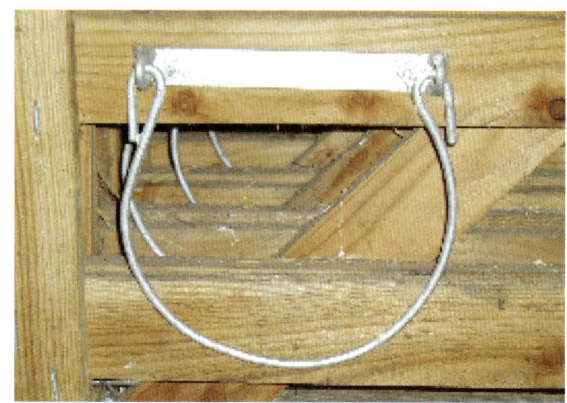

Costings	
Home Made	
Timber for 68 hurdles & 10 hayracks	£180
Eye bolts (272)	£80
Bucket rings (22)	£60
TOTAL	£320
Purchased	
24 x 5′ hurdles	£415
44 x 3′ hurdles	£660
22 bucket rings to hook on metal hurdles	£138
10 pannier style hayracks	£211
TOTAL	£1424
SAVING	£1104

The Shepherd's Year
Spring - lambing time

The gestation length of the ewe is 147 days (give or take a day or two), yet no matter how well prepared you are, and how carefully you've counted the days, the first lamb will catch you unawares! In all likelihood you'll walk into the shed one morning, to be confronted by a proud and possessive mother, with a couple of knock-kneed babies nuzzling at her udder. Well, from now on it's all hands to the pump! From the day that our first lamb is born, until the bulk of lambing is over, we don't leave the shed. We divide up the time into reasonable shifts, but whatever the hour of the day or night there is always one of us on duty. There are many shepherds who believe that night supervision is unnecessary, and there certainly is a strong argument for leaving the flock untended overnight, yet in our case we are adamant that it is this round-the-clock shepherding that has enabled us to really drive down mortality rates. On average, nationally, the number of lambs lost in the first week of life is 15%, although there are many vets who believe that the true figure is closer to 20%. That's a lot of dead lambs. 10% is good, and anything less than this is exceptional. (And if you're thinking that that still sounds like a lot, just bear in mind that just one lamb may represent 10% of a smallholding flock!). In recent years, we've succeeded in cutting our overall lambing time losses (including stillbirths) down to only 3.3%, with a death rate of less than half a percent! It's not that much happens overnight, in fact often there are no lambs born during the night at all, but we spend many hours just watching the sheep. It is by watching in this way that you'll learn to recognise the signs of imminent birth, long before any visible evidence can be seen; you'll learn to identify potential problems, before they even become a problem, and be able to prevent them occurring; and, above all, you'll learn to really understand your sheep. It is this understanding that makes all the difference.

Remember the old saying: "There's plenty of folk that looks at sheep and never sees" and make damned sure you're not one of them!

The bare necessities

EQUIPMENT CHECKLIST

Every book on sheep keeping includes a list of "essential items the shepherd must have at lambing time", and a similar list is faithfully reproduced annually in many magazine articles that deal with the subject of lambing. My own contribution will be no exception, although it does concern me somewhat that no-one appears to have updated, amended, or re-prioritised the familiar inventory for about 40 years! To this end, I hope that my list will be a little different, at least in order of importance, if not in content!

Our sheep shed is divided into two areas: the main part of the building where the ewes are housed and give birth, and the attached lean-to that contains all the individual bonding pens. Each of the two parts of the shed has its own set of procedures and protocols, and its own box of indispensable bits and pieces. You'll notice that some items appear on the lists for both areas, and this is indeed the case

– it's better to have two of something, than to be forever having to stop what you're doing in order to fetch what you need from the other part of the building.

The Bare Necessities

Together with a notebook and pencil, these first seven items are my absolute "must have" list

Time And Patience

Remember this: the safest place for a lamb to be is inside its mum! If a ewe needs 2 hours to lamb, then give her 2 hours. If she needs longer, give her longer. "Maximum observation, minimum interference" should be the lambing shed motto. I've known a young ewe take more than four hours to give birth, from the time that I could first see the nose of the lamb peeping through the vulva. I kept her under observation throughout, from the corner of my eye, content to let her get on with it. When eventually the lamb (a big single) was born, the ewe was on her feet immediately, and mothering well. The lamb was up and suckling within minutes. No fuss, no stress. I dread to think what the outcome would have been, had I acted on impulse and rushed in to assist with the delivery after only an hour or two of labour! Undoubtedly I'd have risked damaging the ewe, by attempting to deliver the lamb before she had fully dilated; subsequent infection could have led to retention of the placenta, metritis, or eversion of the uterus; large doses of antibiotic would have been required; perhaps the ewe's milk would have dried up, or she'd have refused to mother the lamb; the lamb itself would, in all likelihood, have sustained damage, and, if it survived the birth trauma, would remain stunted throughout its life. All in all, untimely intervention is probably the second

most significant cause of deaths at lambing time, though many people do not realise this as they don't associate, say, the loss of a week old lamb with the fact that it was dragged from its mother's womb far too soon.

If you are in any doubt about whether a ewe in labour requires assistance, then sit down and have a cup of tea while you think it over – she'll probably have her lambs while you are waiting for the kettle to boil!

Thermovite

This is the nearest thing to a miracle that I've ever come across! Available from the company Tithebarn, thermovite is a high energy paste containing colostrum, pro-biotics, vitamins, minerals and allicin. We administer thermovite at birth to all lambs that are compromised in any way, and to triplets. Its effect is almost immediate, and long lasting. I wouldn't consider starting the lambing season without it.

Disposable Shoulder Length Obstetric Gloves

A lot of people seem to think that there's something rather "cissy" about wearing gloves to lamb a ewe, or that there's no time to go fiddling about with things like that. What nonsense! The average shepherd has hands like a bit of coarse sandpaper at this time of year – just put yourself in the ewe's position for a moment…

Keep a couple of gloves in the pocket of your overalls, where they are always handy for any eventuality. In addition to using them to assist deliveries, the gloves should be used to contain contaminated material such as aborted foetuses – pick up the dead lamb, then roll the glove down off your arm and over the hand holding the lamb. It's a simple matter, then, to tie up the mouth of the "inside-out" glove, sealing the potentially infectious material within.

Disposable gloves can be used to contain potentially infectious material

Lubricating Gel

Use plenty, even for minor examinations. It is more effective if slightly diluted. If you always use gloves and gel, then you'll seldom have to inject antibiotics after an assisted delivery (assuming that there were no major complications).

1m Length Of Plastic Coated Clothes Line

Tied into a loop with a great big knot, this is simply the best lambing aid I've ever used.

Bactakil

After trying out all manner of sprays and things, I've come to the conclusion that Osmonds Bactakil is the most effective product for applying to the navels of newborn lambs.

An Outdoor Hot Water Supply

By "outdoor" I mean not in the house. If you've got outbuildings adjoining your dwelling then it'll be a simple matter to run a pipe through from the domestic hot water supply, otherwise fit an immersion heater or geyser in an outhouse or in the lambing shed itself. Keep a squirty tub of anti-bacterial handwash close by, and an old towel. You'll be amazed at how much cleaner you keep everything, if you don't have to go to the house every time you need hot water.

Additional Items For The Lambing Shed

• *Lambing ropes x 3* – clean, and in a sealed container.
• *Calcium borogluconate* – together with a 60ml syringe and wide bore (16 gauge) needles, for subcutaneous injection.
• *Penicillin* – e.g., Duphapen, together with 10ml syringe and medium bore (18 gauge) needles, for intra-muscular injection.
• *Oxytetracycline* – e.g., Tetroxy LA, together

with 10ml syringe and medium bore (18 gauge) needles, for intra-muscular injection.

• *Marker crayon* – the sort with a twist base, small enough to fit in your pocket.

• *Prolapse restrainer* – keep a spoon type restrainer handy in the shed, but for persistent cases you may also need to apply a truss.

And For The Night Shift…

A torch, a flask of tea, a packet of tobacco, and the ability to sleep standing up with one eye open!

The Box Of Bits And Pieces In The Individual Penning Area

The following list of equipment is always to hand in the lean-to. I store the whole lot in a deep hook-over manger, which can be hung on the hurdles of whichever pen I'm working in. Don't be tempted to use one of those rather pathetic little plastic "tack trays" to keep your stuff in, or you'll be forever picking it all up off the floor!

• *Dagging shears*
Not a great long bladed clumsy pair, designed for shearing whole sheep! Get some really small ones, 3½″ blades, with a double bow.

• *Hoof clippers and foot spray*
Serrated clippers, and plenty of spray.

• *Spray markers (at least two colours)*
For spraying identification numbers onto ewe and lamb couples at turnout. Use different colours for twins and singles.

• *Penicillin (such as Duphapen)*
Together with 10ml syringe and medium bore (18 gauge) needles, for intra-muscular injection. Also some finer (19 gauge) needles,

Numbers should be bold and clear. Use a different colour for twins and singles

in case it's necessary to inject any lambs. I also have some extremely fine, short needles (21 gauge x ⅝″), for the treatment of entropion (turned in eyelids) in young lambs. It's a bit tricky, and involves an injection of 1ml of penicillin into the eyelid. If you encounter the problem, I suggest you ask your vet to treat it for you. If left untreated the lamb will be in constant pain, and eventually go blind. It occurs more frequently in some breeds, and is probably hereditary.

• *Oxytetracycline (such as Tetroxy LA)*
Together with 10ml syringe and medium bore (18 gauge) needles, for intra-muscular injection.

• *Ear notching pliers*
By the use of a system of small, discreet notches, a sheep can carry a permanent record of her whole history in her ear! In particular, a notch should be used to identify lambs that should not be retained for breeding, for example if they've suffered from a hereditary condition such as entropion.

• *Tailing rings and applicator*
Only required for some breeds. Keep them in a narrow necked bottle. It's easy, then, to tip out one ring as required, but if the whole bottle falls over the rubber rings jam in the neck and none fall out!

TAILING AND CASTRATION

Most lowland breeds and crosses have their tails docked, and this is best carried out within the first 24 hours, using an elastrator. This places a tight rubber ring around the tail which then withers and drops off. The tail must be left long enough to cover the anus (and vulva, in ewes) which, in effect, means that the ring should be placed where the area of bare skin on the underside of the tail ends. Some semi-upland breeds are left with a ¾ length tail. Many years ago it was common practice to dock the tails of down breeds very short – at the base of the spine, almost – but this is now illegal. Unfortunately, "fashion" dictates that some pedigree breeders persist in the practice.

The elastrator can also be used to castrate male lambs, but you should ask yourself whether this is really necessary. Most people do not bother nowadays as the modern taste is for smaller cuts of meat – ram lambs will be slaughtered long before they become a nuisance. Entire lambs will produce a better conformation carcass with more muscling and less fat. If you do decide to run a few on as wethers for mutton, these can be castrated in the autumn using a "burdizzo" or bloodless castrator. The burdizzo crushes the spermatic cords and the blood vessels running to the testicles, whilst leaving the scrotum undamaged.

It is illegal for anyone under the age of 17 to carry out tailing or castration by any method.

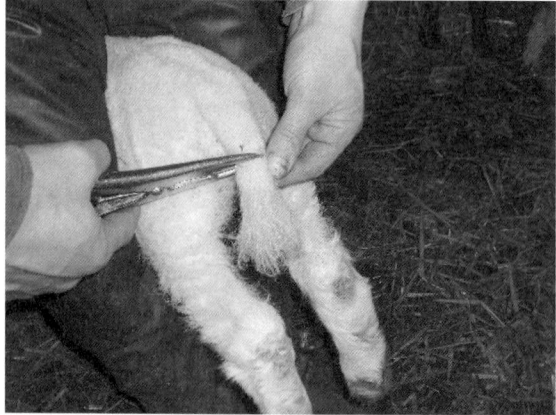

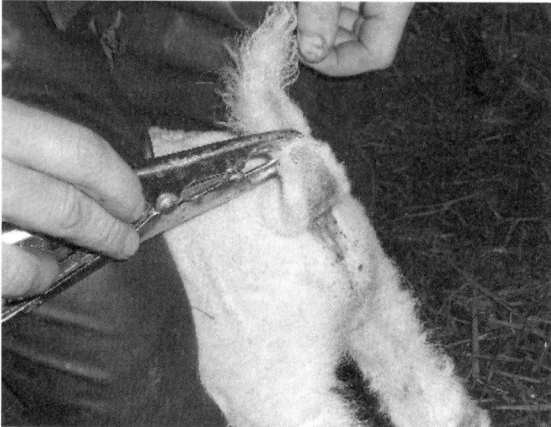

Tailing with an elastrator and, below, castrating an older animal using a burdizzo

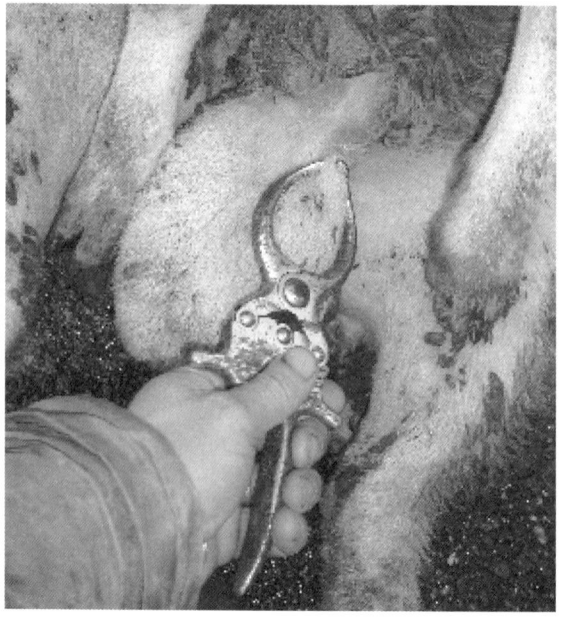

• *Combined Fluke & Worm drench*

Although adult sheep generally have a high degree of immunity to worms, they will shed high numbers of worm eggs shortly after lambing. Dosing ewes at turnout reduces pasture contamination, and therefore reduces the challenge to young lambs.

• *Crovect (or similar pour-on product)*

Although we treated the sheep for ectoparasites in the autumn, housed conditions are ideal for the spread of lice. We find that crovect is the most cost effective treatment at this time of year.

You'll Also Need...

Hydrated lime and a small trowel

Keep it in a bucket with a lid, so it won't get damp. Use to disinfect individual pens between occupants.

and Perhaps...

A spring balance

If you are performance recording your flock you may need to make a note of lamb birth weights. Newborn lambs can be momentarily hung on a spring balance by their front legs, using a loop of rubber inner-tube.

and Of Course...

A notebook and pencil

The notebook that I always have in my pocket at lambing time has its pages divided into columns, in which I record the following:
- Ear tag number of ewe.
- Date lambed.
- Number of lambs born, and whether they are male, female or dead.
- The sire of the lambs.

Newborn lambs can be momentarily hung on a spring balance using a loop of inner-tube

- Birth weights (if required for performance recording).
- Notes, such as whether a lamb needs to be bottle reared, or fostered to another ewe.
- The identification number sprayed on the ewe and lambs at turnout.
- Ear tag numbers of lambs. (I enter this at a later date. We don't tag the lambs straight away, as their ears are so small and delicate).
- Deaths.

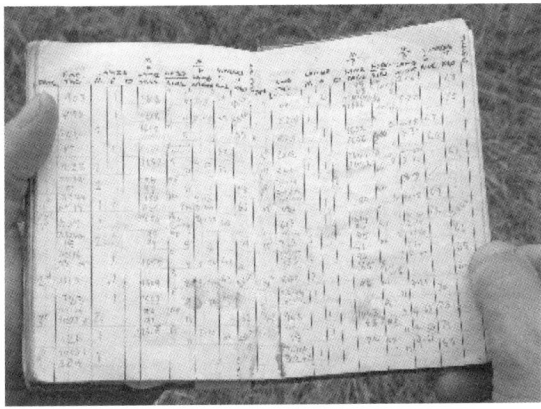

A notebook to record lambing data is essential, even in non-pedigree flocks

Additional items

Somewhere about the place, probably in the vicinity of your hot water tap, you should have the following:

Stomach tube

Always use a stomach tube in preference to a bottle and teat when giving colostrum to new born lambs. Only resort to bottle feeding if it becomes apparent that the lamb will need to be artificially reared. The pipe on some makes of stomach tube seems to be far too long, making them awkward to use. About 14" is best. Remove the plunger from the syringe part of the stomach tube and throw it away – it's nothing but a menace. In use, pour warm colostrum from a jug into the syringe, allowing it to flow gently into the lamb's stomach at its own pace.

Powdered colostrum (together with a mixing jug and whisk)

The beauty of powdered colostrum is that it is always available and can be ready for use in an instant. Take care to mix at the right strength. After feeding reconstituted colostrum by stomach tube, the lamb should be re-united with his mother (if strong enough), and encouraged to suckle her. Suckling is a very important aspect of the ewe / lamb bond – lambs that do not suckle will soon be abandoned.

Some flock owners prefer to keep a stock of frozen goat's colostrum for use in emergencies, which is fine provided that the goat received a full course of clostridial vaccinations. Avoid using cow's colostrum, as some may contain an anti-ovine factor.

In addition to the vital antibodies that it conveys, colostrum contains a powerful laxative to ensure that newborn lambs pass their

HOW TO INSERT THE STOMACH TUBE

(Instructions for a right handed person). Stand with your left leg raised by placing your foot on one of the lower rails of a hurdle, or on a bale of straw, and rest the lamb on your thigh (facing right) with his hind legs dangling over the edge. Hold him against your body using your left forearm, with his head cupped in your left hand. He should be pretty well stretched out in this position. Use your right hand to insert the stomach tube into the corner of the lamb's mouth and over the back of his tongue (pretty much as you would do with a dosing gun), and carry on slipping it in until within an inch of the ferrule at the end of the tube (i.e., 12 – 13 inches of tube will be inside the lamb). If it won't go in this far, or if the lamb is showing signs of distress, then clearly it's going down the wrong way, in which case pull it out and start again. When you're happy that all is as it should be, move your left arm forward a bit, so that the lamb's head is supported by your wrist, and hold onto the attached syringe with your left hand. Use the other hand to pour colostrum from a jug into the syringe. If it doesn't go down, rotate the pipe a bit, or withdraw it by half an inch or so. When the required quantity has been given, pinch the tube and remove it carefully.

Do not stomach tube any lamb that is too weak to hold up its head.

INTRAPERITONEAL INJECTION

Intraperitoneal injections of between 20 and 40ml of warmed glucose solution (20%) are used in the treatment of severely hypothermic lambs, particularly those over 5 hours old. Typically these would have a body temperature of less than 37°C, and be unable to hold up their heads.

The correct site for injection is approximately 2cm to the side of and 2.5 cm below the navel, into the body cavity. The lamb is suspended by its front legs, and the needle inserted to its full length in the direction of the opposite hip. Inject the solution slowly. Don't be alarmed if the lamb urinates copiously as the fluid runs in.

meconium (first dung). It is believed that lambs that do not pass their meconium soon after birth are more susceptible to watery mouth (page 118).

Heat lamp (together with some old cardboard boxes)

Don't hang the lamp too low. 3′ – 4′ above the lamb should be fine. A thermometer is handy to monitor rising body temperature. Avoid overheating. Normal body temperature is 39 - 40°C. Any lamb with a temperature less than 39°C will require warming. Use a separate cardboard box for each poorly lamb, and burn after use.

Glucose solution (together with appropriate syringe and 19 gauge x 1″ needles)

For intraperitoneal injection of very weak hypothermic lambs. When treating a very weak, cold, lamb, which is more than 5 hours old, never place it under the heat lamp without first providing energy in the form of glucose, or it'll die in a low-blood-sugar fit as it warms up. Ask your vet to demonstrate the correct technique. You may never need to do it, but you should know how, just in case.

Also, if you are expecting a lot of multiple

births, it may be wise to prepare for the possibility of having to artificially rear some lambs, so keep bottles, teats, and ewe milk replacer in stock.

THE LAMBING TIME ROUTINE

Whether you're lambing a large flock, or just a few ewes on the smallholding, it is most important that you adopt a well thought out routine, capable of coping with all eventualities. The saying "a place for everything, and everything in its place" is nowhere more appropriate than in the lambing shed! It should never be necessary to go and look for things – all equipment must be to hand, exactly where it should be, and must be replaced immediately after use. Furthermore, all the procedures employed must be fully understood by everyone involved – it's no good one member of the family putting a spray mark on a ewe to indicate some specific treatment she may require, if no one else has a clue what that mark means!

It is well worth writing out a checklist of procedures, together with a simple set of identification marks, and pinning this list up on the wall in the shed. Make sure everyone involved with the flock is familiar with the list, and accept no excuses for deviation! Even when working with small flocks, there is a very fine line between busy-ness and chaos – attention to detail can prevent everything from degenerating into frightful muddle. It saves lambs' lives, too.

Lack of space will prevent me from going into quite as much detail here as I would like. Nevertheless, I'll do my best to talk you through the basic routine of the lambing shed, and try to include information relevant to a range of different breeds and systems:

Approaching The Lambing Shed

Approach the lambing shed gently, and stand quietly for some minutes in the doorway. There is no need to say anything to the flock to announce your presence – communication between a shepherd and his sheep is largely unspoken and telepathic. Most of the ewes will be lying down contentedly, and your appearance in the doorway should not disturb the peace. Remain standing where you are until you are sure that all the ewes know that you are there (even if you can see that lambs have been born in your absence), and in the meantime let your mind and your senses go in ahead of you. Sight, sound, and even scent, are invaluable here.

This moment of calm is of particular importance if you are about to feed the sheep, as they'll all jump up and start shouting for their breakfast as soon as you walk in! The quiet period of observation allows you to suss out exactly what is happening, and where, and enables you to keep track of individual ewes (perhaps requiring attention) that may otherwise go un-noticed in the feeding time melée.

However, for the sake of our example here, let's assume that the housed sheep were fed

Top: Learn to recognise the very early signs of labour
Above: She throws back her head and curls her upper lip

some hours ago, and that you've just returned to the shed after feeding the later lambing ewes that are still outdoors...

Step 1

From your vantage point in the doorway, you notice one ewe pawing at the ground and "turning" in a corner. Several times she gets up and lies down again, occasionally throwing back her head and curling her upper lip. As soon as you are sure that all the ewes in the shed are aware of your presence you enter the building quietly, and get into a position from where you can easily observe the ewe in question. The remainder of the flock should remain undisturbed, mostly lying down chewing the cud.

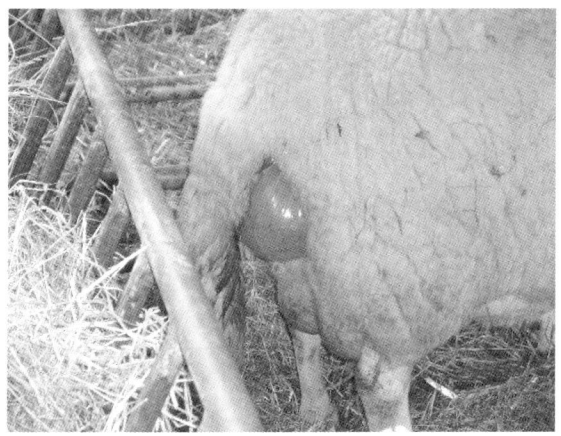

Now that you can see the ewe more clearly you notice the water bag protruding from her vulva, and as she turns to get a better look at you it bursts and falls away. Suddenly, apart from a damp patch under her tail, she looks no different from her flock mates! Had you hurried into the shed without patience, this lambing ewe may well have escaped notice.

Step 2

The ewe will probably spend some time licking up the fluid that spilled from the bursting water bag, and, while she's occupied, you've got a few minutes to decide what to do next. You should be able to tell, at a glance, whether she's an old ewe or a young one, and how many lambs she's carrying (assuming that the sheep were scanned). An experienced shepherd may also be able to spot potential complications, even at this early stage, by an almost intuitive interpretation of the animal's posture and body language, but I'm afraid that an explanation of these deeper mysteries falls far outside the scope of this book. (I'm not sure that I could put it all into words anyway!)

If the ewe is a young one, a first time lamber, and expecting a single, then it is likely to be some time before the lamb is born. Don't be alarmed if it's an hour or more. In this case you can be getting on with some other jobs in the lambing shed, or with the ewes and lambs in the individual bonding pens. Just keep a watch on her from the corner of your eye, until she begins to make more progress. However, if the ewe is an older animal, then the birthing process is likely to be far quicker, and if she's expecting twins then you should remain alert – the second lamb of twins is the one commonly lost at birth, particularly if the birth has been a relatively easy one.

Step 3

Let's assume, for the sake of simplicity, that the ewe in question is an older animal expecting a single lamb. She shouldn't take too long (although a single lamb, by virtue of its larger size, will be slower coming than one of twins), so having double checked that there is no other ewe in need of observation, and made sure you've got an empty individual pen ready to receive the new family, you settle down to watch the whole process.

Shortly the ewe will lie down again, and begin straining in earnest. You'll probably notice the lamb's nose and the tip of one front hoof each time she pushes, only to see them slip back inside between each contraction! However, with each subsequent push, the lamb comes a little further into view, and you'll see the other front hoof slightly behind the first. A big lamb may remain in this position for some length of time, with only the tip of its nose and hooves visible, to all appearances stuck like a cork in a bottle! Don't worry. If you can see a nose and two front feet then everything is fine, it just needs time. Sometimes the lamb's tongue may be protruding and slightly swollen, but again, don't be alarmed – it'll quickly return to its proper size after birth.

Eventually, a more drawn out contraction

will result in the whole of the lamb's head coming through the vulva, together with (hopefully!) the forelegs up to the knee joint or elbow.

Step 4

At this point the ewe will often roll onto her other side, or get to her feet and turn round a few times. You may even notice the lamb shaking his head! Again you'll see the ewe straining hard, to push the lamb's shoulders through, and sometimes, if it's a big lamb, she'll cry out as the shoulders emerge. From here on it's fairly straightforward, and a few good pushes should see the lamb fully delivered.

Step 5

There is no hurry for the ewe to get to her feet after the birth is completed, and no need for you to do anything for a few minutes, either. A single lamb is very rarely born with a membrane covering its nose, as the pressure caused by the passage of so big a lamb through such a small hole will usually have rubbed it away. As long as the ewe remains lying down, and the lamb remains behind her, the umbilical cord connecting the two will continue to transfer oxygen and other essential substances from the ewe's bloodstream into the lamb, and, when finally it is broken, as the ewe turns to lick her offspring, it will break in a natural way with very little bleeding and no risk of rupture.

Sometimes, of course, it will be necessary

Why is the second lamb of twins the one which is commonly lost at birth?

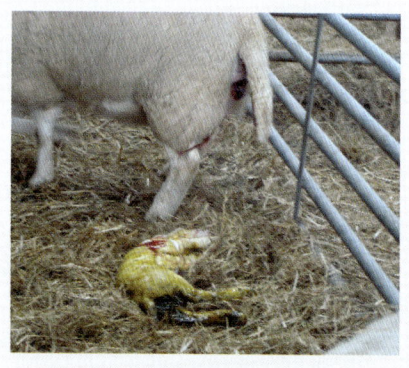

This is the usual sad scenario:
Having spent a few minutes mothering and bonding with the first lamb, the ewe lies down to give birth to the second. While she's thus occupied, her firstborn is taking a few tottering steps, and begins to follow the wall (or another ewe) until he's a couple of yards away from his mother. She is looking anxiously after him, and as soon as a final contraction expels the second lamb, she jumps to her feet and hurries to the rescue of the wanderer. By the time she turns back to lick the newborn twin, it's too late – the sudden parting of the umbilical cord, as the ewe got to her feet, caused him to draw his first breath and he suffocated in the birth fluids and membranes surrounding him. Had the lamb been bigger, for example a single, or had the birth been more prolonged (single or the first of twins), the membranes would probably have been rubbed away from the lamb's face during the birthing process.

If you are not able to be on hand to observe every birth you can help avoid these losses by placing straw bales at intervals against the wall of the lambing shed (at right angles to it), forming small corners where ewes will choose to give birth. The bales will prevent the first born lamb from wandering out of reach.

Leave well alone until you are confident that the ewe has accepted her offspring

for you to clear membranes from the nose of a newborn lamb (particularly a twin), but provided that the cord is still acting as a "lifeline" between mother and baby there is no need to rush – if you startle the ewe in your haste, she'll jump to her feet before you get near, breaking the cord too soon.

Step 6

At this early stage, i.e., within the first few minutes of birth, the single thing most important to the lamb's chance of survival is the initial establishment of the ewe / lamb bond, so, once you are content that the lamb is alive and breathing, leave well alone for a few minutes more, until you are confident that the ewe has accepted her offspring. Only then should you think about moving them out of the lambing yard, applying bactakil (or iodine) to the lamb's navel, and transferring them to an individual pen.

A flighty ewe, or a first time lamber, is best penned up for an hour or so on the spot where she has lambed, using a few lightweight portable hurdles, before trying to move her, but, when dealing with a mature, motherly old girl, simply pick up the lamb by its forelegs,

hold it low down where the ewe can see it, and walk off slowly making little bleating sounds. The ewe will follow closely behind, and it's a relatively simple matter to lead her to the area where you've got an individual pen ready and waiting. Keep the bottle of bactakil hanging on the gate of the lambing yard, and give the lamb's navel a squirt as you pass.

If, for any reason, you are unable to pen up a ewe and lamb straight away – perhaps because another ewe is in need of more urgent assistance – then use your crayon marker to put a small dot on the head of the lamb, and a corresponding mark on the mother. In this way, should any mis-mothering occur while you're busy elsewhere, it's easy to sort out the mix-up.

Step 7 In The Individual Penning Area

The sequence of events at this stage will vary according to breed / type of sheep involved. To avoid getting in a muddle, carry out the tasks in the order that I've listed them:

Smaller, Hardier Breeds

Put the lamb in the individual pen, leaving the ewe standing outside. Restrain her and turn her over into a sitting position as per my instructions on page 47 (and following). Your box of bits and pieces should be hanging on a hurdle within reach. (Use a sturdy container that's deep enough to hold everything, and heavy enough not to get knocked off). Now:-

1. Check that the ewe has plenty of milk, that her teats are not blocked, and that her udder is healthy.
2. Clip all the wool off her tail, around her backend, and in front of her udder.

3. Trim all her hooves, and squirt each one with foot spray.
4. Check her mouth.
5. Read her ear tag.
6. Turn her back up the right way, pop her in the pen with her baby, and write everything down in your notebook.

Heavier Lowland Breeds And Crosses

Put both the ewe and lamb together in the pen, and then:-

1. Restrain the ewe with your left hand under her chin, crouch down beside her, and check her udder with your right hand by using your fingers to draw a jet of milk from each teat into your cupped palm. Examine the milk for flakes, clots or blood.
2. If necessary, clip some wool from in front of her udder.
3. Read her ear tag, and write everything down in your notebook.

TIP! If you struggle to restrain a heavy ewe with one hand, while you check her milk or trim away a bit of wool, then loop an old fan belt around a rail of the hurdle and let her run her head into that.

Step 8

The length of time that the ewe and lambs need to remain penned up will vary, but in normal circumstances it will be 12 – 24 hours. The afterbirth should come away cleanly in the first hour or so – if it doesn't, the ewe should have antibiotic cover until it does. Never try to hurry it up by pulling on the protruding end.

Fill the hay rack, and provide fresh clean water in a small bucket. Replace the water regularly. Always top up the hay rack **before** refilling the water bucket, or you'll drop dust on the surface of the fresh water. Feed at the normal times, and take care not to give too much.

Step 9 Turnout Time

Each morning, after feeding round, you'll need to make an assessment of all the ewes penned up with their lambs, in order to determine which are ready to be turned out to grass. Look for alert, active ewes and contented, well fed lambs. And they shouldn't just look right – they should feel and smell right, too! If you have even the slightest of doubts about any couple, keep them in for a bit longer.

Step 10

Having identified a ewe and lamb(s) that are ready to turn out, remove their water bucket, hang your box of equipment within reach, and enter the pen.

(Again, this is a situation where the sequence of events will vary according to the breed or type of sheep involved.)

Read the ewe's ear tag, and look in your notebook to see if you've recorded any particular reason why she shouldn't be turned out. Also check what spray number should be used (i.e., next consecutive number to the last single or set of twins turned out), and write it down alongside her entry in the book. Then:-

Smaller, Hardier Breeds

1. Ear notch lambs, if required.
2. Spray numbers on lambs.
3. Administer fluke & worm dose to ewe.
4. Spray same number on ewe.
5. Apply crovect (or similar pour-on) to ewe.

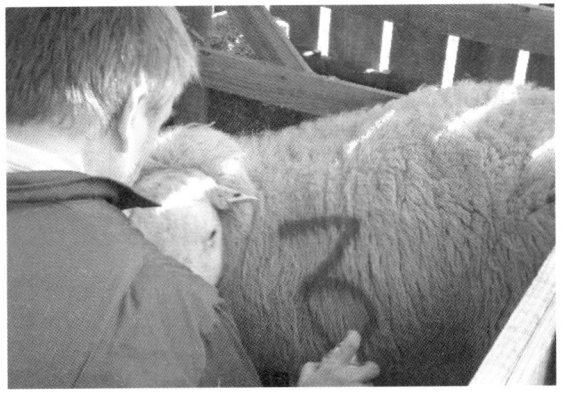

Spraying numbers at turnout

Heavier Lowland Breeds and Crosses

1. Turn the ewe over into a sitting position (see page 47 and following). (If the pen is very small it is quite a good idea to pop the lambs into an adjacent empty pen while you're doing this or, if there are two of you, your helper can be busy ringing the lamb's tails while you deal with the ewe).
2. Check udder thoroughly.
3. Check mouth.
4. Trim all hooves and apply foot spray, then allow the ewe to stand up.
5. Administer fluke & worm dose to ewe. (**Never** do this with the animal in a sitting position).
6. Pick up each lamb in turn and ear notch (if required), apply tail ring, and spray number.
7. Spray same number on ewe.
8. Apply crovect (or similar pour-on) to ewe.

TIP! Well fed, contented lambs, when rising to their feet after a snooze, will always give a luxuriant stretch. When determining which lambs are ready to turn out, go around the pens and chivvy them all up a bit. Look for the stretch. When you see that, you know all is well.

TIP! All numbers should be sprayed on the same side of all ewes and lambs. (The left side, if you are right-handed). It's a simple matter to encourage a group of sheep in a field to walk across in front of you, making all numbers visible at once. Any problems will be immediately apparent.

Step 11

If the turnout fields are close to the lambing shed, ewes and lambs can simply be walked out, or lambs can be carried (by the forelegs) with the ewes following. For longer distances, put one or two couples at a time in a small trailer or link box. Ensure that all ewes pair up with their lambs in the field – in the case of twins, you should return to the field in an hour or so, just to make sure. Where possible, turn out twins and singles separately. Group sizes should be kept small initially, and should not contain lambs of mixed ages. For the first few weeks make use of all available fields, putting just a few ewes and lambs in each, rather than squashing the whole lot into the nearest handy paddock. Regularly push stronger couples through into more distant fields, so those closest to the buildings are always available for the youngest lambs.

High magnesium licks should be provided for each group.

Step 12

Back in the shed, the bedding in the newly emptied pens should be liberally dusted with hydrated lime, and a fresh layer of straw put down ready for the next occupants.

Now return to step 1…

If the turnout fields are close to the lambing shed, ewes and lambs can simply be walked out, or lambs can be carried (by the forelegs) with the ewe following

Feeding At Grass

In early lambing flocks, ewes will require supplementary feeding after turnout. It is best not to feed for the first few days (i.e., don't feed in the first field) as lambs easily become separated from their dams when they come rushing to the troughs – it pays to keep everything as calm as possible where very young lambs are concerned. By the time they've been out for a week or so they'll be well able to find their mothers in a group, and won't be harmed by periods of temporary separation.

The lambs themselves may need to be "creep" fed. A creep feeder usually consists of a low, portable shelter, with a restricted entrance that can be adjusted as the lambs grow (provided that the ewes can still be kept out!). Inside the shelter there's a trough, above which is a hopper sufficient to hold a few sacks of feed. Access to the hopper for filling is from the outside, via a lifting flap in the roof. In some designs there is no shelter as such, and ewes are prevented from feeding by narrowly spaced bars across the front of the trough, through which the lambs have to put their heads. This type of feeder is no good for creeping lambs of horned breeds.

Creep feeding is only likely to be justifiable in "high input / high output" systems, where it is intended to finish lambs as early and as quickly as possible. It's also commonly carried out in pedigree flocks. Start by offering small amounts of fresh feed daily when the lambs are about a week old, discarding any that isn't consumed (feed it to the ewes). Very soon the lambs will get the hang of it, from which point you simply keep the hopper topped up.

SOME POST-LAMBING PROBLEMS

Uterine Prolapse

This is a far more serious condition than the vaginal prolapse mentioned earlier (page 94). In this case the entire uterus is everted, and it doesn't look nice, particularly if damaged or swollen. I am utterly convinced that 99% of these cases are a direct result of unskilled and unnecessary intervention during parturition, particularly if someone's been diving in up to their elbows in an attempt to extract the second lamb of twins, when really the ewe should have been left to get on with it in her own time. Almost without exception, every uterine prolapse I've had to deal with has been in a ewe that lambed at night under the supervision of an impatient shepherd, who simply wanted to get the job over with as quickly as possible so he could go back to bed! It's also perhaps worth pointing out that I've never had a uterine prolapse occur in any flock over which I have had total responsibility for management and staffing.

If you decide to replace a prolapsed uterus yourself you'll ideally need an assistant to hold the ewe up by the hind legs. A straw bale wedged under her rear will help take some of the weight. Any gross contamination (bits of straw, dung, etc.) can be carefully picked off, but don't go overboard on trying to clean up the messy organ or you'll damage it and make matters worse. Smother the whole thing in lubricating gel, and, starting from the base (where it exits the ewe), cup your hands around it, squeeze, and push! Little by little it should go back where it came from. Follow it up by inserting your arm (well lubricated) to ensure that it's fully restored to its rightful place. You'll need to fit a restrainer harness to ensure she doesn't push it all out again (make sure you do it up really tight), and give long term antibiotic cover. Affected ewes should be culled before the next breeding season.

If you get your vet to replace a prolapse he'll be able to administer an injection to prevent the ewe from straining, which makes the whole job a lot easier, but there is the added risk factor of delay, or damage to the uterus while transporting the patient to the surgery.

In its worst extreme, a prolapsed uterus can rupture, resulting in the intestines of the ewe spilling out through the tear. If this occurs, the ewe should be euthanased immediately. Don't stop to think about it. Just do it.

Retained Placenta

This is most likely to occur in ewes that have aborted, particularly if the foetuses had begun to decompose, and in ewes that have suffered from calcium deficiency. Inexpert assistance in the delivery of lambs, and subsequent infection, also increases the likelihood. Provided that affected ewes are given long term antibiotic cover they can safely carry a retained afterbirth for some considerable time, with no ill effects. It gradually breaks down and comes away piece by piece. Never try to hurry it along by pulling on the protruding part. Without antibiotic

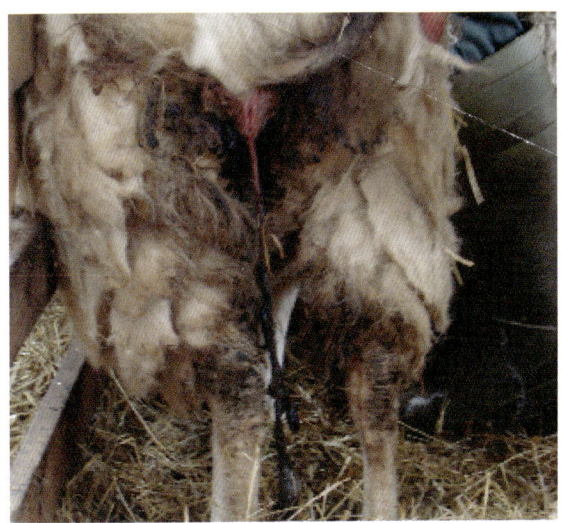

Retained placenta, seen here hanging from the vulva several days after parturition

cover the ewe would become very ill indeed. Often their milk dries up too, so do keep a close eye on the lambs.

Mastitis

Mastitis (lit: inflammation of the mammary organ) is yet another condition where I believe that prevention lies in management. If you selectively breed for good shaped udders, always crutch ewes after lambing, and use hydrated lime to disinfect individual pens between occupants, I don't think you'll see many cases of mastitis in your flock. It will tend to occur more often in years when orf (see page 133) is a particular problem, as the infection can transfer from the lamb's mouths to the ewe's teats. In its mildest form mastitis may simply be seen as a reduction in milk production leading to stunted lambs, but in more severe cases the udder is swollen, hot, and discoloured (purplish red). No milk can be drawn from the teats, only serum, pus, or blood. As the condition progresses the udder becomes cold and large areas will slough away. Once the gangrenous

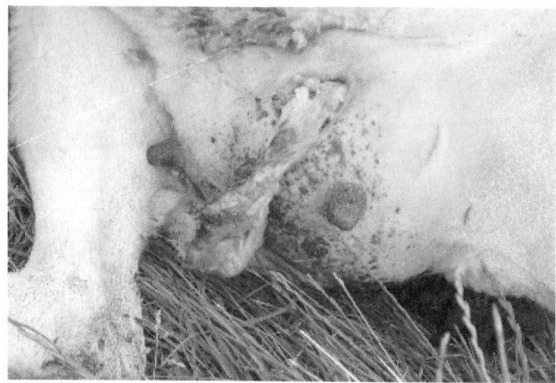

An advanced case of mastitis, where part of the udder has begun to slough away

tissue has fallen off (assuming the animal lives this long) ewes will often recover quite rapidly, provided that suitable antibiotic cover is given and that the wound is protected from fly strike until healed. Obviously these ewes should be culled. If detected and treated in the early stages mastitis can be successfully cured, although it is likely to reoccur in future lactations, so I'd recommend culling even minor cases.

Incidentally, if you ever find a sick ewe standing in a stream then suspect mastitis – affected ewes seem inexplicably drawn to water.

Magnesium Deficiency

Rapidly growing spring grass is notoriously low in magnesium, hence the need to provide high magnesium licks at turnout (see page 115). Magnesium can also be administered to each ewe in the form of a slow release bolus, but I've found that these never provide cover for as long as they should.

Magnesium deficiency is generally characterised by sudden death, which makes treatment rather tricky. Dead ewes are usually found flat out on their sides, and there are often marks on the ground to show that they've had frantic convulsions before death. In the event that you do find a hypomagnesaemic ewe

HUNGRY LAMBS

A hungry lamb tends to stand with its back arched and all four feet fairly close together. The head is drawn back to the shoulders and the neck curved in a rather camel-like posture. If lying down he'll invariably have his head tucked in his flank. When roused he'll follow his dam in a shambling fashion, with neck outstretched and head held low. His sides look clapped, and, when handled, he'll feel very "loose-limbed". It may simply be that this lamb didn't "mother-up" very well after turnout (perhaps the ewe ran off down the field with one of her twins, leaving the other standing pitifully alone just inside the gate), in which case all you'll need to do is pen them up again overnight, but there could, of course, be some underlying problem such as mastitis in the ewe. Either way, if you spot these signs, bring in the whole family.

while she's still alive, on no account attempt to move her before treatment or you'll trigger the final fit that signals the end. Treatment is by a subcutaneous injection comprising of 40 - 50ml of calcium borogluconate mixed with the same amount of magnesium sulphate.

Scours, Watery Mouth And Joint Ill

These 3 significant causes of losses in young lambs are, to my mind, largely attributable to shoddy husbandry at lambing time. If you maintain good levels of hygiene, appropriate supervision, and strictly adhere to a straightforward set of basic management guidelines, they're unlikely to occur in a small flock:

"Gunged" Lambs

After passing the meconium, (which is usually dark greenish black in colour) lambs will produce thick, dark yellow, sticky faeces, as a result of the rich colostrum they're feeding on. This can stick to the wool and the skin around the lamb's anus, fastening the tail firmly down, and effectively preventing the passing of any more dung. Not surprisingly the lamb feels pretty uncomfortable. If the muck has hardened don't try simply pulling it off or you'll tear the delicate skin. Clip it away carefully, perhaps having softened the toffee-like lump with warm water first.

While on the subject of bunged up lambs, I should just mention that one is occasionally born with an imperforate anus. Usually the first thing you'll notice is a lamb a couple of days old blowing up like a balloon! Apparently, in some cases, the condition can be surgically corrected, but I have never known this to be successful.

• Keep lambing yards well bedded.
• Administer thermovite to all compromised new born lambs immediately (ie. within minutes of birth). Ensure that this is followed in due course by an adequate feed of colostrum, either from the ewe or by stomach tube.
• Treat the navels of all newborn lambs, as soon as you are certain that the mother has accepted them.
• Don't keep ewes and lambs confined to individual pens for longer than is absolutely necessary.
• Dry and disinfect bedding in individual pens using hydrated lime, and put down fresh straw, between every occupant.
• Any pen that's held a poorly animal should be thoroughly mucked out, disinfected, and rested.
• Keep stocking densities in turnout fields low.
• If ewes are trough fed after turnout, move the feeding area regularly.

Hypothermia

Hypothermia is the single biggest cause of losses in young lambs in the UK, with maybe more than a million deaths occurring each

TIP! The most effective treatment for digestive upsets in young lambs is natural live yoghurt! At the other extreme, an enema of warm soapy water will soon get a constipated lamb going – definitely worth a try if the meconium hasn't been passed

lambing season. However, except under really extreme weather conditions, (such as we are not likely to experience on a regular basis), well fed, healthy lambs with dutiful, milky mothers will not suffer from hypothermia. Therefore, it stands to reason that the key to avoiding these losses lies in early identification of hungry lambs, and swift action to forestall consequent deterioration.

Treatment Of Hypothermia

Successful treatment of hypothermia depends upon you knowing the age of the lamb. Lambs are born with body reserves of brown fat which provide them with the energy they need to get to get to their feet and search for the teat.

By the time a lamb is 5 hours old it will have

ARTIFICIAL REARING

In an ideal world, all lambs would be under ewes, and all ewes would be rearing lambs, so really, if you have to artificially rear any lambs, then that's a sign of failure! It's also rather costly, so should be avoided if at all possible.

Initially, while there's still hope that you might be able to foster a spare lamb onto another ewe (see page 126), it should be fed by stomach tube, in order that it doesn't get too attached to its two legged "parent". Bottle feeding should be seen as a last resort, if a suitable foster mother isn't found within a day or two.

Inevitably, in some years, there'll be some lambs that need to be artificially reared. We use a powdered ewe milk replacer for this (which should always be mixed up slightly weaker than the instructions on the bag suggest), although goat's milk is ideal if you've got an adequate supply. A moderate sized lamb is going to need in the region of 1 litre of milk in each 24 hour period. Initially, this should be divided equally into 6 evenly spaced feeds, but the frequency of feeding can soon be reduced to four feeds of 250ml each. By the time lambs are a week old the total amount of milk fed may need to be increased to as much as 1½ litres per day. At the earliest opportunity reduce the level of inconvenience by getting lambs onto a cold ad-lib feeding system. This simply consists of a couple of teats mounted on a back plate, which is fixed to the side of the pen. A pipe, fitted with a non- return valve, runs from behind each teat into a lidded bucket, which you fill with a whole day's supply of milk. Lambs will naturally adopt a "little and often" feeding routine, and are much healthier as a result. Offer high quality creep pellets from the very start, and provide either coarse hay or leafy barley straw to stimulate good rumen function. It goes without saying that fresh, clean water should be available at all times.

Wean lambs as soon as is practically possible. The first step in the weaning process is to further reduce the concentration of reconstituted milk, or, if whole goat's milk is being used, water it down quite considerably.

exhausted all these reserves, and will depend wholly on nourishment from outside sources. A chilled newborn lamb (i.e. less than 5 hours old) may simply require warming, but an older lamb must be provided with sustenance before any attempts are made to raise its body temperature, or, as its metabolism speeds up, it'll simply die in a low blood sugar fit (flat out, legs paddling, neck craned back).

Rules Of Thumb:

• Lamb under 5 hours old, body temperature between 37 and 39ºC. Feed warm colostrum by stomach tube.

• Lamb over 5 hours old, body temperature between 37 and 39ºC. Feed warm milk by stomach tube.

• Lamb under 5 hours old, body temperature below 37ºC. Place under heat lamp. Monitor rising temperature. Give warm colostrum by stomach tube when body temperature is between 37 and 39ºC and lamb is able to hold up head.

• Lamb over 5 hours old, body temperature below 37ºC. Give warm glucose solution by intraperitoneal injection (see page 109). Place under heat lamp. Monitor rising temperature. Give warm milk by stomach tube when body temperature is between 37 and 39ºC and lamb is able to hold up head.

OBSTETRICS

What to do when things go wrong:
1. Don't panic.
2. Put the kettle on.
3. Re-read what I wrote about time & patience.

The vast majority of births (94 – 96%) should not require any assistance, and in all likelihood you're worrying unduly. However, at times things do go wrong, and malpresentations sometimes occur – you must be able to determine the nature of the problem (if, indeed, there is a problem at all), without exacerbating the situation. On no account should you attempt to deal with any complication that you feel may be beyond your level of experience. If the initial examination leaves you puzzled, then seek professional help. Don't go fumbling around inside the ewe for ages, expecting enlightenment – you'll be causing untold damage to the poor animal, seriously reducing the chances of a successful outcome.

My advice to anyone who is new to keeping sheep, (and to quite a few who are not so new to it), is to attend a lambing course, such as the one we hold in March each year, where it will be possible for you to spend time learning to correct malpresentations using an artificial ewe. In this way you will be able to develop your obstetric skills without putting any lives at risk.

Above all, remember that obstetrics should be the last resort, never the first line of defence.

Initial Examination

Check that you've got a clean glove in your pocket, and lubricating gel within reach. Restrain the ewe with your left hand under her chin, and lay her down gently on her side by following steps 1 – 4 of my instructions on page 48. At this point, step back a bit and kneel down, pinching a good tuft of wool under your knees as you do so. This has the effect of rocking the ewe back slightly, keeping her hooves off the ground and preventing her from struggling. You are now in an ideal position to carry out an internal examination using your right hand, and, once you've got the glove on and applied plenty of gel, your left hand is free to help hold the ewe. You'll find that this position is far more comfortable (for both sheep and shepherd) than the method that is usually illustrated.

In most cases, (but not always), your examination will reveal the presence of parts of the lamb in the birth canal. You should be able to differentiate between front and rear hooves, and between heads and bums, and be able to determine whether all the bits you can feel belong to the same lamb! If the presentation is normal, (i.e., head and two front feet of the same lamb), then let the ewe get on with it in her own time. If, however, you're faced with something more unusual, then you must make an immediate decision as to whether the situation falls within your capabilities. Don't proceed without help unless you are absolutely sure of what you are doing.

PROBLEM SOLVING

Now, let's assume that you've made an initial examination. Hopefully all is normal, and you were worrying unnecessarily, but supposing that

Learning to correct malpresentations using a dead lamb in an artificial ewe

its not? What can you feel, and what can you do about it?

Nothing At All In The Birth Canal

Firstly (and this may sound a bit daft), but are you sure that the ewe hasn't lambed already, and had her offspring stolen by another expectant mum (known as mis-mothering)? You wouldn't be the first person to be caught out like this! Having ascertained that the ewe does indeed have a lamb (or lambs) still inside her (by feeling her tummy just in front of her udder), consider her age. In the case of a young ewe, lambing for the first time, it may take hours for the cervix to open up and let the lamb pass through. Perhaps you're interfering too soon? When an older ewe is slow to dilate, an injection of calcium borogluconate (60ml, subcutaneous) will often work wonders – a minor case of calcium deficiency will cause poor muscle tone, even though no other symptoms may be apparent.

RINGWOMB AND TORSION

Some books on sheep-keeping seem determined to frighten the newcomer with descriptions of a mysterious condition known as "ringwomb", where the cervix simply will not dilate. The only course of action given is to carry out a caesarean, and subsequently cull the ewe. Well, all I can say is that in nearly 25 years of shepherding, during which I've been responsible for many thousands of lambing ewes, I've never come across the problem, so clearly it isn't a very common one! Generally speaking, time and patience will win in most situations. A uterine torsion will also result in nothing passing through the cervix, but again, this is very rare. I've seen it only twice in my career. If you do come up against it, it's definitely a job for the vet.

There are a couple of malpresentations that may result in you not being able to feel any part of a lamb in the birth canal: firstly, where only the shoulders of the lamb are presented, and secondly if it's coming rump foremost. It can be difficult, initially, to differentiate between the two, as the lamb will just feel like a smooth ball (without any useful identifying features such as legs!) blocking the cervix. In either case, it's helpful if an assistant holds the ewe upside down by her hind legs while you sort things out – this takes the pressure of the lambs off the cervix, enabling you to pass your hand through without pushing. If it's a bit tight, massage the cervix gently with your closed fist until things slacken up a bit. Once you've jiggled everything into position, the ewe can be laid back down on her side for the delivery. In the first case the lamb will be brought out in the normal delivery position, but, where the lamb was presented rump first, you'll deliver it as for a posterior presentation.

Tail Only (Breech)

This is very similar to, but easier to deal with, than the "rump first" presentation described above. Sometimes only the tail will have come through the cervix, but often the hocks can also be felt in the birth canal, and the tail can be seen protruding from the vulva. As always, the lamb should be fully returned to the uterus before attempting to correct the position. It's a fairly simple matter to work your hand down the length of a hind leg, when you've got the lamb's tail and hips present as a starting point. Make sure you keep in contact with the same leg all the way to the hoof. When you reach the hoof, cup your fingers under it, and begin to draw it upwards, while pressing forward on the hock with the heel of your hand. The hoof will come up easily to begin with but, just as you think you're nearly there, you'll run out of space! At this point, rotate your hand slightly, so that your thumb pushes down on the hock, and use your fingers to guide the hoof under your wrist. You'll feel a "click" sensation as the leg straightens and pops through the cervix. Now repeat for the other hind leg, then deliver the lamb in the posterior position. Never attempt to turn around a lamb that's coming backwards.

Hind Legs Only (Posterior)

Firstly, are you sure that they're hind legs? It's easy enough to tell the difference. Hind legs are identified by the fact that the first and second joints (the pastern and the hock) bend

in opposite directions, and the hoof tips point down towards the ewe's udder. In the case of front legs, the two joints (the fetlock and the knee) bend in the same direction, and the tips of the hooves point upwards towards the ewe's spine. Having ascertained that you are indeed faced with a posterior presentation, you must determine whether both the hind legs belong to the same lamb. Feel your way up one leg, then down the other, ensuring that your hand remains in contact with the lamb at all times. If both limbs do belong to the same lamb, then proceed with the delivery. If not, attach lambing ropes as identifiers before pushing both lambs further back into the uterus. Using the techniques already described (see "breech", left) locate and manipulate the second hind leg of the more progressed lamb. Deliver this lamb before attempting to correct the position of the other.

In many cases, ewes will naturally give birth unaided to small lambs in the posterior position, particularly the second lamb of twins. However, if you do need to give assistance during a posterior delivery (perhaps because you've had to manipulate a breech presentation), this is the one and only time where there is a need for some haste. This is because the umbilical cord will become pinched between the lamb's body and the pelvis of the ewe while the head of the lamb is still inside the fluid filled uterus. The immediate response to this, on the part of the lamb, is to attempt to draw breath, whereupon he drowns. So, to be sure of a successful outcome, once you've got the lamb into position, pause for a moment in order to make sure that nothing will impede a smooth and rapid delivery. You'll need a clear space behind the ewe, as the whole lamb needs to be drawn out in one movement, so if she's too close to a wall or hurdle swing her backend round a bit. Now, make sure that the lamb's legs are fully extended, insert loads more lubricant, and go for it! The direction of pull, in one smooth steady movement, should follow the curve of the animal, although larger lambs will need to be rotated slightly in an attempt to lessen the pressure on the umbilicus. Assuming that you're using your right hand to draw the lamb, have your left poised ready to clear the mucus from the lamb's head as soon as it comes into view. You may need to dangle it upside down for a few moments to drain fluid from the mouth and nostrils. Hopefully, the lamb will cough and splutter and begin to breathe immediately. If not, take a firm grip on the hind legs (careful, they're slippery!) and swing the lamb around a few times. And I mean really swing it – full circles, arm's length, up and over your head. Four revolutions should be enough, then drop the lamb on the straw with a bit of a jolt. That usually does the job! If he's still slow to breathe, try poking a bit of straw up his nostrils or squirting cold water in his ear. Sounds bizarre I know, but these old tricks really do work.

Front Legs Only (Head Back)

Using the methods described above, make sure that they are front legs you can feel, and that they both belong to the same lamb. Attach lambing ropes to both legs above the fetlocks, and push the lamb further back into the uterus. Now locate the head. Generally this will be lying to one side or the other so, starting from the lamb's shoulders, pass your fingers over the bend in the neck and slide them all the way down the far side 'til the head is reached. What you do next will depend upon the size of the lamb. If it's a small one, simply cradle the head in your hand and draw it up into position, while maintaining a gentle tension on the lambing ropes (with your other hand) to ensure that the legs are in the correct position too. Now, with your hand over the lamb's face (his nose poking into your palm), spread your fingers a few times to stimulate the ewe into straining. With a bit of luck, the lamb will be delivered straight into your hands, with no effort on your part at all! If the ewe doesn't respond by straining, you may be able to pop the lamb's head through the cervix by sliding your fingers over the crown of his head and applying a bit of pressure from behind. A bigger lamb presents more of a challenge. There just won't be room to cradle the head and lift it up as I've described – you'll need to actually get a grip on it. Never, on any account, be tempted to hook your finger under the lamb's lower jaw to raise the head. It will probably break. The only safe way to get hold of the head is by using your thumb and / or fingers in the eye sockets. Don't be squeamish – the eyes are remarkably flexible, and will not be damaged by this. Once you've got everything lined up for delivery, it'll not be possible to guide the head of a large lamb through the cervix using your hand as I've described above – in fact you may be wondering how it's going to fit through at all, with the front hooves apparently filling most of the available space. If you try pulling on the leg ropes, the head will either stubbornly remain where it is (with the neck appearing to retract like a telescope!) or, worse still, fall back into its initial position alongside the body of the lamb. Now, cast your mind back to my list of essential items for the lambing shed (page 103). Remember that loop of plastic coated clothes line? This is where it comes into play. The stiff yet flexible nature of the clothes line enables you to fairly easily work the loop over the back of the lamb's head and behind his ears. (While you're doing this, keep a slight tension on the leg ropes with your other hand, or you risk inadvertently slipping them off as

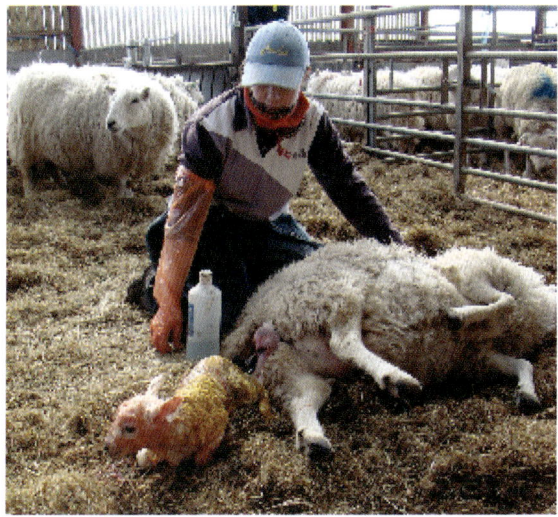

A student on one of our lambing courses puts theory into practice

you move your hand about.) Once it's in place, gently take up the strain, and it should stay put. Draw the legs forward slightly, then steadily pull on the clothes line. It's rather like that trick using a bit of string to remove a cork that's been pushed down the neck of a wine bottle! Once the head has popped through the cervix the rest of the delivery is straightforward. You'll be able to dispense with the ropes, and get hold of the forelegs with your right hand (one between thumb and first finger, and the other between first and middle fingers). As the birth progresses, you'll also be able to use your left hand to grip the lamb behind the head, and, working in conjunction with the contractions of the ewe, ease the lamb into the world.

Head Only (Front Legs Back)

If this problem is identified while the head is still in the birth canal, then it's relatively easy to correct. Often, however, the head will be visible outside the vulva, and may be swollen to quite alarming proportions. In either case, the head must be returned to the uterus before any attempt is made to retrieve the legs. When the

head is grossly enlarged this may seem to be an utterly impossible task, but I tell you, it **will** go! Use absolutely loads of lubrication, and, with your thumb in one eye socket, the joint of your first finger in the other, and the rest of your fingers curled gently under the lamb's chin, push hard between each contraction. Once the swollen head is back in the birth canal hold it there for a few minutes to allow it to deflate a bit before attempting to push it through the cervix. Once everything is back in the uterus, locate each front leg in turn by running your hand all the way down from the shoulder. On reaching the hoof, cup your hand under it to draw it into the correct position. You may like to attach a lambing rope to each front leg, above the fetlock, then proceed with the delivery as outlined above, making use of the clothes line if necessary.

Sometimes, in what appears to be a "head only" presentation, the forefeet may be found tucked up under the lamb's throat. This occurs when the elbows of the lamb have not straightened properly, and are stuck in the pelvis of the ewe. Always check for this, before starting to push the head back in, by running your fingers around the lamb's neck. If the hooves are found, it is usually possible to bring the legs forward into the correct position (without returning the head) by gently pulling alternately on one then the other.

Head And One Front Leg

In the case of small lambs, and in particular the second lamb of twins, it's quite acceptable to deliver lambs with one leg back. However, this should never be attempted with a big single lamb, or when lambing the first of twins from a young ewe. The correction procedure is as for "head only" (see previous), although of course there's only one missing foreleg to locate.

TIP! After an assisted delivery, always turn the ewe around to where the lambs are. Don't drag the lambs round to the head end of the ewe – firstly, that's not where she expects to find them, and, secondly, it'll result in the cord breaking off far too close to the lamb's body with associated risk of haemorrhage or rupture. Turn the ewe by rocking her back further onto her shoulder and simply swivelling her round. Provided that her feet are kept clear of the floor this is very easy.

If the ewe is hesitant about mothering the lambs, rub some of the birth fluids onto her gums, or cover her mouth and nose with a piece of membrane – she'll automatically lick this off, which is usually sufficient stimulant to trigger her mothering instinct.

Two (Or More) Lambs Coming Together

I think it's probably the thought of being faced with a jumbled up mixture of unidentifiable body parts that gives the newcomer the most nightmares! Other people's stories do nothing to alleviate these fears, as the average raconteur cannot help but embellish the complexity of each situation he's overcome! In reality, these multiple mix-ups are usually fairly simple to resolve, due to the small size of twins or triplets. Once everything has been returned to the uterus the lambs can be delivered one by one, using all the various techniques outlined so far.

IMPORTANT NOTE

Always use shoulder length obstetric gloves and plenty of lubricant, even for relatively minor examinations. There simply isn't any excuse for cutting corners. And, if you call in a more experienced neighbour to help with a difficult lambing, insist that they do too. Remember, attention to detail saves lives.

FOSTERING ON

By far the best way to rear a spare lamb (perhaps one of triplets) is to foster it on to another ewe, and, in the case of a milky ewe that's lost her own lamb, it's better to send her out with one of someone else's, rather than none at all. If no surplus lambs are available, consider splitting a set of twins. Always take the biggest or strongest lamb, preferably a female. Choose foster ewes with care. Look for good motherly types. Don't try fostering onto a first time lamber or a flighty ewe, and remember: always, always, always check the udder of a potential foster mum **before** giving her an extra lamb. You'll feel really daft if you've successfully fostered a spare lamb onto a ewe with a single, and subsequently discover that she's only got one teat!

Fostering is made much easier if your ewes have been pregnancy scanned. With really maternal breeds like Poll Dorsets it's often possible to simply pick up a newborn triplet and plonk it under the nose of a ewe that's just had a single, whereupon she'll accept it as her own. This will only work if the lambs are really newborn i.e., within a minute or so of birth. Otherwise you'll need to resort to cunning:

Rubbing on

The most successful fostering method is to catch a ewe that's in the process of giving birth to a single lamb, and deliver her lamb over the spare lamb that you want to foster on, such that it is saturated in her birth fluids. This is the **one** *and only* time that it is acceptable to assist a normal birth. Efficacy is much improved if the spare lamb is first dunked in a bucket of warm water. A further refinement is to place the spare lamb in a plastic feed sack or fertiliser bag, which is then placed under the backend of the ewe. This is made easier if you cut a slit in the bag for about half its length, starting from the open end. The ewe's own lamb is then delivered into the bag complete with all the birth fluids. Mix everything up a bit then give the whole contents of the bag to the ewe. It might be necessary to loosely tie together the front legs of the older lamb for a while. When you apply bactakil to the navel of the newborn lamb put some on the older lamb as well, and spray a little bit on the nose of the ewe, just to make sure everything smells the same.

Dustbin fostering

This method is worth a try if you're just too late for rubbing on. The ewe's own lamb and the one you want to foster on are placed together in a narrow container (such as a small dustbin) in a corner of the pen containing the ewe. She should just about be able to reach the lambs with her nose when they're standing up, but not harass them in any way. Periodically lift both lambs out together and put them to suckle, (under supervision), before returning them to the bin. A couple of days of this regime will usually do the trick.

Skinning

Where a ewe's own lamb has died it can be skinned and the skin put on the spare lamb like a baby's romper suit!

The skin should be removed in one piece, starting with a small incision at the belly. Cut around the leg joints, and around the head in front of the ears, and pull off the whole thing inside out. Cut through the tail bone on the

inside so that the tail also remains attached to the skin. Turn it back the right way and dress up the lamb to be fostered. If you've carried out the skinning correctly you'll not need any string to hold it on.

Generally speaking, once a ewe has allowed the fostered lamb to suckle a few times all will be well. Ewes have a scent gland either side of their udder (have a look next time you turn a sheep over to trim its feet) which oozes a waxy yellow substance – this finds its way on to the head of the suckling lamb and enables her to recognise it as her own.

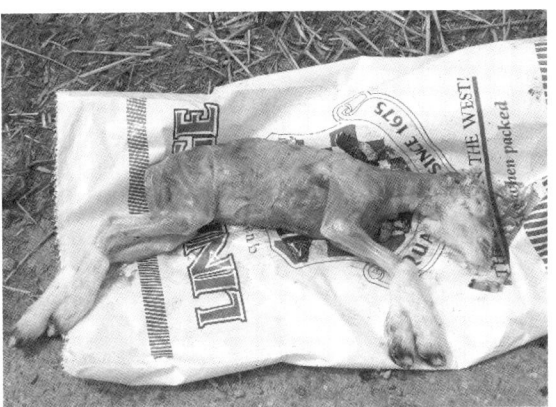

Above: One lamb dressed in the skin of another
Below: Skinned lamb

The Shepherd's Year
Summer

*Regular exercise on a hard surface is good for the hooves,
and helps keep the handler in good shape, too...*

On most traditional livestock holdings, May 1st marks the end of the winter regime. It's the day for turning out the cows after their long winter confinement, closing off the hay fields (and perhaps spreading a bit of fertiliser), and moving the sheep to their summer pasture. For us, this means taking the flock to the mountain.

PREPARING FOR SUMMER

Throughout April, the ewes and lambs should remain dotted around the holding in small groups, probably in the fields that you're about to close for hay. Now, on the 1st of May (or thereabouts) you'll need to gather those groups together. We prefer to handle each small batch separately, usually by setting up a portable pen of hurdles in the field, before merging the various groups. However, depending on the size of your flock and the layout of your holding, it may be simpler to walk all the sheep to a central handling system, which can be permanently set up near your buildings (see page 135). Also, we handle the twins and singles separately, as we'll not be sending the twins to the roughest part of the hill. The fact that we used different colour spray markers on the twins and the singles at turnout avoids mix-ups now.

IDENTIFICATION

Provided that your sheep are identified in accordance with the legal requirements (see Appendices) there is no need to record individual breeding details, except in pedigree flocks (many breed societies will have their own system of ID, which will need to be applied in addition to the compulsory Defra approved tags) and / or performance recorded flocks. However, if you intend to retain any of your

Provided that spray numbers were applied with care at lambing time, they'll still be clearly readable now

female lambs as flock replacements, even if non-pedigree or crossbred, it is well worth keeping accurate records. We have done so for many years, and it is fascinating to be able to trace back the lineage of our ewes to our original foundation stock. Detailed records prove invaluable when selecting replacement breeding stock, or deciding which ewes should be culled.

We do not eartag our lambs when they are very young as we feel that their ears are just too small. Instead, we rely wholly on the spray mark for identification purposes, hence the need for clear, bold numbering at turnout. By May 1st, however, the majority of the lambs are at least six weeks old, and their ears are well able to carry the weight of a tag. Provided that the spray numbers were applied with care, they'll still be clearly readable now, so it is a simple matter, when applying an eartag, to cross reference the tag number with the number sprayed on the side of the lamb, and thus complete the record in your notebook.

Tagging

There's a vast range of tags available for

Eartags come in many different sizes, shapes and colours

sheep, in all sorts of sizes, colours, and shapes. Choose a size and shape proportional to your breed. Many farmers use different colours to denote the year of birth, whereas others (like us) always stick to the same. In our case, the fact that all our sheep have the same colour tags makes recognition of stray animals much simpler. Whichever type of tag you choose to use, the basic method of application remains the same for (almost) all. Each tag consists of a "male" and a "female" half, the male having a sharp pin, and the female a hole. Often the two parts of the tag are separate, (two piece tags), but in some makes they're joined at the ends forming a continuous "one piece" tag. Generally, one part of the tag will be printed with your UK flock number, and the other will be printed with an individual number issued by Defra, but on some of the larger "flag" style tags, both numbers may be printed on one part, leaving the other side clear for you to record management information (such as the animal's name, or a year letter / number) using an indelible pen.

For insertion, the tag is held in an applicator that resembles a pair of pliers. The male part of the tag is inserted from the back of the ear. When the pliers are squeezed together, the sharp male pin pierces the ear and locks into the hole in the female.

Tattooing

Although the current rules for identifying sheep are described as "double tagging", it is acceptable to use a tattoo in the ear as the secondary identifier, rather than another tag. Tattooing has been a recognised form of identification for pedigree animals for many years, and, where accurate breeding records need to be maintained, is far more reliable than tagging, as tags are inclined to get lost. If the lost tag was the only form of ID used, then the animal's identity (and her history) is lost too. Although tattoos are usually applied to the ear, some breed societies suggest tattooing in the groin or under the tail.

The equipment required is fairly expensive to buy (£150 - £200 for everything you'll need), but it should pay for itself in time as you'll only have to purchase one set of tags each year. What's more, if you also keep pigs and goats, the same piece of kit can be used for the legal identification of all three species, representing a further saving.

(NB. Tattooing is not a permitted form of ID in animals intended for export).

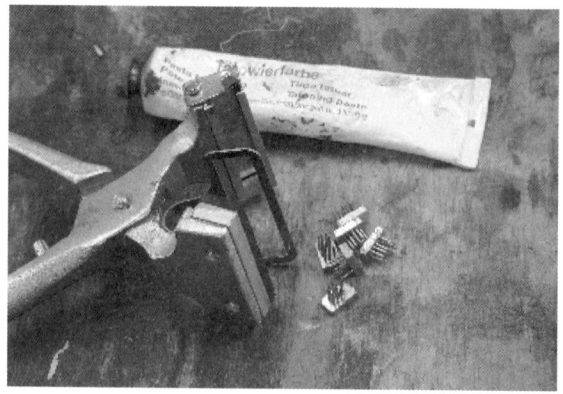

Tattooing equipment

MEDICATION

Having invested so much time and effort in the sheep over lambing time, it would be foolish to undo all that hard work by compromising on the health of the flock now. A few timely treatments ensure that lambs enter the summer grazing period well protected against the perils they'll encounter:

Parasite Control - Internal

Despite the fact that the ewes were dosed at turnout, lambs will be at risk from worm infestation as soon as they begin to graze. The most significant threat at this stage is probably *Nematodirus*, and, as the eggs of this particular worm will have been deposited on the pasture during the previous season, the situation is not affected by the dose given to the ewes after lambing. Large hatches of *Nematodirus* may occur during May and June, and, as the young lambs have not developed any previous immunity, the effect can be devastating. Lambs should be dosed now against all the major stomach worms, and if a product with a persistent effect is used (such as the one we use – Cydectin) this can reduce the number of times you'll need to treat throughout the season. However, the persistent effect does not extend to the control of *Nematodirus*, so where this is a significant problem you'll need a slightly different strategy.

The old recommended strategy of regular dosing and moving to clean pasture is now outdated, as it has been shown to accelerate the onset of anthelmintic resistance in parasite populations. Ideally we should be dosing lambs as little as possible, perhaps by using faecal egg counts to establish exactly when treatment is required. It's also a good policy to randomly not dose about 10% of the lambs on each occasion. This ensures that there is always a carry over of normal "healthy" worms to dilute the population of resistant ones.

When determining dose rates, always base your calculation on the weight of the heaviest lamb in the group – the use of an average figure would lead to under dosing of some individuals, which is another way that resistance may develop. Always check the calibration of your dosing gun before you use it, by squirting a set amount into a syringe or measuring cylinder.

Parasite Control - External

Ticks and blowfly are the worst pests at this time of year. You may experience problems with either one or the other, or both, depending on your location. On our grazing, flies do not really become a nuisance until June, but there are lots of ticks, so we treat each lamb with Spot-on before turning them to the mountain. Later in the season we use Dysect, which is effective against both ticks and flies. On lowland holdings, ticks may not be present at all, and the blowfly is likely to be more of a problem early in the year – there are a number of pour-on products available, such as Vetrazin and Crovect, for use as a preventative.

Clostridial Vaccination

If the flock lambed early in March then, by now, the immunity passed on to the lambs from their mothers, as a result of the pre-lambing booster against clostridial diseases, will be wearing a bit thin. For full protection, lambs should receive a primary course of vaccination consisting of two injections separated by an interval of 4 – 6 weeks. If the first jab is given now, the second can be done when the flock

Pneumonia

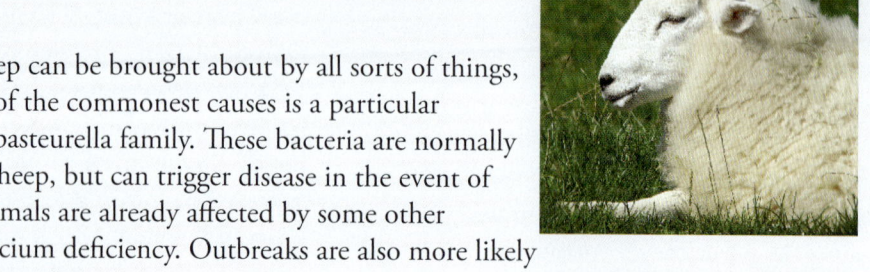

Pneumonia in sheep can be brought about by all sorts of things, but probably one of the commonest causes is a particular bacteria from the pasteurella family. These bacteria are normally found in healthy sheep, but can trigger disease in the event of stress, or when animals are already affected by some other malady such as calcium deficiency. Outbreaks are also more likely in certain weather conditions, particularly when cold nights are followed by warm, misty, still days (e.g. early autumn, about the time of the first frost of the year). Similar climatic conditions occur in poorly ventilated buildings, so newly housed sheep are vulnerable to infection. Avoid stressful handling, particularly in frosty weather, and ensure that sheep are dry at housing. Store lambs are also at risk following a move to new pasture.

Affected sheep are generally found apart from the main flock, looking fairly miserable, with ears down. If the flock is gathered these sheep will hang back, and, if you try to hurry them along they'll start coughing frightfully and find it very difficult to catch their breath. After coughing they'll often shake their heads to clear their mouth and throat of the white froth they've coughed up, and may stagger and go down.

Mild cases can be given oxytetracycline injections, but more severely affected animals may need to be treated by a vet.

is gathered for shearing in early June. For late March / April born lambs, give the first injection at shearing time, followed by another 4 – 6 weeks later, by which time you'll probably be wanting to gather the lambs for a 2nd worm dose anyway. Some of the vaccines available also provide protection against pasteurella pneumonia, and it is well worth paying a bit extra for a product that does.

Bluetongue Vaccination

Bluetongue is a notifiable disease that has fairly recently appeared in the UK (see appendices). All sheep in the flock should be vaccinated prior to the high risk period (mid-summer – early autumn). Lambs must not be vaccinated at less than one month of age, and full immunity occurs some 3 weeks after vaccination, so it appears that early May

is an ideal time to bluetongue jab the whole flock. Unfortunately the vaccine cannot be administered concurrently with any other injectable product, so you won't be able to tie it in with doing the clostridial vaccinations.

SUPPLEMENTATION

Clearly the ewes won't need any extra feed now that the grass is growing, but supplementary minerals and trace elements may well be required throughout the grazing season, in the form of blocks, licks, powders, boluses or drenches. Local knowledge should give you a good idea of what deficiencies are significant in your area, so seek the advice of your established farming neighbours.

We also provide lumps of rock salt for the sheep to lick, which they seem to enjoy.

TRANSPORTATION

For many of you, moving the sheep to their summer pasture will be a simple matter of opening a gate and shooing them through. However, there are lots of smallholders who rent or own additional grazing land some distance from their main holding, and in this case a trip in the trailer will be required. Remember: handle lambs with care! It is all too easy for lambs to receive injuries such as broken legs, or to be smothered, in an overcrowded vehicle. In our own case, we use a twin deck livestock trailer, with the lambs carried on the top deck and the ewes run in underneath. If your trailer is a single decker, then use a partition to separate ewes and lambs in transit, or put the lambs in the back of the towing vehicle. Keep the stocking density in the trailer fairly low – it is better to have to make two trips, than to arrive at your destination and find dead lambs on board.

When moving sheep between parcels of land separated by a distance of more than five miles,

ORF

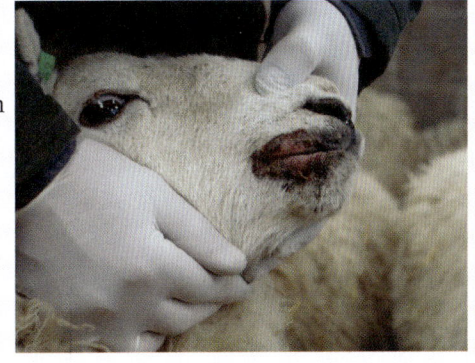

Orf (contagious pustular dermatitis) is very common in lambs in early summer, though also occurs at other times of year (usually autumn) in previously unaffected animals. It first causes small pustules, but, generally, by the time it is noticed, it looks like particularly nasty cold sores around the lambs' mouths. It also occurs at the tops of the hooves (causing lameness) and can sometimes spread from lambs' mouths to ewes' udders, leading to mastitis. Orf is caused by a virus, and cannot be treated directly, as it does not respond to antibiotics. However, it is very important to treat the lesions against secondary infection, as it is this that can result in the disease being prolonged, painful and unsightly. Purple foot spray or bactakil can be used to treat the sores, or even the sort of cream you'd use for putting on a baby's bottom at nappy changing time! Sudocreme is particularly good, but make sure you have separate pots for the bathroom and the sheep shed as orf is transmittable to humans. (For the same reason you should wear gloves when handling infected stock). Very severe cases of secondary infection will require antibiotic injections. Provided that secondary infection is kept under control, the disease will clear up after about 2 weeks.

The orf virus gains entry to the lamb via small wounds, such as may be caused by grazing amongst thistles, so regular topping of pasture to control thistles and remove coarse stemmy growth can limit the spread of the disease (see page 230). The virus can survive for long periods in dry places where sheep like to congregate, such as field shelters, and re-appear year after year. Animals that have been affected develop an immunity. There is a vaccination that can be used against orf, but, as it is a live vaccine, sheep that have been treated may become carriers. Buying in vaccinated stock is one way that orf can be introduced to a flock where it's not been seen previously.

We put young lambs on the top deck, and run the ewes in underneath

it is necessary to submit a completed movement licence to your local authority, even if all the land is owned or managed by yourself. All the animals moved must be correctly identified, and you'll need to obtain a separate holding number for the outlying land. Movements within the same holding, and less than five miles, do not require a licence. Movements between holdings, for example seasonal grazing rented from a neighbour, must be accompanied by the correct licence, even if the distance travelled is only a matter of a few yards!

Irrespective of whether a licence is required, all movements between parcels of land must be recorded in your flock register (see appendices). Movement of sheep, cattle or goats onto a holding triggers a standstill that prevents any animals being moved off the holding for 6 days, except for stock being moved directly to slaughter. It doesn't stop more animals being moved onto the holding within that period. Movement of pigs onto a holding triggers a 21 day standstill on pigs only, and six days for all other classes of stock. If you are intending to transport your sheep on a regular basis, perhaps exhibiting members of the flock at various events throughout the summer, it may be worthwhile creating an approved isolation

facility. This would mean that stock returning from a show needn't result in the rest of the holding being on standstill. It can also be very useful at times to have two holding numbers, so animals moved on using one number won't prevent animals being moved off the other. The extra holding number could relate to a single field.

While on the subject of transport, it is worth mentioning that EU rules governing the welfare of animals during transport came into force in January 2007, and since 5th January 2008 a competence certificate has been a legal requirement for anyone transporting farm livestock, if the journey exceeds 40 miles. These rules apply to all vertebrate animals being moved in connection with an economic activity, and when you come to think about it, there are very few movements that are *not* connected, in some way, with an economic activity. Taking a sheepdog bitch to be mated, for example, if she's a working dog and you intend to sell the pups. Or taking a couple of porkers to the abattoir, or rare breed sheep to a breed society show and sale! And be warned, if you take lambs to a market less than 40 miles from your holding, but fail to sell them, so bring them home, it's the combined journey distance that's counted, despite the fact that the animals were unloaded, rested, and the trailer washed and cleaned before re-loading. But if you do sell at the mart, the purchaser can then move them up to 40 miles without a certificate of competence, regardless of the distance you travelled bringing them to the sale! Crazy! And you have to pay for this piece of paper! You have to pay for, and sit, a separate test for each species of farm animal you may wish to carry, including horses and poultry. Good luck!

EWE LAMBS

One sector of the flock that I have not mentioned much so far is the ewe lambs (now yearlings) that were selected as flock replacements last year. Rearing good quality replacements can be difficult on a limited acreage, as pregnant ewes need to be given priority during the winter. Hopefully you'll have kept the ewe lambs growing on nicely, but, if you've struggled to find enough grazing for them, then perhaps you should send them away "on tack" in future years, (i.e., to spend a few months on another holding that has surplus grazing due to the fact that its own stock are housed for the winter), or consider whether it is worth rearing your own replacements at all.

Towards the end of lambing time, when the shed is nearly empty, we bring all our yearlings in, where they'll remain for a month or so. The principal reason for this is to teach them to eat concentrates, and to get them used to the housed environment. Some of them will stubbornly refuse to touch the feed, and may become quite thin, but eventually they give in. This may sound like a rather harsh regime, but it pays dividends in the following year, when these ewes will be lambing for the first time – they suffer no stress at housing, take readily to the feed, produce strong healthy lambs, and milk well.

Also, while they are housed, there is a fair chance that they'll come into contact with the protozoan parasite *Toxoplasma gondii* (spread by cats), and the mild infection that ensues in the non-pregnant animals renders them immune to toxoplasmosis for the rest of their lives. This is known as "vaccination by exposure".

While the yearlings are in we give them their booster jab against the clostridial diseases; dose them; check, trim, and spray all their

We house all our yearlings at the end of lambing time

hooves, and shear them. When we turn them out sometime in May we do the bluetongue vaccination, and apply a pour on against ticks and flies, then they're ready to spend the summer on the mountain, with very little interference from us. Their growth rate seems to accelerate after early shearing. By late summer they look fantastic, and so they should – they are the future of our flock.

HANDLING SYSTEMS

Design And Layout

However many sheep you keep, there will come a time when you need to pen them up for routine handling procedures such as dosing, foot trimming or weaning, or perhaps in order to load sheep into a trailer to go to market, or a show, or to move to fresh grazing. Whether you intend to construct a permanent handling system (which could be as simple as one or two pens in a convenient central location), or set up a temporary arrangement of hurdles when required, it is well worth spending a bit of time to get the layout right. Sheep are idiosyncratic creatures, and by taking account of their little quirks and foibles when siting your pens you can avoid much stress to both man and beast!

A Word Of Caution

Although I'll be talking here about gates and hurdles, what I have to say applies equally to pallets and string! Smallholders, on the whole, are great innovators, and the ubiquitous pallet tends to feature highly in many construction projects. However, on numerous occasions, I've seen someone waste a whole day in dragging heavy pallets into place, lashing everything up with yards of twine, with perhaps a bit of fence netting here and there for good measure. Eventually the flock is coaxed inside, whereupon they stream out again through some weak point, and career off down the field, much wiser now! No amount of shouting and running about will get them anywhere near that pen again! Finally a gang of neighbours are enlisted in an attempt to outmanoeuvre the wayward flock, and what should have been a pleasant 20 minute task of dosing a dozen ewes becomes a major operation. But it needn't be like that. Learn to think like a sheep, and set up your pens accordingly, whatever the materials you're using. It's worth it, if only in terms of blood pressure!

The Components Of A Handling System

The Shedding Race

Central to any well set up handling system is the shedding race. In fact, even if you're building your own system using odds and ends of reclaimed materials, I'd suggest that the shedding race (and associated gates) is one component it may well be worth purchasing. Basically, the race is a narrow passage (about 16 inches wide by 16 feet long) with a shedding (or drafting) gate at the exit end, and usually a guillotine gate at the entry. The sides of the

Using the shedding gate

race must be solid (so the sheep can only see ahead of them), and may be sloped in towards the bottom to prevent animals from trying to turn around. The drafting gate, fitted at the end of the race, swings from side to side allowing sheep to be shed off one way or the other, for example when weaning lambs. It should also be possible to close off the end of the race by keeping the drafting gate in a central position (it's a good idea to fit some sort of catch to hold it thus), although the sheep must still be able to see out forwards, as they tend to back away from a solid barrier – just one nervous ewe walking backwards down the race is enough to cause a pile up of literally hundreds of sheep! Some shedding races have a third exit in the form of a small gate in one of the side panels, immediately before the shedding gate. This gate opens inwards to intercept sheep coming down the race, and by using it in conjunction with the main shedding gate it is possible to split the flock into 3 groups in one pass. This is called three-way drafting, and requires a great deal of concentration and pretty good eye / hand co-ordination. Having said that, it really does speed things up – I've found that just myself and a good dog can split a flock of around 300 ewes into 3 specific groups (for example, according to raddle marks) in just over quarter

Dosing store lambs in the race

of an hour.

The guillotine gate, sited at the entrance to the race, operates just as its name suggests. The small gate is raised vertically, controlled by a rope led over a pulley at the top of the frame in which the gate slides. Once sufficient sheep have passed under the gate into the race, it is lowered down to close. (Don't drop the gate too soon, or heads may roll!). The rope controlling the gate should be fairly long (at least as long as the race), and may be operated by someone behind the sheep in the forcing pen or, if working alone with a dog, the rope can be led forward and operated from by the shedding gate.

Sometimes a race is fitted with "anti-backing" gates protruding inwards from the sides, supposedly in order to prevent sheep from backing out of the race. In practice, all they seem to do is impede the smooth flow of sheep through the race, turning shedding into a frustrating stop-start affair. You are much better off without them. In fact, there should be no encumbrances in the race at all, so any other bits of kit, such as footbaths or roll-over crates,

should be located elsewhere in the system or put into the race only when required.

The shedding race is also used to hold groups of sheep for routine tasks such as dosing, checking teeth, applying pour-on, vaccinating, marking etc. – it's a simple matter to walk down the outside of the race, applying the appropriate treatment to each animal, as they are held nice and neatly in a straight line, within reach. A 16 foot long race will hold 8 – 12 ewes, depending on breed.

The Working Race

The working race generally runs alongside the shedding race. Typically it would be 4 – 5 feet wide, and divided into two pens 8′ long (where the shedding race is 16′). The divider should ideally (but not essentially) be a hurdle with a sliding panel, so that space is not wasted by having to allow for a swinging gate. Sheep can enter the working race either from the forcing pen (instead of entering the shedding race), from the side of the shedding race (in the case of 3-way drafting), or by turning back after exiting the shedding gate. The working race is where tasks such as foot trimming and dagging are carried out, and is also used when dosing batches of young lambs, as it is not convenient (or safe) to try to handle very small animals in the shedding race (although they can be run through with their mothers and shed off for treatment).

Handling systems for very small flocks would probably have either a shedding race or working race, rather than both. In the first case, husbandry tasks can be carried out in the forcing pen or one of the exit pens, and in the second case sheep would need to be sorted manually using the sliding hurdle dividing the working area.

Exit Pens

Two pens are required at the exit end of the shedding race, in order to hold the respective groups of sheep. In some cases, for example when weaning lambs, it may appear simpler to have one side exiting straight back to the field. However, it only takes a momentary lapse of concentration to allow a lamb to slip the wrong side of the shedding gate, resulting in a lengthy chase to rectify the muddle! Even if you're short of hurdles, it's generally better to make a small pen that has to be regularly emptied, than try to cut corners.

Holding Pen

The holding pen is the area that the sheep are initially collected into. It could be a section of the farmyard that can be closed off with gates, a yard in the lambing shed, a small area fenced off at the end of a field, or simply a temporary ring of hurdles. The holding pen should be large enough to contain all the sheep to be handled, without having to squash them in. At peak times, for example when the whole flock with lambs at foot needs to be brought in, it's a good idea to open up other gates within the handling system allowing the sheep to draw through and fill all the available space. This simple tip will make gathering the flock much, much easier! Once the flock is in off the field, and the gate securely shut, they can be pushed back into the holding area.

Forcing Pen

The forcing pen is located at the apex of the holding pen, where the sheep are funnelled into the race. In more elaborate fixed handling systems it may consist of a circular pen with a gate hung on a central post. For most simple applications, however, it will be sufficient to have a gate that closes across the neck of the funnel (a "forcing gate"), or a few hurdles that can be pulled around behind a small group of sheep. As I mentioned earlier, the person operating the forcing gate can also control the guillotine gate at the entrance of the race.

Gates and Hurdles

Gates, gates and more gates! I've said it before, and I'll say it again – you can't have too many gates about the place! Galvanised metal gates are pretty good value for money, and are very versatile. For the construction of a permanent or semi-permanent system, I'd recommend building more or less the whole thing (except the shedding race) with gates. Once you've got posts with hinges set in appropriate locations it's easy enough to fiddle around with different combinations of gates to give the most appropriate layout for the task in hand. Also, it facilitates re-circulation of sheep through the system if any section can be opened to allow stock to pass. The same gates could be re-located and used for yard divisions at lambing time, as you'll not be needing the handling system for a while, once the flock is sorted and housed.

For very small set-ups (or for temporary handling facilities in the field) the same can be achieved using 4´, 6´ and 8´ sheep hurdles (of the type often used for individual lambing pens), though it is still a good idea to knock in a few posts here and there to give the structure some rigidity.

A very well designed weigh crate, manufactured by Modulamb

Weigh Crate

Although the weigh crate is perhaps not technically part of the handling system, it seems appropriate to mention it here.

Regular weighing of growing livestock should be part of every stockman's regime, but most particularly in the case of anyone new to stock rearing. Without the benchmark of experience, the beginner has no other way of knowing whether his animals are thriving or not. On many occasions I've seen very poor examples of animal husbandry on smallholdings, where the owner was blissfully unaware that anything was amiss. Regular monitoring of weights, and comparing the results with target figures, would have identified the problem at a very early stage, enabling management practices to be amended before welfare became an issue. Regular weighing can also be used to assess the effects of a change of diet, possibly allowing for more efficient use of lower cost or more readily obtainable feedstuffs. Recently, for example, I had to change the diet of my tup lambs, as the ram mix I had been using (the best available, I thought) was temporarily unavailable through my local merchant. The substitute feed did not look nearly so appetising, and I would have quickly reverted to the former had I had no means to monitor the effects of the change. To my surprise, the weights of all the sheep in the group increased considerably, after having been more or less static for some time!

Dip Bath And Foot Bath

These items can be considered as "optional extras".

Dip Bath

At one time, not so long ago, twice yearly dipping against sheep scab was a compulsory part of the shepherding year. Almost all fixed handling systems would have included a dip bath. In many cases this would be located beneath the shedding race, with a sturdy wooden cover enabling the race to be used for its usual purpose at other times of year. The forcing pen would become the catching pen, and the two exit pens would be used as alternate draining areas, allowing surplus dip to run off the sheep and back into the bath, before returning the flock to pasture. Larger handling systems would have a designated dipping area.

Nowadays it is not worth the additional expense of incorporating this feature into a small scale handling system. Dipping is no longer a legal requirement, there are serious environmental and human health concerns relating to the chemicals used, disposal of used dip requires an expensive permit and, what's more, there are a whole range of injectable and pour-on products now available which are capable of doing the same job with less hassle.

Foot Bath

Personally, I'm not a fan of footbaths, and would not bother to include one as a permanent fixture in any handling system. If you must footbath your sheep, a removable bath can be placed in position as and when required – a long, narrow "walkthrough" type bath (usually used with formaldehyde based solutions) can be temporarily located in the shedding race, and the flock run straight through, or one of the wider "stand-in" baths (for use with zinc sulphate) could be placed in part of the working race.

A permanent footbath, if deemed necessary, should be located in its own race. This could run alongside the shedding race, on the other side from the working race, avoiding the need for a second forcing pen.

SETTING UP

It is a well known fact that sheep will follow one another like…well…like sheep! In shepherding parlance this is called "drawing", and the phenomenon can be used to good advantage in the design of handling systems – sheep will "draw" through a well designed setup virtually unaided. Indeed, I knew one farmer whose flock of several hundred ewes would footbath themselves, while he went to have his lunch! (And if he hadn't returned by the time they'd finished, they'd all go through again!). Compare this situation to that of a smallholder of my acquaintance, who had to physically drag each and every ewe through his race, shedding gate, footbath etc., turning even the simplest task into a back-breaking nightmare! Yet with a few simple tweaks to the layout of his system, the one worked almost as well as the other.

SOME TIPS

• Sheep "draw" most happily uphill, so ideally the whole system should be on a slight slope. This is particularly important for the shedding race. Trying to get sheep to enter a race sloping downhill can be very frustrating!

• Sheep will move away from, rather than towards, buildings etc. However, it is helpful to have the handling pens in the vicinity of the sheep shed, with the race exiting close to the building, so that ewes can be shed off into the building if required. A suitable compromise is to have the race running parallel to the length of the building. The side of the building then forms one side of the handling system, saving hurdles and gates (see diagram page 142).

• Sheep will move from a dark area to a light one. This is one reason why they will run readily through a solid sided race, towards the see-through shedding gate. If setting up a handling system in a building, have the race running towards the open doorway.

• If your sheep regularly outwit your attempts to pen them up, try locating a temporary handling system at the gateway between two fields. Leave the field gate and the pen gates open for a few days, allowing sheep to move from field to field *through* part of the handling system. Now it's your turn to outwit them! Close a few gates and hurdles, then casually pretend you're simply moving the flock from field to field. Hey presto…gottem!

• Right angle turns in confined spaces, for example when entering or exiting the race, are to be avoided at all costs. This was one of the main faults with the smallholder's handling system I mentioned earlier.

• Some very up-to-date handling systems have a gently curved race, as it has been found that sheep are more likely to keep moving through. Presumably this is because they are trying to keep up with the sheep in front, which always seems to be disappearing around a bend!

• Make the holding pen roughly "L" shaped, or leave part of the handling system open when gathering the flock, in order to create

This semi-permanent arrangement in the corner of a field becomes a useful handling system with the addition of a few gates and hurdles

Fixed handling system at Glynllifon college

this effect. This will prevent the flock swirling around and pouring out again before you've had a chance to shut the gate!

• If using a temporary pen of hurdles when gathering sheep off the field, make the entrance narrower than the width of the pen, with a lead-in funnel if necessary.

• When erecting temporary pens in the field, it's common practice to set up alongside the boundary fence, making a few hurdles go a long way. Be aware, though, that handling sheep up against stock netting often results in lost tags and ripped ears.

• If you're behind a group of sheep trying to push them forward, and the foremost ewe decides she's not going to enter the race or pen, then no amount of shouting, whistling and arm waving will get things moving. Just back off a bit, and let them "draw". Remember that your voice will rebound off the side of a building or even the sheeted panels of the race – to the sheep it sounds like there's a second person in front of them pushing them back! A bit of peace and quiet will work wonders.

• Use your position and body language to work with, rather than against the sheep. So, for example, rather than standing behind a mob of sheep and trying to push them forward, try standing at the front of the group, close to the race entrance. Use your body to create a gap through which the sheep will draw, and step forward and back, or from side to side, in order to speed up or slow down the woolly torrent.

SUGGESTED BASIC LAYOUT

The following diagram gives a suggested layout for a permanent or semi-permanent handling system. Basically, it's the layout we use. It's centred on a 16′ shedding race, and the whole thing (apart from the race) is made of standard 6′ sheep hurdles and conventional "off the peg" galvanised gates, which do service elsewhere on the farm at other times of year. In our case, the holding pen is an area of the farmyard that can be closed off with gates, but a small fenced paddock at the end of a field would do just as well. From time to time we dismantle the whole thing and take it to another part of the farm, or set it up inside the building. The same basic layout can be scaled up or down according to requirement, and still be based around a 16′ race. A shortened race is not nearly so effective.

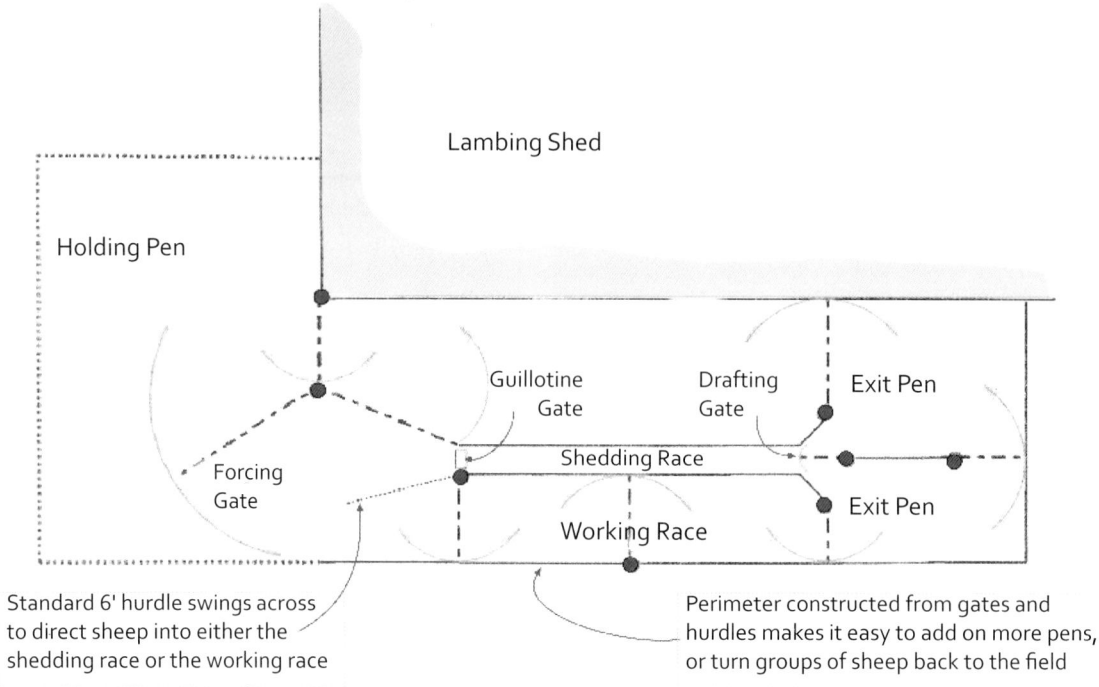

Lambing Shed

Holding Pen

Guillotine Gate

Drafting Gate

Exit Pen

Forcing Gate

Shedding Race

Working Race

Exit Pen

Standard 6' hurdle swings across to direct sheep into either the shedding race or the working race

Perimeter constructed from gates and hurdles makes it easy to add on more pens, or turn groups of sheep back to the field

ANOTHER WAY

So far I've mostly referred to fixed handling systems, or temporary arrangements that may remain in one place for some time. However, both farmers and smallholders are increasingly finding themselves running "fragmented" holdings, with sheep grazing parcels of rented land dotted around the countryside, often some distance from home. The fact that the flock is unlikely to be all in one place at one time, together with the short-term nature of many tenancies, effectively precludes any investment in a fixed system. A temporary setup of gates and hurdles wouldn't be altogether appropriate either – it's not a quick or simple matter carting a load of farmyard gates around from place to place, yet you can be sure that if you leave them set up in one field you'll almost immediately wish you had them somewhere else! Clearly what's needed here is a truly mobile handling system, and thankfully just such an item exists!

There are several models on the market, but for the purpose of research I paid a visit to sheep farmer John Hughes of Llanllechid, to see his Alligator in action (and I bet you thought shepherds only had working dogs!)

The Alligator

Pull into the field towing a lightweight trailer, and within 10 minutes you've got a fully equipped handling system, complete with holding pen, forcing pen, shedding race and drafting gates. And when you've done whatever shepherding task you came to do, the whole thing simply folds up like a concertina and becomes a trailer once again. That's the Alligator.

The basic unit is the race, which becomes the trailer when lifted onto its wheels (a simple process).

The drawbar is also detachable. The lightweight 5´ hurdles (really lightweight –

The Alligator handling system

Alligator race and drafting gates

they can be lifted above head height with one finger!) are carried, linked together and folded zigzag fashion, on the outside of the race, and all the other paraphernalia stows inside. The drafting gates are an integral part of the race. The system usually starts off with 15 hurdles, but in reality every Alligator is a custom build to the customer's own requirements, so a smallholder could begin with a very small setup, and add to it bit by bit as his flock develops. The biggest systems may carry up to 40 hurdles, together with extras such as shearing pens. Sheeted hurdles are also available in a range of lengths – these can be used to extend the shedding race, which at only 10′ is rather on the short side. The sheeted hurdles can also be used, together with gates and spreaders, to construct a sturdy working race. Small access gates can be added in here and there, making life much easier for anyone not nimble enough to step over into the pens. These would be particularly useful if you've taken up the option of having higher hurdles.

I am amazed at the lightness of the components – certainly a far cry from heavy wooden pallets, – yet all parts appear to be strong and well constructed. The sledge-like profile of the bottom of the hurdles allow them to glide over grass or uneven surfaces, so

the whole ring of 15 (or more) hurdles can be pulled shut behind the flock with the minimum of effort.

In use, I found that the sheep moved well through the system, and entered the race readily, and although the race was too short to be as useful as it could be, it would be fine for dosing small numbers of fairly docile ewes (see note earlier about lengthening the race). The drafting gates caused some problems as their use interfered with the smooth flow of sheep through the race. Again, maybe not an issue when handling small numbers of docile sheep, or when using the race for dosing, but certainly not an ideal arrangement. The problem could

The Alligator system folds up and becomes a trailer in minutes

be rectified by making the side exit drafting gates solid, rather than barred, so that each sheep being shed off can only see the outlet that you wish it to take, rather than the outlet its friends have just taken. (Remember what I said earlier about the need for the race to be solid sided). I've spoken to the chap in the technical department at CoxAgri (Alligator distributors) and suggested that they address this. I also found the drafting arrangement rather clumsy to use, but Mr. Hughes assured me that it's just a question of getting used to the way it operates.

In conclusion, I would say that a mobile handling system such as the Alligator would undeniably be a boon to anyone with sheep dotted about on short term rented grazing, though perhaps difficult to justify financially for a small flock. Having said that, it would be a simple matter for a group of smallholders to club together to buy the basic unit, with individual members owning any additional bits of kit needed to tailor the set-up to their own requirements.

As for me, well, I've been putting it on my Christmas present list for years, but clearly Father Christmas' budget doesn't stretch that far!

SHEARING

Come all my jolly boys, and we'll together go
About with our Captain, to shear the lamb and ewe,
In the merry month of June, of all times in the year,
'Tis always the season the ewe and lambs to shear;
And then we must work hard, boys, until our backs do ache
And our master he will bring us beer whenever we do lack!
(Old Sussex sheep-shearing song)

It seems that no sooner has the flock been turned away to its summer pasture, than it's time to fetch them home again for shearing! To the novice shepherd, shearing may appear almost as daunting a prospect as lambing time, and you may decide to have your sheep shorn by a contractor (certainly that's the easiest option!). However, I hope that this chapter will encourage you to tackle the job yourself. Don't despair if you struggle at first – it's bound to take a while to get the hang of it! I remember my own early attempts clearly; my father and I sheared a sheep between us, taking it turn and turn about to wield the clippers, while the other read out instructions from a book! It took us so long that we had to stop halfway, and allow the ewe to stand up to suckle her lambs!

Believe it or not, though, once you've got the knack, shearing becomes quite enjoyable, and if you take the time to become a reasonably skilful shearer, contract work can provide a useful sideline to running a smallholding. Professional shearers are often loathe to take on small flocks, due to the rigamarole involved in setting up all the gear just to shear a handful of ewes, so provided that you don't mind taking a bit of time, and doing a bit of travelling, you'll find that there are plenty of small scale shearing jobs available, where care and patience are more important than speed.

Shearing Gear

"Gear" is the name universally given by all shearers to their equipment, and, more specifically, the word is used when referring to the cutters and combs used in machine shearing.

By Hand Or Machine?

The choice is yours! If you've only got a few sheep to do, then hand shearing is fine. It's quieter, more portable, and a pair of clippers costs a lot less than all the gear required for machine shearing. And you can sharpen them yourself, too. There's no doubt that hand shearing is slower – it takes me almost five minutes to shear a ewe by hand, but as little as 1min 45 secs using a machine (and I'm a pretty slow shearer by professional standards!) – so if you intend to carry out any amount of shearing you'll need a machine. Having said that, for very small flocks, or where sheep are going to be shown later in the season, hand shearing remains a viable option, even for contract work. Ideally you should be able to do both, using the machine for the bulk of your shearing, then the hand clippers to finish off the odds and ends at the end of the season, such as ewes that had gone astray, late lambers, "sticky" sheep, and anything else that for some reason you'd been unable to shear when you had the machine set up.

Hand Shears

Hand shears can be purchased (for under £25) in a whole range of sizes and styles, so make sure you're using a pair appropriate for the job you're doing. My own experience of the different styles, and their uses, is as follows:
- 3½ inch blade, double bow, straight, for dagging.
- 6½ inch blade, double bow, cranked, for show trimming.
- 6½ inch blade, single bow, cranked, for show trimming, where fleece is matted or coarse.
- 6 inch blade, single bow, straight, for hand shearing.

You'll find that the single bow shears require a firmer squeeze, but, if you regularly hand milk a cow, goat, or ewe, then all the right muscles will be well developed! The advantage of using the single bow for shearing is that the blades spring open more crisply, so don't snag in the fleece as you slide the shears forward for the next cut.

Hand shears can be sharpened using the fine stone wheel on a bench grinder – cross the

TIP! Never shear wet sheep!

Quite apart from the risk of electrocuting yourself and the unfortunate animal, damp wool deteriorates in storage. If you've set a specific date for your shearing, perhaps because you've arranged extra labour for that day, then you may have to house the sheep overnight. This can present problems in keeping the animals, and their fleeces, clean. If you provide feed and bedding, such as hay and fresh straw, the wool will be heavily contaminated with vegetable matter by the morning, reducing its value and blunting your shears. But if you don't provide them, then the sheep will be hungrily paddling about in faeces and urine all night, and everything will be filthy. The answer is not to muck out the deep litter of the lambing shed 'til after shearing. Any loose material can be swept off the surface, leaving a hard packed absorbent bed, ideal for the purpose. If shearing day dawns bright and sunny then turn out the ewes to graze for an hour or two before you start – they'll be much easier to shear with nice, full, rounded bellies!

blades, then draw the cutting edge of each in turn across the grinding wheel, fairly briskly. Never try to grind the face of the blades, just clean them up with emery paper or wire wool. If you keep the shears well oiled when not in use, they should last you for years.

Machine Shears

The shearing machine is made up of three parts – the motor, the dropper (or shaft), and the handpiece. These components are not cheap (about £750 would be needed to buy a basic set of equipment), but can be purchased individually, so it is possible to build up a full complement from a mix of new and secondhand gear, depending on availability. Spare parts for most makes are readily available, and elderly machines can be reconditioned. The old Lister 3 speed shearing machine remains one of the most popular, due to its longevity, I suspect, and these often crop up at farm dispersal sales, where even a tatty specimen may fetch a nearly new price! Look out for a secondhand Lister single speed machine – these are less popular with contractors, so represent a more affordable option. There are also some good 12 volt machines available now, which may be worth considering if you'll be shearing out in the field. These simply connect to the terminals on your car battery, and give perfectly adequate performance for shearing smaller flocks. There are a number of machines – both 12 and 240 volt – that have the motor in the handpiece. These are often sold as being ideal for small flock shearing, but in reality they are clumsy and awkward to use.

Rummage about in the various boxes of miscellaneous junk at any farm sale, and you're quite likely to unearth an old Lister handpiece – after being properly serviced and re-set, with perhaps a few small parts replaced, it'll last for years. My own is over 40 years old and still going strong!

You'll also need cutters and combs. At around £15 per set these aren't cheap either,

Lister handpiece

but by buying a couple more at the start of each season, you will eventually build up a reasonable stock of gear. Always use a comb with a width proportional to your level of experience, and to the type of sheep being shorn. At one end of the scale there are narrow, rounded combs, used for dagging and crutching or for shearing sticky ewes, and at the other extreme are the really wide, hollow combs, with flared outer teeth – only to be used by experienced shearers shearing good sheep. The Lister "Maverick" (88mm wide with a 6mm bevel) is a sensible general purpose comb, although the total beginner should start with something just a little narrower.

Sharpening gear is even more complex than shearing sheep, but as a grinding machine costs in excess of £500 it's hardly the sort of thing you'll buy on a whim anyway! Many of the companies that sell shearing equipment also offer a grinding service, so send all your gear away to be done at the end of the season. Newly ground combs will need to be "dressed" – use a piece of fine emery paper to round off the tips and remove any burrs, then drag the points back and forth through a piece of soft wood to polish them – or you'll leave unsightly stripes on every ewe!

Clothing

Footwear

Moccasins are best. Shearing gear suppliers sell felt or leather moccasins, but you could easily make a pair from a piece of hessian. Failing this, carpet slippers, soft plimsolls or just bare socks will do. Wellingtons and work boots are a definite no-no, I'm afraid.

Trousers

The best shearing trousers are made from a double layer of finely woven, soft, denim. They are fairly close fitting, and come high up at the back. A double row of loops enables two broad belts to be worn, giving extra support to the shearers back. Belt buckles should be done up at the side, not in front, to avoid discomfort when bending over all day. Ordinary trousers are OK for a limited amount of shearing, but without the double layer of cloth, grease from the sheep soon works its way through, and causes irritation to the shearer's legs. In some people this may develop into a nasty rash or small boils.

Top

A vest or singlet is usually worn. This must be long enough to tuck well into the shearer's trousers, keeping the small of the back protected against draughts. Always have a warm top handy to pull on whenever you stop for a break – even on a hot day it's all too easy to get a chill when you've been sweating.

Accessories

Oil Can

As well as fully lubricating the hand piece and dropper a couple of times a day, you'll need to give a quick squirt of oil to the cutters and combs between each sheep.

Screwdriver

For changing combs.

Stiff Scrubbing Brush

For cleaning accumulated grease from the underside of the comb, the top of the cutter, and from the forks that press the cutter onto the comb.

Broom And / Or Light Rake

To keep the shearing area tidy by frequently sweeping up loose snippets and oddments of wool.

Sweat Rag

Shearing is hot work, but you must resist the urge to wipe the sweat from your face with the back of your (greasy) hand – to do so would result in an irritating rash. Keep a small face towel hanging up within reach, and periodically mop your brow with that.

Tally Counter

To keep a count, when contract shearing.

Antiseptic Spray

Any small cuts on the sheep (and there's bound to be a few!) should be treated immediately with an antiseptic spray, powder, or ointment.

First Aid Kit

Hopefully it'll not be needed, but even relatively minor cuts and scratches are best covered up.

Circuit Breaker

A circuit breaker must always be used when shearing with a 240volt machine.

Wool Sheets Or Clean Paper Sacks

For packing fleeces.

SETTING UP

You'll need a nice clean area for shearing. If you're working indoors on a sound floor, then just have a good sweep-up, but if the surface is poor, or if you're working outside on the grass, then spread a large tarpaulin over the area. For actually shearing on, you'll need a sheet of plywood. This must be on level ground, just outside the gate of the pen in which the waiting ewes are contained. The pen gate should open outwards to the left (looking in), and the machine should be hanging a couple of feet to the right of the gate, at such a height that the dropper just "kisses" the board. Make the pen fairly small – just big enough for half a dozen ewes, quite tightly packed – and have a larger pen behind to hold the remainder of the flock.

Now, clearly it won't always be possible to lay everything out as I've described above, as you'll be working within the constraints of whatever buildings or handling facilities are at your disposal, but if you can get the relationship right between the relative positions of the board, the gate, and the machine, it'll make the whole job go far more smoothly. You'll see this layout replicated on all the mobile shearing trailers used by contractors.

Getting Started

Adjusting The Handpiece

Turn the handpiece over, and you'll see two large headed screws. These are to hold the comb. Slacken them off and slide a comb into position (with the ground face away from you, as the handpiece is upside down). Tighten the screws thumb-tight only. Turn the handpiece back up the right way, slacken off the tension screw, and place a cutter (with the ground face downwards) under the forks. Screw down the tension just enough to hold the cutter in place, ensuring that the pins on the forks are properly located in the holes in the cutter. Now set the "lead" and "throw".

• *Lead*

You'll notice that there is a small unground area at the tip of each tooth of the comb. This is called the "scallop". The cutter must not be allowed to run over the scallop, and the correct distance between the tips of the cutter and the scallop is called "lead". Adjust the lead by moving the comb in or out. Early in the season, or when shearing sticky sheep, allow only a little lead (1.5mm), but on nice, easy shearing, open woolled ewes, increase the lead to about 5mm.

• *Throw*

This is the side to side movement of the cutter over the comb, and is also adjusted by moving the comb. When at either extremity of its arc, the cutter must not project over the edge of the comb.

After setting the lead and throw, fully tighten the screws holding the comb, and adjust the tension screw until it is just possible, with firm pressure, to reciprocate the cutter by using your thumb to turn the cogs at the back. Fully lubricate the handpiece and connect it to the dropper, then place it flat down on the shearing board, pointing towards the gate of the pen.

Now you're ready to start shearing.

Style

There are many regional variations of style used in hand shearing, but for machine shearing the "Bowen" style has become universally adopted, more or less the world over. The rhythm and pattern of the Bowen style can equally well be followed by the hand shearer, the principal difference being that when hand shearing, you can, to a certain extent, shear round the sheep, whereas when machine shearing it is necessary to be constantly moving the sheep around, in order for the shearer to remain in roughly the same position relative to the machine. It is this sequence of co-ordinated movements, worked out by New Zealander Godfrey Bowen some 60 years or so ago (and described by him as "…a graceful slow waltzing movement") that gives the style its name.

Shearing Your First Ewe

Enter the small pen, and select a ewe standing facing away from the gate. Turn her over in such a way that you are exiting the pen as you lift her into a sitting position. Walk backwards out of the pen, pulling the ewe along on her bottom, and swing round so that you are facing down the board with your right shoulder alongside the shearing machine motor, and the ewe's right hind leg is alongside the handpiece. If the pen gate swings freely, fit a piece of bungee cord to pull it shut after you, or have an assistant to open and close the pen as required (the same assistant can then stand by to switch off the machine if things get out of hand). With the ewe sitting comfortably between your knees (in pretty much the same position as is required for foot trimming, but with the weight of the ewe resting on her right hip), pick up the handpiece in your right hand, reach across with your left to switch on the machine…and you're off!

1. After clearing the brisket, I then remove the belly wool. First I make a blow down the right hand side of the belly, ending on the bare patch to the right of the udder. The sheep's right front leg is then pushed behind my right leg, and I raise her left foreleg with my left hand, before making another blow, down the left hand side of her belly this time. I break open the fleece with my left hand, then proceed to clear the remaining belly wool using downward strokes of the handpiece. The belly wool comes off in one piece, and is tossed aside.

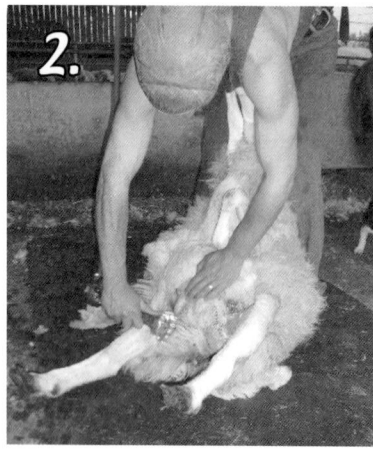

2. In shearing the crutch I run the handpiece down the front of the right hind leg, then clear the crutch from right to left, using my left hand to protect the ewe's teats. Then, with my left fist pressed into the ewe's left flank (in order to straighten her leg) I make one blow down the front of the left hind leg followed by a return blow up the outside of the leg.

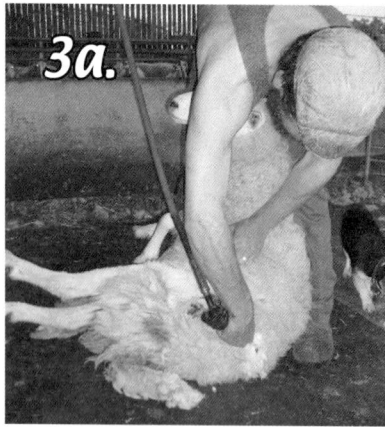

3a. Using my knees and feet I turn the ewe towards the machine, and clear the wool from her left hind leg by making a series of blows towards her backbone, with each successive blow being closer to the tail.

Here (3b), I've broken open the fleece to show you the area covered. Next, I'll shear her tail, followed by a short longitudinal blow either side of the backbone, to clear the rump.

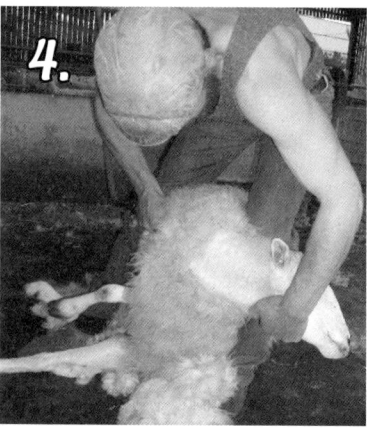

4. After shearing the left hind leg, I move my left leg forward while lifting the sheep with my left hand, then step forward with my right leg, placing it between the sheep's hind legs. I'm now correctly positioned to shear the neck. With my left hand, I stretch the ewe's neck around my knee and shear the right side of her neck. The blow commences just above the brisket, and exits the fleece just under the jaw, and then I break open the wool with my left hand. Many beginners are nervous of making this blind stroke, but provided the sheep's neck is properly stretched out, smoothing any wrinkles, everything will be fine.

7. As I make the last stroke of the long blows, I bring my right leg forward as far as the sheep's shoulder. Now, with the fingers of my left hand hooked under her jawbone, I begin to clear the right shoulder. I gradually lift the ewe, shuffling my feet backwards as I work, and put her head between my legs.

5. I turn the sheep's head, so her nose points upwards, in order to shear the left side of the neck and the back of the head. From this position, I start clearing the wool from her left shoulder, while my right foot moves her back end into the correct position for the long blows.

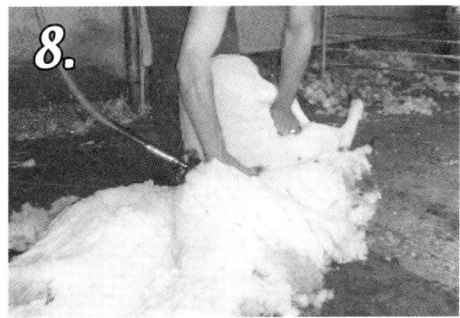

8. I continue shuffling backwards, and, after shearing the shoulder, I make a blow downwards towards the flank, following the edge of the area cleared when I removed the belly wool. The next blow runs parallel to the last, but continues out along the sheep's hind leg. I now press on her right knee in order to keep her hind leg straight, and continue shearing parallel strokes, until the whole of the leg and hip are shorn. And that's all there is to it! I switch off the machine while allowing the sheep to rise to her feet between my legs, whereupon she runs off behind me, leaving her fleece lying neatly on the board.

6. When the shoulder is done, the ewe is lowered to the board and I commence the long blows. These blows run the length of the sheep, parallel to the backbone, starting with a full comb width, and running out to a half, or less, on the neck. Initially, my right leg remains between the ewe's hind legs, but after the first couple of blows I step over to the back of the ewe, using my foot to roll the ewe into a position that enables me to shear a comb's width over the backbone.

ROLLING THE FLEECE

Shearing follows a set sequence of movements, so it stands to reason that each and every fleece will be left lying on the board in the same way, and you'll know which part is which. It's a simple matter, then, to scoop up a whole fleece in your arms and toss it out on the board with a flick of the wrists such that it lands neatly spread out, with the skin side underneath, looking for all the world like a sheepskin rug! If you do this correctly, the hind legs will be closest to you. (Blackface, Herdwick and Rough Fell fleeces should be thrown out skin side up.) The belly wool, which was put to one side during shearing, should be placed in the centre. Pull off any daggings and other obvious contamination, and fold in the sides of the fleece. Start rolling from the hind end, and finish at the neck. The neck wool is then loosely twisted and pushed into the rolled fleece in order to hold the whole thing together.

It used to be the practice that the neck wool would be twisted into a tight band that was wound right around the rolled fleece before tucking in under itself, making a very secure bundle. If, for any reason, you aren't able to pack your fleeces immediately, it would be worth using this old method of rolling.

Fleeces that you're keeping for your own use can be stored individually in paper sacks (empty feed bags), but any wool that will be sent to the BWMB should be packed into proper wool sheets. Hang up an empty sheet by the two top corners and fill it evenly, starting with a fleece in each of the bottom corners. Squash in as many as possible before closing with the skewers or tape provided. A floppy, loosely packed sheet is much harder to handle than a well packed firm one, despite the difference in weight.

Registration

If you keep more than 4 adult sheep then you will have to register with the British Wool Marketing Board (BWMB), and, in most cases, you'll have to sell your wool to them. Certain rare breeds are exempt from this requirement. At the time of writing the list of exempt breeds is as follows:

Balwen, Boreray, British Merino, Castlemilk Moorit, Galway, Hebridean, Leicester Longwool, Lincoln Longwool, Llanwenog, Manx Loaghtan, Norfolk Horn, North Ronaldsay, Portland, Soay and Teeswater.

This list is revised and updated periodically, and registered wool producers will be advised of the current ruling each season. A producer can also request an exemption for the purpose of retaining their home produced fleeces for artisan craft purposes, up to a limit of 3,000kg annually, or up to 15,000kg annually for non-textile uses such as insulation. Applications for exemption are considered on a case by case basis.

The board provides producers with wool sheets in which to pack fleeces, and arranges for collection of wool, either from the farm or from an authorised collection point (in which case there'll be a charge made for each sheet), or, alternatively, producers can deliver their own wool to their nearest grading depot or intermediate centre. Organic wool should be packed in green sheets, and kept separate. Once your wool has been weighed and graded an advance payment will be made, based on a percentage of its expected value. A balance payment is made later, according to the price that each grade fetched at auction. Usually the balance payment is made at the same time as the advance payment for the following year's clip. All producers have a right to be present when their wool is graded, should they so wish.

BWMB Training

The British Wool Marketing Board also organises training courses for shearers in the UK. In some areas, funding may be available through LANTRA. Courses can be held at an agricultural college or hosted by a large farm, and cater for all levels, from total beginners, up to the highly coveted "Gold Seal" standard. In addition to learning correct shearing techniques, students are taught about the proper maintenance of the shearing gear.

Many contract shearers attend these courses at the start of each season, in the hope of adding another seal to their certificates (blue for beginners, then bronze, silver, and finally gold), so it's an excellent opportunity to pick up tips from the professionals.

Buying British Wool

The following information was kindly supplied by Andrew Todd, Purchasing Manager for Brintons Fine Carpets, who gives us a buyers view of British wool.

"Brintons have been making carpets in Kidderminster since the 1780s, and still purchase over 4000 tonnes of British wool each year.

Climatic and geographical conditions allow differing breeds of sheep to prosper across the whole of the British Isles, and each breed produces its own individual style of fleece. Management and husbandry can make the fleeces higher or lower in quality, but the breed controls the overall character of the fleece.

British Wool has a great advantage over the wool grown in New Zealand, as their wool is predominately from the Romney breed of sheep, and, although this is a high quality fleece, the

FLY STRIKE

Although generally a mid-summer problem, I've found maggots on sheep as early as April, and as late as November! Any sheep with a mucky bottom in warm weather is at risk from fly strike. Growing lambs are most susceptible, due to the effects of lush pasture or worm infestation, but, in muggy weather, even clean sheep will be attacked. And it doesn't always occur on the rear end of the animal, either – ewes with footrot will often get maggots in the affected hoof, and also on their side and belly as a result of lying on the bad foot. Rams may be struck where wool is urine soaked around the pizzle, and a surprising number of cases begin in clean wool on the middle of the back.

Above: This ewe has been struck on her side. Note the dark patch of wool and the characteristic behaviour. Below: Clipping away the wool from the maggot infested area

The best protection you can give your ewes is to shear them fairly early. Leaving shearing 'til late is simply asking for trouble, particularly as the first signs of fly strike can be difficult to spot in a heavily woolled animal. Shearing time is also a

good opportunity to clip the bums of any dirty lambs, and apply a preventative pour-on to the whole flock.

To treat flystrike, clip the affected area down to bare skin, and clip a good way into the clean fleece surrounding the area, too. Sometimes it is simpler just to shear the whole animal. Minor cases will need no further treatment, but where the damage is extensive or deep you'll need to treat the area with a pour-on, or proprietary "Maggot Oil", and possibly administer antibiotics.

lack of cross breeding limits the NZ wool to one product type.

When selecting fleeces to create our blend, firstly I look for a high bulk fibre, which will give resilience and bounce to the yarn and to our carpet. Next, I add strength of fibre, to allow the fibres to be spun into a strong yarn, which will avoid fibre breakages in our production processes and reduce the amount of shedding of fibre in our carpets. Finally, I look for fleeces which are crisp to

handle.

In addition to these 3 main attributes, I aim to avoid contamination from marking fluids, and try to avoid contaminations such as baling twine within the fleece, and excessive vegetable matter.

The cost to the industry caused by contamination needs to be reduced by improved handling and by the use of registered marking fluids, which can be washed out in the commercial scouring process. A single producer's action, by marking with a product such as bitumen / oil based liquid, can cause many thousands of pounds of loss to the end user.

White kemps do not take dye and therefore excessive amounts of these dead fibres needs to be avoided. Red kemps give the yarn dyers a major problem as they tint the wool with their redness, resulting in variable dye uptake. When a kilo of dyed yarn costs between £3.00 and £4.00 the loss to the industry your product supports is excessive in failed dye batches.

Excessive amount of grey fibre is also a problem, and should be engineered out within the breed selection process wherever possible. This will allow wool to be used in paler yarn colours, and therefore can increase the demand for your product."

SHOWING

I'm not really suitably qualified to write about showing – we've only exhibited our sheep on a handful of occasions. However, those few outings have secured us 18 first prizes, 5 Reserve Champions, 3 Champions and 1 Supreme Interbreed Champion, together with many other lesser placings and a couple of trophies, so somewhere along the line we must be doing something right! Here I'll attempt to pass on some of what we've learnt over the past few years, which will hopefully provide a useful starting point for anyone considering launching themselves (and their sheep) onto the show circuit.

Where To Go

Some enthusiastic showmen will enter every event within reasonable distance, and are fully occupied throughout the whole of the summer. However, maintaining sheep in show condition for extended periods, without having them "go over", or stale, is an art in itself, and best left to the professionals. My advice to the beginner would be to set your sights on one particular event, and gradually build up the sheep so that they look at their best on that day, rather than trying to hold the sheep in show condition for several outings. After that event the sheep can be roughed off, and, if you've enjoyed the experience, you can begin to plan a more ambitious timetable for the following year.

For most would-be competitors, the obvious place to start will be at a local agricultural show. These small events are usually "un-affiliated", meaning that it is not necessary for your sheep to be registered with a breed society, as there are no society awards on offer. Having said that, there may well be classes or prizes specifically for members of local clubs or organisations, particularly in areas where one specific type of sheep (usually a local breed or sub-type) is very popular. The local shows tend to be very friendly affairs, where everyone knows each other, but there may be long established hierarchies – everyone knows that old so-and-so always wins such-and-such a class – so don't be too dismayed if you're beaten by the judge's wife's second cousin, with a sheep that's not a patch on your own!

Our own "target" event is the RWAS Smallholder & Garden Festival, held at Builth Wells each year in May. I think it would be fair to say that this is the only large show of its kind, where the amateur takes centre stage. One of the beauties of this event is that there are no classes for Maedi Visna (MV) accredited sheep, so the weekend is not dominated by professional showmen with commercial breeds.

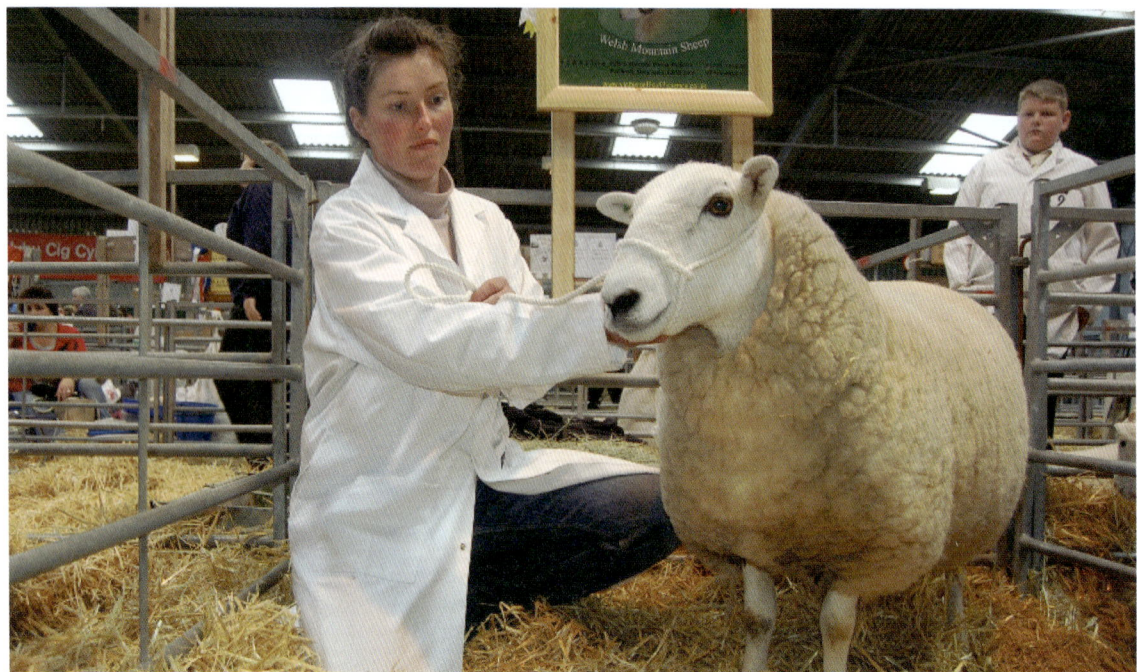

Dot with "Lucy", Supreme Champion at the RWAS Smallholder and Garden Festival, 2007

M.V. STATUS

Consider that in some breeds, most flocks have MV accredited status (see page 59). It may be financially unjustifiable to attempt to attain such status in a small flock, particularly where a mix of breeds are kept, yet competing with non-MV sheep of these breeds may place you at a disadvantage in the show ring, and could prevent you from exhibiting at the breed society's own events.

Choosing A Breed

The best thing to do would be to get out and about a bit, visit some shows, and learn to identify the various breeds, some of which are very similar in appearance (woe betide you if you put your foot in it by calling someone's Kerry Hill a Speckleface!). If it's your intention to attend local shows with a local breed, then be aware of regional variations within that breed – you may have lovely Welsh Mountain sheep, but if they happen to be the type from the other side of the hill, you'll not get a look in! Speak to the breeders and exhibitors of the types that take your fancy, and find out about the relevant breed societies – some are very open and welcoming, whereas others may be cliquier. Most showmen are happy to help a newcomer, and will gladly guide you as you take your first tottering steps into the show ring, seemingly very proud of their protégée. Be warned though: the supply of goodwill may cease abruptly when you start winning prizes – from that point on, you're no longer a newcomer, you're a rival!

Picking The Team

At all shows there will be classes for aged rams, shearling rams, aged ewes (which must have reared lambs in the current season), and shearling ewes. Events later in the year will also include classes for male and female lambs. In addition there may be specials such as a group of three (one male / two females), a ewe with her own lambs, and for a pair of butcher's lambs. I suggest that you don't try taking too many sheep to your first show, but by selecting a suitable team you may in fact be able to enter plenty of classes. One ram, together with a ewe and her twin lambs (one male, one female) would enable you to compete in 7 of the above mentioned categories, whilst allowing you to devote all of your attention to just a few animals.

Begin making a tentative team selection well in advance, perhaps even in the previous year, and make sure you pick a few extras to safeguard against mishaps – it would be a shame to pin all you're hopes on one ewe, only to find her barren! You can gradually thin down your selection, depending on how each individual develops – some exhibitors won't make a final choice until the very morning of the show!

It goes without saying that the sheep that you choose to show should be correct in terms of breed characteristics, but equally important, if not more so, is the animal's conformation. In some breeds, obsessive selection for particular markings or other characteristics of appearance has been detrimental to the overall strength of the breed, and the show ring is largely to blame for this trend. In my opinion, a really well put together animal shouldn't be moved down the line just because it happens to have a little bit of black where there should be white, or *vice versa*.

What To Look For

- **Teeth**

  Having a poor mouth is probably the most common reason for a sheep being placed down the line. The incisor teeth must bite directly onto the dental pad, not in front of it. It is a well known fact that housing and heavy feeding in the run up to a show are inclined to bring the teeth forward, so, where possible, keep the show team out at grass 'til the last minute. This is particularly important with yearlings that are cutting their new teeth. It is not uncommon for a good mouthed sheep to have gone over by the end of a long show season as a result of trough feeding throughout the whole of the summer.

- **Shoulder**

  The shoulder should be strong and flat. A weak, pointed shoulder can, to a certain extent, be masked by trimming the fleece, but you'll not deceive the judge's hand!

- **Back**

  The back should be long, strong, wide and level. In particular there should be no dip behind the shoulder, and it shouldn't slope off sharply behind the hips.

- **Chest and Neck**

  In a ram, the neck should be well set onto the shoulder, and the chest broad, flat, and deep, giving a strong masculine stance. The neck of a ewe should be more slender, without being weak, and the chest more rounded, giving a less aggressive, more maternal appearance.

- **Body**

  The body should be deep, though take into account that some breeds are characteristically more "leggy" in appearance than others.

- **Legs and Feet**

  Any sheep being considered for showing must stand well, with "a leg at each corner". The front legs (when viewed from in front) should fall straight from the shoulders, and, when viewed from the side, should not appear over or back at the knees. The back of the fetlock should not appear to sag – the sheep must stand well up on its hooves. The back legs must have a short pastern, or the sheep will be inclined to go down at heel, particularly after a long trailer ride. The part of the leg between the fetlock and the hock should be strong and straight. Hind legs should be set well apart, in-line with wide hips, and the sheep should naturally stand with its legs slightly back, not tucked in under its body. Action must be straight. Thickness of bone should be good but not excessive, proportional to breed, age, and sex.

Hooves should be trimmed to a good shape well in advance of show day, and kept in sound condition by regular exercise on a hard surface.

To Shear Or Not To Shear?

That is the question! Many breed societies specify that sheep should be shorn bare after the first of January in the year in which they are being shown. This is of course rather an open ended ruling, and liable to deliberate misinterpretation – some exhibitors may shear at the end of the show season, say October, claiming (correctly, of course) that October *is* after January! However, this is not the spirit in which the rule was intended! Other breed societies specify a period within which sheep should be shorn, for example *"…all sheep must be shorn bare between 1st January and 15th July of the year in which they are being shown"* which very sensibly closes the loophole. And some individual shows have their own ruling – the schedule for Cerrigydrudion show, held in September in North Wales, clearly states that *"All sheep must have been really and fairly shorn bare after March 1st [of the current year] and judges are directed to disqualify any exhibit if they*

are of the opinion that the same have not been properly sheared."

If you aim to exhibit at just one show in the season, then you can shear at the time that will have your sheep looking at their best for that event, and still remain within the rules. For a show in May, June or early July, shear the sheep late the year before, then shear them again immediately after the show. For shows later in the year, shear in early January. (Unless you're going to Cerrigydrudion…)

Feeding

This is one area of show preparation where there are almost as many trains of thought as there are shepherds! There are no hard and fast rules here, and everyone tries different combinations of ingredients and regimes in the hope of giving their sheep an advantage over their rivals! I once asked a fellow Welsh Mountain breeder what he was feeding his rams, as they were so much bigger than my own, whereupon he looked me straight in the eye and said "only grass and clover!" which, I think, was a polite way of telling me to mind my own business!

A Few Feeding Guidelines

• Start feeding at "tit bit" level well in advance of the show, so that the sheep are used to coming up to the troughs. Hopefully this will result in them taking readily to the increased feed level in the final run up, and should mean that less feed will be required to put a final edge on their condition.
• Whichever feed you choose to use, it must be something that your sheep like! There are a number of "ram mixes" available that are very palatable.
• As you increase the level of feed in the final

few weeks before the show, divide the total amount into several feeds per day.

• Include plenty of sugar beet pulp or nuts in the diet, in order to maintain sufficient fibre intake. Either feed dry as part of the ration, or soak and feed separately. Feeding sugar beet is also said to "bring out the head" of show sheep, particularly rams, but I've got no idea why it should have this effect. If anyone's got an answer, I'd love to hear it!

• The occasional minor digestive upsets that may be caused by high levels of cereal feeding can often be countered by dosing the animal with live yoghurt.

• Provide high energy licks, of the sort usually used for ewes in late pregnancy. Tithebarn's "Triple Energy" blocks are excellent, and give the sheep a real bloom.

• Rams in particular should have a lump of natural rock salt to lick.

• Ensure plenty of fresh water is always available.

• When the sheep are housed in the final few days before the show, provide roughage in the form of straw rather than hay. If they're allowed to bulk up on hay they will have a reduced appetite for the trough feed, and may develop a rather pot-bellied appearance.

• Don't feed the sheep immediately prior to departure, or immediately on arrival at the showground. Whilst away from home, reduce the level of concentrate feed, and provide coarse hay and / or soaked beet pulp.

Washing

Whether or not the fleece should be washed prior to a show varies between breeds, so find out what is the norm for the type you'll be showing. Even then you may be caught out – I once took a group of carefully washed rams to a sale, only to find that everyone else had applied a yellow bloom to theirs! My snowy white animals stood out like a sore thumb, and looked rather daft. Needless to say I did not receive a good price. The following year I bloomed my sheep, only to find that everyone else had washed theirs! Sometimes you just can't win!

If sheep are to be shown looking really white they'll need washing quite close to show day, using some kind of shampoo, but for most breeds a quick run over with the hosepipe a few weeks prior to show day will suffice. Some exhibitors do not wash their sheep at all. A "bloomed" appearance can be achieved by applying a weak solution of Jeyes fluid using a knapsack sprayer, then turning the sheep out in hot sunshine to dry.

Halter Training

Most people's halter training methods seem to be based on the principles of "fight & fright", so the poor animal is utterly defeated and plods along docilely at heel, without an ounce of spirit left. Usually this is achieved by tying the haltered sheep up tight to a gate and leaving it for a few hours until it stops struggling, or dragging it about on a short lead until eventually it simply gives in. Clearly this is not ideal, and if you want a sheep to really stride out and present himself well in the show ring then halter training needs to be approached from a rather different angle. This was pointed out to me by an American lady on one of our lambing courses, who had considerable experience of training horses. We challenged her, then, to halter train our roughest, toughest, un-handled Welsh Mountain ram, which she did, in an open field, in under half an hour! Needless to say, we've applied her principles ever since. She started with a very long lead rope attached to the halter – at least 15′ long, maybe more – as this immediately removed some of the inevitable conflict caused by the handler being too close to the animal. In effect, this meant that the ram's behaviour was a direct response to the restriction of the halter, rather than a response to the intimidating presence of the handler. With sheep that are already moderately tame we've found that the method works fine with a lead rope of about half that length. Initially, while the ram was leaping about like a mad thing at the far extremity of his available range, Debi simply leaned back on the rope, applying plenty of pressure, preventing him from turning away. The instant the ram stood still, the rope was allowed to go slack. Then, gently, she backed off a bit, applying a steady tension to the rope. If, as a result of this, he started monkeying about again, the pressure came on – hard! If, on the other hand, he responded by taking a step in her direction, the rope was immediately slackened. It didn't take many repeats of this sequence for him to realise that taking bold steps in the direction of the handler meant no pressure on the rope, and within 20 minutes or so Debi was able to walk down the field towards us with the ram proudly strutting along close behind, with all his eye-catching character intact.

Preparation

The final preparation – trimming, brushing, etc. – takes place in stages over the last fortnight before the show. The down breeds of sheep, such as the Hampshire and Southdown, are heavily trimmed, with the fleece being meticulously sculpted into a shape that enhances the animal's meaty conformation. At the other extreme are the really hardy hill breeds, where washing of the face and legs and a quick brush down are all that is required. Some breed societies (the Lleyn, for example, and the Texel) specify that sheep must be shown untrimmed. Preparing longwool breeds, such as Wensleydales, is another matter altogether, and something I know nothing about. Our own breed, the Improved Welsh, takes a middle-of-the road approach, so the following sequence should provide a useful basis for most breeds:

Trimming stand in use

Tools Of The Trade

• Trimming stand
Many people hold their sheep on a trimming stand for show preparation, the big advantage being that the animal can be raised to a

convenient working height. We only use the stand when washing (so the animal isn't standing in a puddle of dirty water), preferring to trim on a halter.

• Carding combs

One coarse blunt comb; one finer blunt comb; one fine, sharp pointed, heavy backed comb.

• Shears

2 pairs of cranked hand shears – a double bow pair for the majority of the trimming, and a single bow pair for using wherever the fleece is particularly dense or matted.

• Patting board

A flat piece of board, about 10″ square, with a handle. Some showmen prefer to use a badminton racquet.

• Hand spray bottles

For misting the fleece with water or setting lotion.

• Cloths, brushes, etc.

An assortment of scrubbing brushes, cloths, towels, buckets, shampoo and so on, for washing faces and legs.

• Halters

You'll need nice clean white (or black) ones for the show ring, but any old halter will do when trimming.

• Sheep coats

To keep your carefully prepared show team in pristine condition during transit.

Preparation Schedule

1. (Starting approx. 2 weeks prior to show day) After picking out as much debris as I can from the fleece (hay seeds, gorse prickles, bits of bramble, etc.), I make a start by thoroughly brushing the sheep all over with the coarse carder. This raises the wool in tufts and lumps all over the place, making the poor animal look frightful!

2. Next I decide whereabouts I want the top of the animal's shoulder to appear. This can be quite far back on a ram, to enhance the masculine appearance of a powerful neck and forequarters, but not so far back that the shoulders appear to slope, or the body appears short. Move the shoulder further forward in a ewe, to give a lighter, more feminine neck.

3. After trimming the shoulder down to the required level, I work towards the rear end of the animal, trimming the back flat, in line with the shoulder. Take care not to let the back slope off behind the hips – keep everything fairly square at this stage. Some people like to lightly mist the fleece with water when trimming, as it stops everything getting too "fluffy".

4. Now I return to the front end of the animal, and again using the shoulder as my starting point, I trim up the back of the neck.

5. Next the chest. This should be big, bold and square in a ram, but for a ewe make the chest more rounded and taper it up into the neck.

6. Having trimmed the back of the neck and the front of the chest I now join these two areas together, and trim down the front leg. The whole of the forequarter is completed.

7. At the back end of the sheep again now, I tidy up the base of the tail, trim the sides of the hips, and down in front of the stifle. I leave the whole of the thigh area untrimmed, then give it shape by combing and fluffing up the wool. However, this is only my personal preference – most people do seem to trim a bit more than that on the hind legs, and in some breeds the hind legs may be heavily trimmed into shape.

8. Next I trim the sides of the sheep, joining together all the areas already covered. Great care must be taken to ensure that the animal doesn't end up looking pot-bellied. I do the sides in three stages: first over the ribcage, then lower down, along the edge of the belly, and finally the small strip along the top, joining the side to the back – this last part must be done carefully, to avoid narrowing the appearance of the back

or making it seem too rounded or droopy.

9. Over the next ten days or so, I use the finer, blunt carder, and repeat the sequence of brushing and trimming every couple of days. With each successive brushing fewer tufts of wool will be lifted, and the fleece will become progressively more even and dense in appearance after each clipping. Also at this stage Dot gives the sheep's face and legs a preliminary wash using warm water, fairy liquid, and a soft scrubbing brush.

10. A couple of days before the show, I start using the sharp pointed carder. This is used with firm patting and lifting motion, and has the effect of consolidating the fleece whilst at the same time raising a fine fluff, which is trimmed off.

11. (On the morning of departure) After a final light trim, I mist the fleece with a lacquer or

setting lotion, then pat the wool firmly all over with the patting board. A coat made of some breathable material, such as an old linen bed sheet, is put on the sheep to keep the fleece clean and tidy in the trailer.

12. With the sheep held in the trimming stand, and towels positioned to keep the neck and chest dry, Dot gives the face and legs a good scrub with a whitening shampoo. She also uses a small stiff brush to clean the horns, which can be dressed with linseed oil or hoof oil when dry. Some people sandpaper the horns, making them smooth and rounded, but personally I think that looks a bit silly.

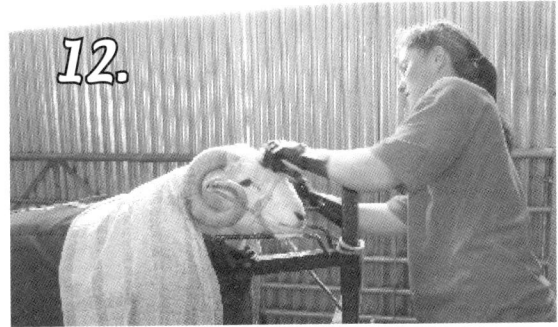

13. (At the show) We leave the coats on the sheep until shortly before our classes, although we must make sure we allow time for a final tidy up before going in the ring. We scrub any travel stains from the sheep's legs, and freshen up their faces with a cold wet cloth.

Success in the ring as a result of our efforts

Any irremovable blemishes on faces and legs are disguised using tennis shoe whitener. After taking off the coats, any marks they've caused in the fleece are removed by lightly patting the squashed areas of wool with the spiked card, and trimming off the resultant fluff. The whole fleece is again flattened down with the patting board. A bit of lacquer is misted onto the thighs and tail, which are then thoroughly fluffed up, giving a nice full appearance to the animal's back end.

And now we've done all we can. Shortly the steward calls our class, so we put nice clean halters on the sheep (remembering to lead from the nearside), slip on our white coats, and enter the ring. Now it's all in the hands of the judge. With luck, we'll win. If not, well it's only the opinion of one man on one day isn't it? There's always next time…

FULL CIRCLE

This section concludes a season by season review of the key aspects of management involved in shepherding a conventional spring lambing flock. In order to complete the full circle, we must turn our attention to the next generation of potential breeding stock, by selecting our home bred flock replacements,

and also begin to think about preparing the ewes for the forthcoming breeding season.

WEANING

There comes a point during the summer when lambs and ewes will be in direct competition for available grass. The ewe's milk quality will also have declined considerably, so leaving lambs with their mothers for longer than is necessary will actually have a detrimental effect on their performance – put simply, it holds them back. The ewes also need to be given time to dry off and regain lost body condition before the next breeding season – delayed weaning may have a serious knock on effect on the output of the flock in the following year, with lower conception rates leading to more barren ewes and less lambs.

Where lambs have been creep fed it will be possible to wean them early in the summer, and push them ahead on better pasture, but for most conventional spring lambing flocks, following a less intensive regime, any lambs remaining with the ewes (i.e. any that have not already been sold for slaughter) should certainly be weaned in early August. By this time there should be some nice aftermath grazing available for the lambs, and they'll suffer no setback at all.

Using the shedding gate at weaning – ewes go one way, lambs go the other

The act of weaning simply involves separating the ewes from the lambs, permanently. But of course, there's a bit more too it than that, in management terms! You'll need good fencing, for a start, as they can be quite determined to get back together for the first day or so. Ideally, they should be out of earshot of one another, but, failing that, try to leave at least one empty field between them. Put the ewes into a really bare paddock for the first couple of days, to ensure that their milk dries up quickly. Alternatively, if your facilities allow, house or yard the ewes for a while, with only straw for fodder. Do not restrict access to water though.

After sorting through the lambs, any that will be remaining on the farm should be dosed if necessary, (or if it is part of your flock health / management regime to do so at this stage), and moved to the best pasture available. (Not forgetting, of course, that you'll need to save some good grass to flush the ewes in a month or so).

SORTING THE LAMBS

Apart from any that will be retained for breeding (we'll discuss those in a minute), there are two (or maybe three) other groups that can be selected at this stage:

Fat Lambs

On handling, a number of the lambs will be found to be "fat". This is something of a misnomer, as the modern taste is for a far leaner carcass, but the old terminology lives on! Really we are looking for fit, well fleshed lambs, with good muscling and only a little subcutaneous fat. These lambs can be sold off the holding on the day of weaning. In fact, if you handle your

sheep regularly, you'll probably have marketed a fair number of lambs by now, anyway. If you are new to shepherding it would be well worth asking the grader in your local mart to show you what he's looking for, and try to get a bit of hands on experience. Choose your moment though! Early in the morning would probably be best, just as the first few sheep are arriving on site – later on he'll be far too busy, and won't appreciate your request.

If you've got a lot of lambs to sell, and you think that a fair few are ready to go, you can arrange for a fieldsman from one of the larger abattoirs to come onto your holding and make a selection. This may be easier for anyone trying to juggle running a smallholding with other commitments, as there's no need to take the lambs into market at a certain time on a certain day – once the fieldsman has made his selection he can arrange for the lambs to be collected, and you'll be paid on a deadweight basis.

Even in very small flocks, where most of the lambs may sold in freezer packs to friends and family, it is well worth learning correct selection methods – your customers won't thank you for over fat or super lean (OK, I mean thin,) lambs.

Fat Lamb Selection & Carcass Classification

Selection of "fat" lambs is fairly similar to condition scoring of ewes (see page 167), although it's necessary to handle lambs over the rib cage and at the dock, in addition to the loin. In a well finished lamb, the loin will feel similar to CS 2.5 – 3.0, and there'll be a firm covering of muscle across the ribcage (but without rolls of soft fat). The dock (base of the tail) is the last place to flesh out on a fattening lamb, so it should be possible to detect the individual vertebrae. If the gaps between these bones are indiscernible then the animal is too fat.

Carcass conformation is classified using the EUROP scale (see table, below), with E being excellent and P being poor. The poorest may be split into 2 groups, P+ and –P. Each category is subdivided according to fat class, on a scale of 1 – 5, with 1 being leanest and 5 fattest. Carcasses at the fatter end of the scale may be penalised, as may carcasses of poorer conformation. Fat classes 3 and 4 are further sub-divided into light (L) and heavy (H). Optimum classification would be E2 or E3L, although the majority of conventionally produced lambs are classified R3L. Hill breeds and primitives are more likely to be around the O2 mark, although female mountain lambs are rather inclined to go over fat at low weights.

Well finished lambs of good conformation may have a killing out percentage (KO%) of around 48% (i.e., the carcass will weigh approximately 48% of the lamb's liveweight (LW)). This could rise to as much as 50% in early finished continental X lambs, and fall to as little as 43 – 45% later in the year, or in lambs of poor conformation. Remember that some males are horned, and that testicles can add a fair bit to liveweight in the autumn! The target weight at slaughter will depend upon the market into which you're hoping to sell. We've found that early in the year there's a good trade in light lamb carcasses of 8 – 11kg for export,

so we try to sell a few in late May / early June (shearing time) at around 20 – 25kg LW. For direct marketing of whole boxed lamb, a carcass weight of 12.5 – 14kg (27 – 31kg LW) seems to be ideal, or, if selling half lambs from larger breeds aim for 18 – 22kg deadweight (DW). For conventional marketing e.g., direct to abattoir, aim for a carcass of 14.5 – 18kg (or 32 – 40kg LW if you're taking them to market). Outside the normal marketing period, heavier carcasses generally fetch higher prices, which is worth bearing in mind if you're running store lambs on over the winter to sell in the New Year. Liveweights over 50kg wouldn't be excessive, provided that they've not gone too fat. Smaller hill breeds and primitive types simply aren't going to reach these sorts of weights and classifications, but they do lend themselves to niche marketing, or for running on to produce mutton as a specialist product.

Store Lambs

This group consists of the remainder of the lambs that you wish to sell, but which are not yet sufficiently well fleshed. These can be kept on the holding for fattening, either quickly ("short keep") on the best aftermaths available, or held over the winter ("long keep") for finishing in the New Year when prices are

	1	2	3L	3H	4L	4H	5
E		Pure continental					
U		Continental crosses					
R		Crossbreds & traditional lowland					
O	Hill & primitive						
P+	Hill & primitive						
-P							

The EUROP scale used for carcass grading. Shaded areas show achievable mean classifications for different types of sheep.

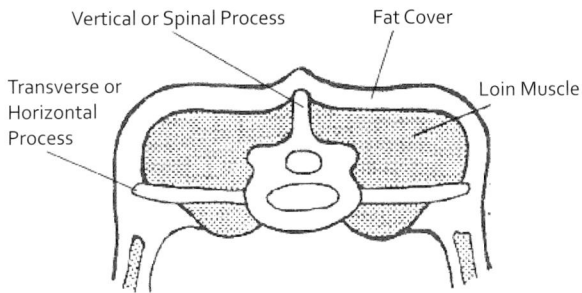

generally higher. If you do intend to keep them on you'll need to castrate the ram lambs, if you haven't already done so, or run them in single sex groups. Really though, on the smallholding where space is limited, you must ask yourself whether it is worthwhile keeping lambs on into the autumn, when ideally the ewes should be given priority for available grazing – hanging onto them in the hope of getting a better price later could be a serious false economy. A better plan may be to sell all unfinished lambs as stores by the end of August.

The possible third group in this category will consist of all the misfits – pet lambs, poor doers etc. These are the ones to put aside for home consumption. Remember the motto – sell the best, eat the rest!

SORTING THE EWES

This is an appropriate time to sort out the ewes according to their condition score – doing it now gives you plenty of scope to correct any problems before the critical autumn period. Some ewes will be very fat, perhaps because their lambs were sold earlier in the year, so will need to be thinned down a little on poorer pasture. Others may be extremely lean, having milked heavily all summer, and will need to build up condition steadily over the next six weeks or so, in order to be within half a condition score of the target for their breed in time for flushing at the end of September. Most ewes will need to gain up to 1½ units of body condition during the period between weaning and tupping – this may take up to 8 weeks on fairly good grazing. Any very thin ewes will require some supplementary feeding.

Body Condition Scoring

Body condition scoring is a universally recognised method of assessment used to determine the level of body condition of breeding sheep. Sheep are scored on a scale of 0 – 5, using half grades where appropriate, following an assessment made by handling the animal over the backbone in the loin region. Although condition scoring follows a standard scale, there is of course a degree of subjectivity – a ewe that is given a score of 3 by a hill shepherd may scarcely make 2.5 when assessed by his lowland counterpart – but having said that, provided that it is always the same person on the holding that makes the assessment you'll find that regular condition scoring is a very valuable aid to management.

Score 0 – Basically this means emaciated and at point of death! The skin will be stretched tightly over the spine, and it will not be possible to detect any muscle or subcutaneous fat. Hopefully you'll not see this condition often, but it will occur from time to time in even the best managed flocks, due to extreme old age or illness.

Score 1 – This is a very lean ewe. The horizontal and vertical processes of the spine will feel prominent and sharp, and you'll be able to easily feel below the horizontals. The loin muscle will be very thin, with no fat.

Score 2 – The spinal processes will feel prominent but smooth, and the horizontals feel rounded. It is still possible to feel below the horizontal processes, but only with pressure. There'll be a moderate amount of muscle, but very little fat.

Hill ewes, primitive breeds, and lowland ewes rearing twins are likely to be CS 2 at weaning time. By tupping time they'll need to have gained 0.5 – 1 unit in the case of hill breeds, or 1 – 1.5 for lowland sheep.

Score 3 – A sheep at score 3 is in good condition. The spine will feel rounded and smooth, and the individual bones can only be felt with pressure. The individual ends of the horizontals can be detected with difficulty. The loin muscles are full, with a moderate covering of fat.

Score 4 – The vertical processes are only detectable as a line, and the ends of the horizontal processes cannot be felt at all. The full loin muscles have a thick covering of fat. In all likelihood this is a ewe that lost her lamb early in the year.

Score 5 – Seriously overweight! Probably a barrener. It is not possible to detect the spinal processes, even with firm pressure – in fact there'll be a dimple in the fat layers over the spine. Neither is it possible to detect the horizontals. There will be a very thick layer of fat over very full loin muscles. Unless this animal goes on a diet she'll be barren again next year!

Selecting Home-bred Flock Replacements

Firstly, consider whether it is really worthwhile rearing your own replacements at all. On a smallholding, with a limited acreage, this non-productive portion of the flock will take up valuable grazing that could more profitably be utilised by the breeding ewes. You may end up overstocked, and have to buy in additional feed or forage, which will really knock the economics of the whole enterprise into a cocked hat! Believe it or not, it is often cheaper to buy in replacements, rather than rear your own. I've discussed the pros and cons of home-bred flock replacements earlier in this book (page 41 and following) so now, assuming that you've decided that you *will* rear your own, read on…

Selecting Female Replacements

I must admit that, for us, the day we select our replacement ewe lambs is one of the most pleasurable days in the whole of the shepherding year! These lambs are the cream of the crop and the future of the flock – there is a huge sense of satisfaction in putting together a nice evenly matched group of ewe lambs, and, quite rightly, we feel proud of them.

With all the female lambs yarded up together, we begin by drawing out the few real "eye-catchers", the very best, the ones we'd spotted months ago and have been admiring ever since. We pen these few up together, then waste a lot of time just leaning on the gate and looking at them – and well worth looking at they are too! Eventually we drag ourselves back to the task in hand, and draw out all the lambs at the other end of the scale – a few poor doers, perhaps, or something with an undershot jaw or wonky legs, anything, in fact, that just doesn't fit in. Taking out the best and the worst leaves the bulk of the ewe lambs looking quite even, so now the selection process becomes more complicated.

With all the female lambs yarded up together, we begin by drawing out the few real "eye-catchers"

Selection Guidelines

• Only pick what you actually need. It is always tempting to keep a few extras "just in case", but rearing ewe lambs is not cheap, and smallholdings can't afford to carry passengers.
• Ewe lambs selected for flock replacements must be at least as good as (and preferably better than) the best ewes in the flock. Don't be tempted to keep sub-standard animals just to make up the numbers – it would be better to keep a smaller group, and do them really well, rather than see your breeding program sliding backwards.
• If your flock is reasonably prolific, and you want to maintain that, don't simply pick the biggest ewe lambs, as they are probably singles. Prolificacy is hereditary, so by selecting twin born lambs as replacements, considerable improvements in flock output can be achieved.
• On the other hand, you may have rough mountain grazing that will not sustain a high output flock, in which case you won't want to breed for prolificacy so should select single born lambs. Use your lambing time records as an aid to selection.
• The points I made earlier (page 157) about conformation when selecting a show team apply equally here, but remember that these are your potential breeding stock, so we're looking for a good all-round robustness that can be maintained without the level of pampering that show sheep receive.
• Look for nice tight fleeces and good skins. If wool production is one of your aims then you'll place a higher emphasis on the finer points of fleece quality, but in general terms avoid selecting lambs with an open, dry-looking, coat.
• If you've decided to performance record your flock, now is the time to begin to make use of the figures. There is only so much you can do by eye, and at best you can only make an assessment of what is in front of you on the day. However, the performance index figures will enable you to take into account a whole range of unseen factors that may impact upon the future breeding potential of each lamb, and enable you to select for very specific aims.

Selecting Ram Lambs

When you attend a ram sale and see the prices some shearlings make, you could be forgiven for thinking that tup sales might make a valuable contribution to the profitability of your holding. Indeed, I know two brothers who sell around 30 rams each year, and who regularly achieve prices of over £1000. I can almost see the smallholder's mind at work here – "Now, if I were to keep 10 ewes…get 16 lambs per year…half would be male…that's eight rams to sell at, let's say, £500 apeice?… mmmm…not a bad return from a little flock."

Now let's put this into perspective:

Those two brothers I mentioned have around seven-and-a-half thousand ewes! They're producing in the region of 10,000 lambs per year, so their thirty sale rams will be selected from a group of 5,000.
That's 0.6 %.
Introduce the figure 0.6% into the sums for a flock of 10 ewes, and you'll see that it may take over 20 years to produce one good tup!

OK, I admit that that's a fairly extreme example, but it's a true scenario and does serve to illustrate the fact that tup breeding is unlikely to be your quickest road to a profitable smallholding. Even if you are producing good quality stock you'll find that there's quite a long "apprenticeship" period, i.e. the length of time it'll take for the regular buyers to stop viewing you as a newcomer, and start to take a serious interest in what you've got on offer. We've been selling rams for 10 years now, but it's only recently begun to have a positive effect on our financial situation. Having said all that, many successful ram breeders are in fact smallholders with small flocks, so clearly it can be done. If

you do want to give it a go I wish you the very best of luck. Just don't say I didn't warn you!

Selection Guidelines

• You must be very hard-hearted in your selection of females, for unless your ewes are correct in every respect you'll never breed good tups.
• Only consider keeping the very, very best male lambs as potential breeding stock. It may cost as much as £200 to rear each one from birth to point of sale at 18 months of age, and you won't want to waste that sort of money on a no-hoper.
• This may sound like stating the obvious, but remember to check that the lambs you select all have 2 testicles!
• Scrapie Genotype. (See the next page).
• Once again, my earlier notes on conformation can be applied here, but do bear in mind that show quality rams may not be the best in commercial terms.
• Look for length, but without a dipped back.
- To the commercial lamb producer (i.e., your customer) length means weight.
• Minor breed characteristics such as specific markings, size of ears, etc., are of relevance only to the pedigree breeder, so if you're aiming to sell pedigree stock to other breeders you'll need to pay strict attention to type. However, there's always a market for good quality commercial purebreds, where the finer points of appearance are of less importance than conformation. Some breed societies organise inspection days, and only ram lambs that are deemed to be of the required standard are eligible for registration in the society's flock book.
• Regularly re-assess your selection, and reject any animal that's not turning out as you'd like.
• If you're selecting a ram lamb for use in your own flock then be aware that a twin (or triplet)

born ram will sire more prolific daughters.

• Don't retain bottle fed ram lambs as breeding rams – they tend to become rather pushy, and are often very aggressive when mature. This is particularly significant if there are small children on the holding.

• If you are not already performance recording your flock, you should seriously consider doing so. Ram buyers, particularly of the terminal sire breeds, are becoming increasingly more aware of the value of performance figures.

SCRAPIE GENOTYPING

Although the National Scrapie Plan (NSP) has drawn to a close, it is appropriate to mention it here due to the impact it's had on the breeding and marketing of rams over the last 8 years or so. Besides, it appears that some form of genotyping scheme will continue into the future, although this time, of course, sheepkeepers will be expected to pay for it!

The NSP was set up following fears of the possibility of links between scrapie in sheep, BSE in cattle and nvCJD in humans. Subsequent research has shown these fears to be groundless (as sheepkeepers believed all along). Over the lifespan of the scheme a huge amount of public money was spent on blood testing and culling out of certain genotypes, and there is a risk that many valuable blood lines may have been lost. All is not doom and gloom though – despite the fact that it has been shown that scrapie is not a human health issue, it remains very much a sheep health problem. Selective breeding for natural disease resistance can only be a good thing given the current move towards reducing the level of drug usage on farms, whilst maintaining high standards of animal welfare. There is no doubt that there has been a significant drop in the number of incidences of scrapie occurring in sheep flocks over the past few years, with an all time low of only one clinically confirmed case in the whole of the UK in 2008. This difference is of course more marked in breeds that have historically been susceptible to the disease. Keepers of the historically less susceptible breeds have not benefited from this improvement, so understandably many consider the whole thing to have been a waste of time and money!

In conjunction with the NSP, a national semen archive has been established to safeguard against the loss of genetic diversity, particularly in rare and minority breeds – I believe that the Rare Breeds Survival Trust has worked closely with the NSP on this, and has now joined forces with the NSA to administer the archive for the future. However, the semen archive really has been "too little, too late" – a genuine case of closing the stable door after the horse has bolted! In fact, it was left so late that a lot of very sub-standard animals have been collected from, simply to make up target numbers.

How The NSP Worked

Membership of the scrapie plan was not compulsory, but as it is now almost impossible to sell rams that haven't been tested it became more or less compulsory by default. There are compulsory genotyping schemes for flocks that have experienced cases of scrapie, and in some cases these farmers are given help towards the cost of purchasing disease resistant rams. Technicians or vets visited members' flocks to take blood samples from lambs selected as potential breeding rams, and at the same time inserted an electronic ID bolus into each animal. In some cases all male lambs were tested, and even a portion of females – when the genotype of ewes is known, it is possible to determine the likely outcomes of mating different animals together, by the use of simple

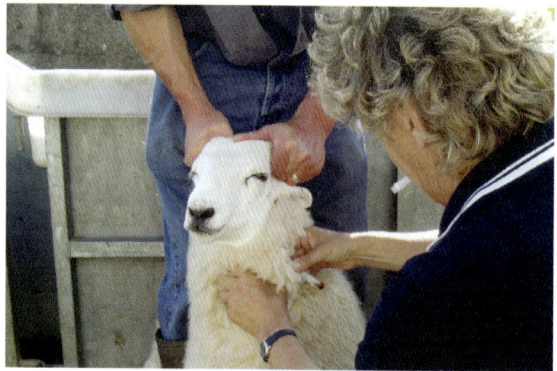

Taking a blood sample for genotype testing

Mendelian genetics. Lambs are categorised into one of 5 groups according to their genotype, with 1 being least susceptible and 5 most susceptible to scrapie, with certificates or slaughter / castration notices provided for each tested ram lamb. Members were obliged to slaughter or castrate any male sheep in groups 4 or 5 within 90 days, without any form of compensation, although an exemption could be made to allow the animal to be used in the semen archive – following successful semen collection the ram was killed and the owner received a payment of £50 plus the slaughter value. 10% of the stored semen remains the property of the owner of the donor ram, for future use in the event of the archive being opened.

Rams in groups 1, 2 and 3 are all acceptable for sale as breeding rams, although earlier restrictions (subsequently abandoned) placed on the use of group 3 rams led to huge discrepancies in prices, with some breeders being very hard hit. Understandably, many farmers set out to breed wholly group 1 flocks, as these rams commanded the highest prices, but with the close of the scheme, these breeders are finding themselves paying the price for having a considerably reduced genetic diversity in their flocks, with many reporting that there is now a lack of vigour in their sheep.

REARING RAM LAMBS

Select more than you intend to keep, initially, and revise the selection as performance data and scrapie test results become available. If the lambs haven't been creep fed it's quite a good idea to put these few aside with their mothers a week or two before weaning, and offer a bit of feed twice a day. The ewes should come readily to the troughs, teaching their offspring to do the same. You can then carry on feeding the tup lambs after weaning – achieving a bit of extra growth now will pay back later. By late November you should have thinned the group down to a manageable number (though always be prepared to make a reassessment, and further reduce the size of the group if necessary – it's no good rearing substandard animals for sale as breeding stock), and may consider housing them for a while – feed conversion ratios will be much higher if they're protected from the wet weather that's so typical of our British winters. Take care over their diet while they're indoors. Plenty of protein will indeed make your lambs grow big and fat, but they'll be flabby and weak, with poor feet and legs, and not much get-up-and-go. High energy feeds will do a better job, so include whole oats and sugar beet pellets. Don't use feeds formulated for ewes, due to their high magnesium content. A mineral supplement containing zinc and selenium will aid skeletal growth, and the provision of rock salt will encourage water intake, reducing the risk of urinary calculi (or "stones"). Good quality forage is a must. Pay particular attention to the lambs' feet, as the hooves of housed sheep don't wear down as they would do outdoors, and encrusted dung and bedding may make them very lame, very quickly.

We always house our chosen few towards the end of November, shear them sometime in

Shorn rams in January

Top: A male lamb selected at weaning to rear as a potential breeding ram, and the same animal at 19 months of age (above). The same sheep can be seen on page 129 (at six weeks old), and on page 35 (at 15 months)

January (check with your breed society before doing this, as some societies specify a date before which sheep must not be shorn), then turn them out to a sheltered paddock at the beginning of March, making way for the ewes coming in to lamb.

Although it's important to keep these yearlings growing on, it's also essential that they aren't given trough feed (or hay) during the critical period that they're raising their new teeth. For this reason we stop feeding them some time in April, by which time, hopefully, there's plenty of grass, and don't re-start until mid-late July. The type of feed used is as for show animals, though be aware that,

for rams intended for sale as breeding stock, growth, muscling and early maturity are more important than fat. Therefore build the feed up to a moderate level quite quickly, certainly by the end of July, and maintain that level right through to the point of sale. Feeding at this lower level over a longer period is preferable to stuffing them with an ever increasing amount of grub in the last few weeks. Buyers want tups that are fit to work, and that won't melt away as soon as the mollycoddling stops.

Preparation of rams for sale (as breeding stock) is also as for show preparation, although any clipping should be carried out well in advance, say 3 – 4 weeks prior to the date of the sale, in order that the fleece can regain a natural looking surface. The aim is to produce an animal with all the sculpted shape of a show sheep, without it actually looking like it's had any pampering at all! The exception to this will be where a breed society holds its annual show in conjunction with an autumn sale, though even here you may see good tups presented "in their working clothes" fetching higher prices than the dolled-up prize winners.

HORNS

In horned breeds there's a risk that rams' horns will curl too tightly, and grow into the back of the head, or puncture the face or an eye.

While the male lambs are still fairly small, in-growing horns can be cut back with an ordinary pair of hoof clippers. Beware of taking them off too short, or you'll cut into the quick which will bleed and be painful. Remember that they'll need trimming again at a later stage, as they keep on growing. The horns of older sheep can be cut using an ordinary wood saw or a butcher's meat saw – don't try using a hacksaw. On mature rams it may be possible to cut a thin slice off the inside face of the horn, giving the necessary clearance without adversely affecting appearance. In the case of show sheep, the horns can be heated and turned (see right) rather than cut. Best of all would be to buy stock rams with good shaped horns, and hope that they pass the characteristic on to their offspring. It really is no joke when you've got lots of lambs with tight horns, as, left unattended, an in-growing horn soon causes a wound which, in summer, becomes flystruck. Even without the flies, the affected sheep would eventually die in misery as a result of the horn's relentless penetration.

Turning The Horns On Show Sheep

Provided that the horns are not ridiculously tight, it's possible to turn them out to clear the face. Ideally, they should be left as long as possible before attempting this, as trying to turn small horns on very young sheep will generally result in them breaking. Usually, by cutting off the tips, it is possible to gain a few months grace. The sawn tips can be sanded

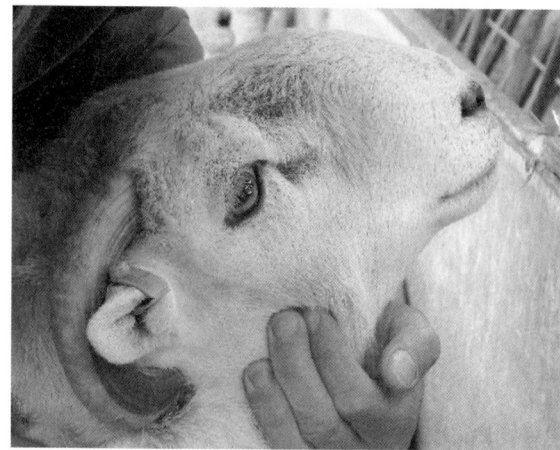

An in-growing horn on a Welsh Mountain shearling ram

back to a point at a later stage.

You Will Need

- Kitchen foil.
- High melting point grease.
- A ring spanner large enough to fit over the horn almost up to the point where you want it to bend.
- Blowtorch.
- Plumber's heat resistant mat.
- Stopwatch.
- Another pair of hands, and maybe thick gloves.

Method

Restrain the sheep in a head yoke. If you've got a trimming stand use that, so you can raise the animal to a suitable working height. Coat the part of the horn that needs bending, and a few inches either side, with grease. Tightly wrap the whole area in several layers of kitchen foil, and slide the heat resistant mat in between the horn and the animal's face. Now use the blowtorch to heat the foil for exactly 2 minutes.

Above: Turning the horn on a Torwen ram
Below right: Job done! The offending horn, which had been forcing one eye painfully closed, is now well clear of the animal's face

Don't hold it too close, and keep the flame playing back and forth to ensure that the area is evenly heated. As soon as the 2 minutes are up, use the ring spanner to lever the horn out to the desired position. Move it out slightly further than where you eventually want it to be, as it's bound to spring back a bit when the pressure's off. Hold it in place until cool.

The Harvest

PURE BRED
WELSH MOUNTAIN
LAMB

HOME SLAUGHTER

Now we come to what, for many smallholders and self supporters, is the ultimate goal – feeding the family with the produce of your own holding. There is a great sense of satisfaction in tucking into a meal that's been wholly bred, reared, killed, butchered and cooked at home, particularly when the whole family has been involved at every step along the way. For us this is the norm, but I appreciate that for many smallholders, new to livestock keeping, it can involve quite a psychological step, this business of eating a piece of mutton that you remember skipping about with the children when it was a lamb! It probably had a name, too! Interestingly, it is always the adults in a household who find this difficult to deal with (though generally, when questioned, they shuffle about a bit and make excuses such as "not wanting to upset the kids"!). Children, on the other hand, tend to be remarkably matter-of-fact about the whole process, have no qualms about being present at (or even helping with) slaughter and butchery, and, at mealtimes, are generally delighted to know the name of the animal they are eating, particularly if it's one that they played a special part in the rearing of, such as a bottle fed lamb.

Why Do It?

Why slaughter stock at home rather than making use of a licensed abattoir? For us the answer is simple: Why not? It's what we've always done, it's appropriate to our lifestyle, and it works well for us. I'm sure that if there were still small local slaughterhouses in rural areas we'd use them, but the fact is there aren't. Of course, any meat intended for sale must go through licensed premises, which in our case

Food production on the smallholding should be a family affair

involves a 120 mile round trip to deliver the animals to the slaughterhouse. The carcasses are then transported, via another abattoir, back to our local butcher, from where we collect the meat for distribution to our customers. So much for food miles! Is this really the way we should be operating, given the impending fuel crisis that the world is facing? The catchphrase "local food for local people" has rather a hollow ring to it when you consider the huge circular journeys that may be involved. And then there's the stress factor. We like to think that our animals lead a fairly stress free life, and we tell our customers so, but is this really the case if their final hours consist of a long trailer journey, a strange and frightening environment, and unfamiliar people? But for home consumption there is a better way – I slaughter the animal myself, on the farm were it was bred and reared, in familiar surroundings and with sympathetic handling. In this way we really do know what we eat, how it lived, and how it died.

I appreciate that there will undoubtedly be readers who have no intention of eating their own stock (although this begs the question why keep them in the first place?) but I would urge everyone to read on, whatever their feelings on the matter – anyone who keeps animals should have the knowledge (and preferably the means) to carry out humane destruction of casualty stock for welfare reasons. To avoid prolonged suffering of the sick or injured it is often preferable to make an on-the-spot decision and carry out the deed immediately. To do otherwise, to leave it to "see if it's a bit better in the morning" amounts to little more than cruelty. It is also worth considering that on a smallholding, where finances are often tight, it may be that "first loss = least loss". In the case of a well grown lamb with a broken leg for example, the most sensible and economical course of action is to butcher it for the freezer, rather than spend on vet's bills for a doubtful outcome. Also consider that many veterinary products will render the meat inedible for some time, so if the treatment is unsuccessful you may be left with an animal you couldn't eat even if you wanted to.

The Law

Various pieces of ludicrous legislation are in place in the UK to ensure that home slaughter within the law is almost impossible. (Strangely enough, pig killing seems to be the least affected). The rules relating to the actual act of slaughter are simple enough, but once the animal is dead things can get complicated!

Slaughter on farm, by the farmer for his / her own consumption is legal. It is illegal to make use of the services of an itinerant slaughterman to kill and dress an animal, as the slaughterer would be deemed to be supplying goods (i.e., the dressed carcass) to the farmer. Were the slaughterman to only kill the animal, leaving the farmer to cut it up, he could be considered to be supplying a service in which case the activity may be held to be legal, although the law is far from clear over this. (Food Standards Agency *"Private Slaughter of Livestock" – guidance for local enforcement authorities in England, December 2003*)

Should you decide to carry out home slaughter, it is necessary to abide by the various welfare codes of practice:-

• No person engaged in the movement, lairaging, restraint, stunning, slaughter or killing of animals shall:-
a) cause any avoidable excitement, pain or suffering to any animal.
b) permit any animal to sustain any avoidable excitement, pain or suffering.
• The animal must be killed humanely, in accordance with *The Welfare of Animals (Slaughter and Killing) Regulations 1995.*
• Animals must be stunned before slaughter.
• No person shall slaughter any sheep by a religious method, or cause or permit any sheep to be so slaughtered, other than in a licensed slaughterhouse.
• All slaughter waste not intended for human consumption or classified as specified risk material (SRM) must be disposed of in accordance with the *Animal By-Products Regulations 2005.* This would include hooves, horns, skins and blood.
• No-one other than the person who actually carried out the act of slaughter may consume any part of any home killed sheep, unless the *Transmissible Spongiform Encephalopathy (TSE) Regulations 2002* relating to the removal and

disposal of Specified Risk Material (SRM) have been correctly adhered to, in which case it could be shared with family members living in the same household.

The issue of pre-slaughter stunning can be neatly sidestepped by the use of a shotgun. According to the Farm Animal Welfare Council, this technique causes instantaneous death, therefore both stunning and slaughter are carried out in one simple action. Our own preferred method is to use a small bore (.410) shotgun – this is most effective in the hands of less experienced users, has simpler licensing / ownership requirements than other weapons (although a license is no longer required for captive bolt stunning equipment) and of course has other uses on the holding, such as pest control. If you don't have a shotgun or shotgun license, it is still possible to use one on private premises under the supervision of the owner / license holder. Provided that someone within your local smallholding community has both gun and license, it should be possible to carry out your home slaughtering under his watchful eye – an offer of home brew may do the trick. Sadly he cannot be paid in lamb chops – this would constitute supply to a third person, which is illegal. If you have a shotgun license but do not wish to keep a gun on your premises, you can borrow one for up to 72 hours without the need to inform your local police or firearms department.

Other legally acceptable methods for the slaughter of sheep include:

Captive Bolt Pistol

Captive bolt pistols are no longer subject to firearms legislation, so are readily available. A captive bolt pistol is a stunning device –

immediately following stunning it is necessary to sever all the major blood vessels in the neck with a sharp knife, in order to ensure that death occurs as quickly as possible.

Humane Killer, 0.32 Calibre Free Bullet

A humane killer can be used to kill sheep of all ages, outright, although bleeding should still be carried out immediately. Generally, a humane killer should only be used by a vet, knackerman, or some other suitably qualified person. There is a very real risk of ricochet should an inaccurately placed bullet happen to pass through the animal, hence the need for proper training in the use of this device.

0.22 Rifle Or Revolver

Only to be used as a last resort, when no other method of humane destruction is available, for example in the case of casualty livestock, in order to prevent further suffering.

There are tight restrictions on what you may do with your home killed meat and who may consume it.

No part of any home killed animal (except in certain cases poultry, rabbit, and game) can be sold, bartered, swapped or given to any third party. This would include guests, friends, and family not normally resident in your own home. In the case of home killed pigs, it is acceptable for the meat to be consumed by the farmer and his immediate family living with him in his house. With sheep, cattle and goats however, the situation is rather different. Due to the Transmissible Spongiform Encephalopathy (TSE) regulations that came into force on 19th April 2002, it is now virtually impossible for a farmer to supply privately killed meat to the rest of his own household. This means that private killing of ovines, bovines and caprines is generally only

TSE REGULATIONS AND SRM

The Food Standards Agency does grudgingly concede that, in certain circumstances, it may be possible to satisfy TSE regulations at home, in which case meat could be consumed by the whole household.

According to the age of the sheep, the slaughterer must remove from the carcass, stain blue, store and correctly dispose of, certain Specified Risk Material (SRM). In the case of sheep or lambs under 12 months of age (or having no permanent incisors erupted), SRM consists of the ileum and spleen. In older sheep, the skull (including brain and eyes), the spinal cord, the tonsils, the ileum and the spleen are all classified as SRM.

In practice, although I know dozens of smallholders who regularly slaughter their own sheep and lambs for family consumption, I am not aware that any one of them takes the slightest bit of notice of this daft piece of nanny-state legislation. I must let you make up your own mind over this. If you decide to invite your friends round for a barbeque, and kill a sheep for the purpose, fine! No-one can actually prevent you from breaking the law; it's just that you may be prosecuted for having done so. In my opinion it is high time that the law included some exemptions for small scale producers supplying home killed meat to relatives and close friends, or within a tight-knit community, provided that there has been no attempt to deceive.

Anyone who would fraudulently endeavour to sell home killed meat to an unknown and unsuspecting customer, as a legitimate product, perhaps by applying a false Meat Hygiene Inspection stamp to the carcass, quite rightly deserves to be punished.

permissible when a farmer slaughters the animal himself, processes it himself, and consumes it himself! Any part of the carcass not actually eaten by the farmer (including the bone left on the plate after Sunday lunch) would classify as waste material under EU animal by-products regulation 1774/2002 and should be disposed of in line with that regulation.

It seems to me to be somewhat bizarre that my wife and family are effectively banned from eating an animal that has been reared on our own farm, has been healthy and well cared for, and has been killed in familiar surroundings, simply because it was me who pulled the trigger! Aren't they considered capable of making up their own minds? Should our children have to eat imported meat of doubtful origin, reared under conditions over which we have no control simply because they are not yet old enough to handle a shotgun? I could, if I wished, legally kill, process, and sell, up to 10,000 chickens each year, yet I cannot feed one home killed lamb to my own family! Where will it all end? And whatever happened to freedom of choice? Maybe you are lucky enough to live in a country where these rules do not apply, or perhaps you are single with no family. Be that as it may, you can rest assured that I, for one, will continue to kill my own stock for my own consumption, and I can't really see the rest of my family being content with a tin of Brazilian Corned Beef whilst I tuck into a roast saddle of mutton, can you?

Mutton Or Lamb?

For us it's mutton, every time! Once you've tasted proper mutton (spiked with garlic, peppered with crushed rosemary, and served with home made redcurrant jelly) you'll begin to wonder why anyone bothers with lamb at all! Sadly, though, mutton seems to have a poor reputation these days, which is not helped by the reality that much of the so called "mutton" that is available is in fact worn out old cull ewes, which are generally sold into the ethnic market for the making of kebabs and such like. Properly reared mutton is an altogether different product, and in terms of flavour and quality it surpasses even the finest of beef. Admittedly it may be rather fatty for modern tastes, but the very fact that you're opting for a degree of self-sufficiency, aiming to produce your own food traditionally and with sympathy, on a limited acreage, implies that your tastes aren't very modern anyway!

Our own policy is to castrate a few of our poorest ram lambs each autumn (i.e., the ones that, for whatever reason, are not saleable as fat lambs, stores or breeding stock), and run them on the mountain until they are 2, 3 or even 4 years old. They lead a very natural semi-wild lifestyle up there, and thrive, existing on a diet of sparse upland grasses, herbs and heather. They are handled only infrequently, for shearing and the like – that is if we can catch them! We kill one from time to time, usually when we're getting fed up of eating pork (or if the level in the freezer looks particularly low), and the flavour is outstanding!

A Place To Work

Unlike killing a pig, where you need lots of hot water, facilities for scalding and a fairly draught proof building, you can handle a lamb pretty well anywhere. Often I've carried out the whole process out of doors, hauling the carcass up on the overhanging bough of a tree (or the front-end loader of the tractor) where, in cold, dry weather, it can safely remain overnight. For the sake of comfort though, (and to avoid alarming the neighbours), an outhouse is the best place to work. No particular facilities are required, but it's handy if you can rig up a block and tackle from an overhead beam, or failing that improvise some sort of frame from scaffold poles or similar. In a large agricultural building the tractor and loader can be brought under cover – this is by far the simplest way to haul up a really heavy mutton beast! You'll not need a lot of space to work, but ideally you'll want room to move all around the animal once he's hanging up. You'll also need somewhere safe to place sharp knives etc., and a small pen to restrain the sheep immediately prior to slaughter. Be aware that there are complex regulations governing the use of firearms in confined spaces, so, unless you are using a captive bolt stunner, you may be better to locate the small pen just outside the shed door.

TOOLS OF THE TRADE

Apart from the block and tackle already mentioned, you'll need a gambrel and a couple of meat hooks, a really good sharp knife with a blade length of about 6″, and a butcher's saw. The gambrel can be improvised from a bit of mild steel bar, (or double up a short length of chain and hold the ends apart with a spreader, forming an "A" shape), and a Jewson's hard point wood saw will do instead of the butcher's saw. (Don't try using a hacksaw – I speak from experience!). Another useful piece of kit, if you can get hold of one, is a small hook-shaped knife used to open the carcass for evisceration

Tools of the trade

– the shape of the blade prevents any risk of puncturing the stomach, which could lead to contamination of the meat.

And of course, you'll need your shotgun. We find that the .410 is ideal for slaughtering all classes of stock, without any fuss or unpleasantness. Shot size should not be less than No.6, with a 2½″ cartridge being OK for most situations. (For very large pigs, or cattle, use a 3″ magnum, but do check that your gun is correctly chambered to take these longer cartridges). Larger bore shotguns may be used equally effectively, but the results can be messy and unappetising.

COMMITTING THE ACT

Get everything set up and organised before you begin. Think the job through carefully, and talk it over with your helpers. Everyone involved must know exactly where to stand, what to do, and when. You can't afford to make mistakes – that's the sort of thing that gives home slaughter a bad name.

Ideally you'll have had your sheep penned up overnight, with only a bucket of water and a little straw to nibble at. This ensures that his guts are reasonably empty (making evisceration much simpler), and also means he's hungry, which is an advantage – it's a simple matter now to place a pan of feed in front of him, and shoot him as he eats. Don't dilly-dally. If you miss the moment he will become suspicious of your behavior, and things will begin to go wrong. The gun should be held 6 – 8 inches from the head, and you shoot him in the middle of the forehead, just above the eyes, in a straight line with the neck. The shot makes only a small hole in the skull but completely destroys the brain, and death occurs immediately. He'll know nothing about it at all.

If your sheep aren't used to hand feeding it may be necessary to put a halter on the animal and tie him to a rail of the pen. Certainly this is the case with our Welsh Mountain wethers. To be honest, I prefer this method as the sheep is held with its head up after having been shot, so it is much easier to make the necessary cut for bleeding.

Bleeding must be carried out immediately, by making a deep cut at the angle of the jaw which severs all the major blood vessels in the neck. As soon as this cut has been made hoist him off the floor by the hind legs, and let him drain for a while.

Bleeding is carried out by making a deep cut at the angle of the jaw

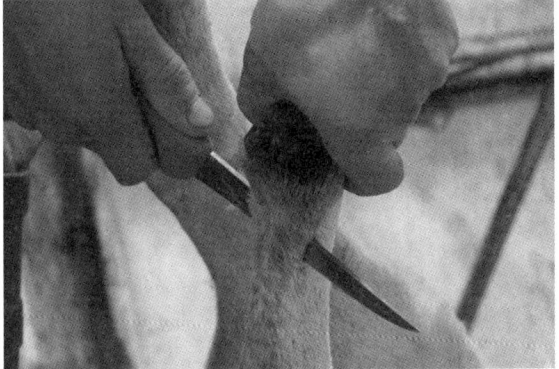

Cut behind the tendons just above the hind hooves (not behind the tendon at the hock) in order to insert meat hooks

Skinning

Generally, I think it is the usual practice for sheep to be partially (or wholly) skinned whilst lying on a low trough shaped bench, but when I started slaughtering my own stock, more than 20 years ago now, I didn't know that. I taught myself, relying on rather vague references in books on self-sufficiency, and got into the habit of skinning the whole animal with it hanging up, as follows:

Start by making a small incision in the skin slightly below, and to one side of, the pizzle (in a male animal), then, with knife blade facing outwards (remember, we're only cutting the

Hoist him off the floor, and let him drain for a while

skin at this stage, and don't want to puncture the abdominal cavity) open all the way up one side of the pee pipe (for want of a better word) to where the clear area of skin starts in the region of the groin. At this point veer off line, cut across the clear skin (still with the blade facing away from the animal), and all the way up the inside front of the hind leg until just past the hock joint. Repeat for the other side. In the case of a ewe, a single cut up the belly is all that is required, before branching just in front of the udder region. Loosen the skin as much as possible on the hind legs before cutting right around and separating the skin from the leg at the hock. Finish skinning the

Open up the front of the hind leg, until just past the hock joint

You may need to use a sharp knife to help separate the skin from the animal

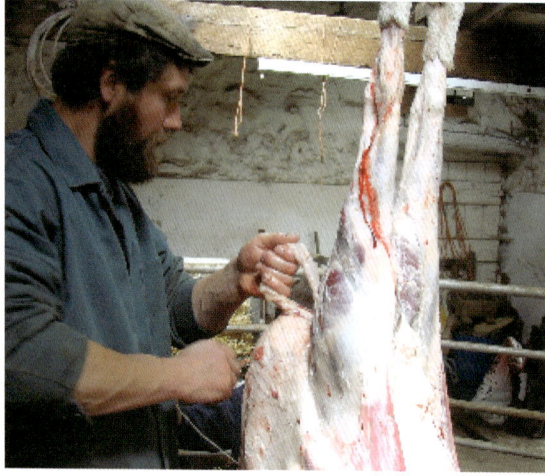

Pull out a few inches of pipework

hind legs by pulling down on the hide while "fisting" with the other hand. This is easy to do in a lamb, but in older animals you may need to resort to the use of a sharp knife to help separate the skin from the animal.

All the various layers of fat and connective tissue should remain on the carcass, not on the skin, as they are necessary to protect the meat whilst the carcass matures.

Now peel back the V shaped piece of skin formed where your original cut divided – pull it over, between the hind legs, to the back of the sheep, until you come to the rectum. Hook your finger around this, pull out a few inches of pipework, and tie it off with a piece of clean string. Cut the rectum away from the skin, then drop the tied off end down into the body of the animal. The next obstruction will be the tail – this can be cut off close to the carcass, and remain attached to the skin. From here on it's fairly plain sailing to remove the skin from the rest of the body, rather in the fashion of pulling off a sweater, using your fist, and sometimes a sharp knife, to loosen the skin as you pull. When you draw level with your original incision, extend the cut downwards, all the way to the throat. Skin out the front legs in the same fashion as the rear – it's very helpful here if you can raise the carcass a bit higher off the floor, and have someone hold each front leg still while you're working on it. Saw off the front hooves just below the knee. Now finish skinning the length of the neck, until you reach it's juncture with the head.

Evisceration

Start by sawing through the breastbone – I wish someone had told me that, years ago! Score a line with the knife, and saw all the way through the full length of the breastbone. Next, use the knife to open up the animal's neck from breast to jaw, exposing the oesophagus and windpipe.

Now, taking care not to puncture any innards (this is where the hooked knife comes in handy), open the carcass from chest to crutch, and all the guts will flop out towards you. Now you can see why it was sensible to cut through the breastbone first! Locate and remove the bladder, without spilling its contents onto the meat, and also pull out the kidneys if you wish – some people prefer to leave them in the carcass. Lift out the liver, which is found to the right of centre, nestled against the diaphragm, and put it to one side for supper, having carefully removed the gall bladder.

Next, a couple of sweeping semi-circular cuts with the knife are used to separate the diaphragm from the inside of the ribcage, whereupon the whole caboodle – the guts and the contents of the chest cavity – simply drop out onto the floor, remaining attached to the carcass only at the head. Saw off the head and the job is done!

Truss up the front legs before the carcass cools, and saw off the hind feet, having re-positioned the meat hooks to the hocks. Don't be tempted to hose down the carcass, as this merely serves to spread contamination into cut surfaces; any visibly contaminated areas should be trimmed off with a sharp knife, later.

Butchery

Before we even begin to think about butchery, the carcass has to hang, so we transfer it to our store-room, where it is allowed to mature for 10 days or a fortnight – the older the animal, the longer it needs. Not only does hanging result in a richer, more complex flavour, but it gives time for enzymes within the meat to soften the muscle fibres, and for connective tissue to relax, giving tenderness.

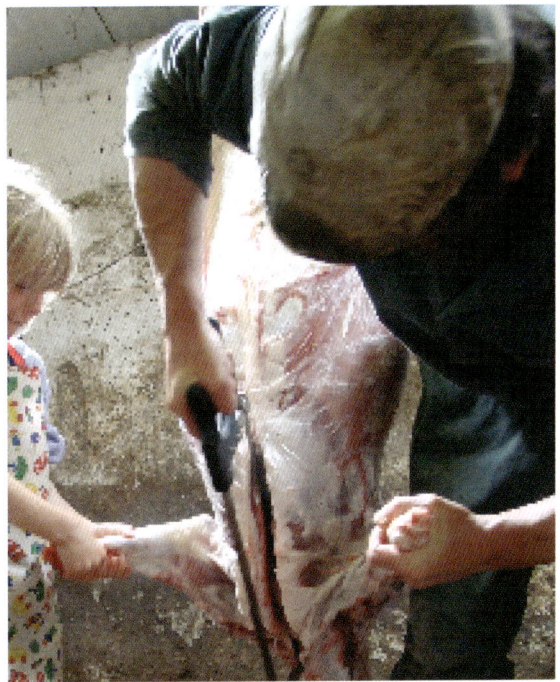

Sawing through the breastbone

Six-year-old Rhian gets to grips with an armful of innards!

In order for hanging to be successful, carcasses must have a good covering of fat – without this protective layer meat may begin to deteriorate rapidly after a few days. The carcass does not need to be refrigerated, but must be kept reasonably cool, so an unheated store-room in winter is ideal. In damp weather, a little mildew may appear here and there, but this is no cause for alarm – we simply wipe it off with a cloth moistened in vinegar.

Butchery is not my strong point I'm afraid. I start by trimming off any areas of contamination, and any particularly unpleasant looking bloody bits, then simply cut the animal up into oven sized pieces! I did once attend a butchery course, but when I attempted to demonstrate my new found skills at home I was told to "stop farting about, and just get on with it!" Apparently Dot needed the kitchen table for something else, and I was clearly holding her up! In the past our butchery has always taken place on the kitchen table, but now we are the proud owners of a butcher's block, which simplifies matters no end (and avoids domestic conflict!). Also, luckily, I have a friend who's very good at cutting things up. He knows exactly where to chop, and knows the names of all the bits and pieces too! So for the sake of getting some pictures for this book, and to christen the new block, I called him in to help.

Disposing Of The Odds And Ends

Just as in killing a pig, when it's possible to use "all but the squeak", so it should be possible to make full use of every part of the carcass of a sheep. The liver and kidneys we've already mentioned. The heart, too, makes a tasty dish, or can be included in mince, where it gives an attractive speckled appearance. A quick flick through two recipe books on our kitchen shelf has turned up seven recipes for

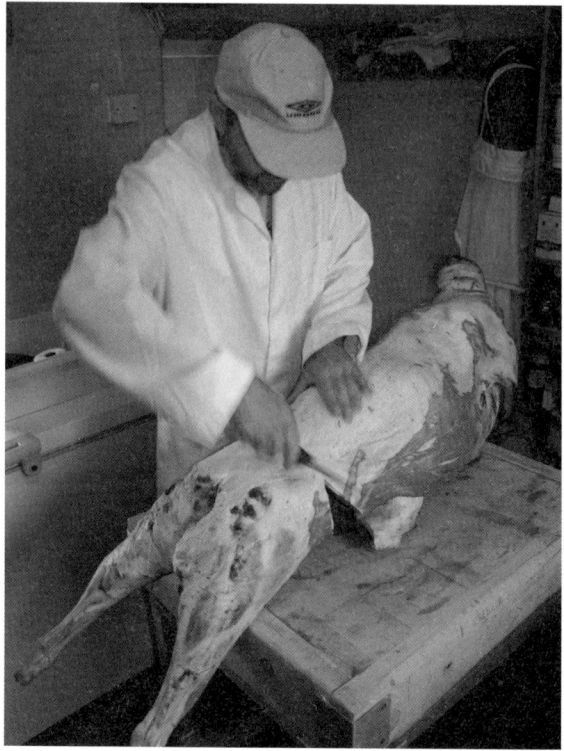

Breaking the carcass down into primal cuts (above and top)

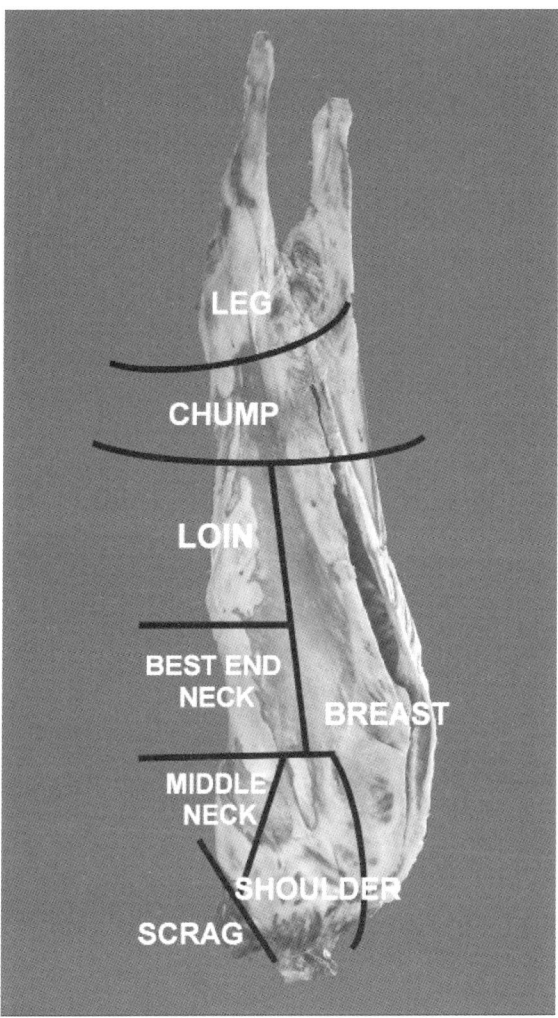

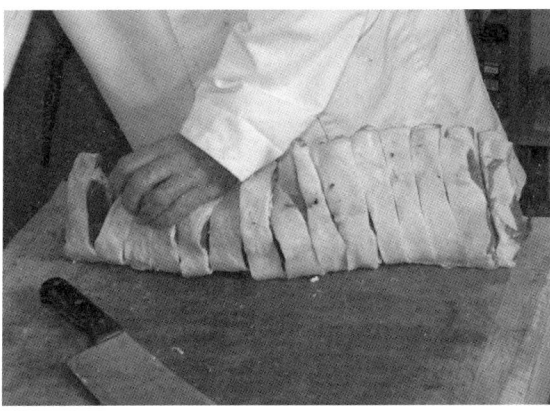

sheeps' heads (and a further 11 for tongues and brains), two recipes for hooves, and one for tails, so that's that lot dealt with! Use the horns for shepherd's crooks, obviously. The hide will make a rug, or slippers, or even a new skin for your banjo! Quite a bit of the intestines can be salted down for sausage skins, and the rest of the innards (together with the lungs and any other bits and pieces) can be fed to the dogs – there's been a huge amount of interest recently in providing more natural diets for dogs and, not surprisingly, they do very well on it. This only leaves one thing – the blood. This should be sprinkled around your raspberry canes, where it'll do a power of good. Now, when an inspector calls, you'll be able to account for every part of every beast!

(Or simply write "missing" in the flock record book, like everyone else does…)

Above right: Scraps can be cubed or minced
Above: Mutton chops!

PRODUCTION OF MEAT FOR SALE

Everything I've discussed so far in this section relates to the production of meat for your own consumption only. Where meat is intended for sale the animal **must** be slaughtered in a licensed abattoir by a suitably qualified person, and butchered in a cutting plant, a butcher's shop or at a registered food premises. Note that not all abattoirs have cutting facilities, so it may be necessary to arrange for transport of carcasses between sites.

Finding your nearest abattoir is easy – just look in the yellow pages – but finding one that'll do private kill might not be so simple! You could 'phone a few and ask, but by far the best thing to do is ask other people locally who they use. Prices are very variable – we currently pay around about £7.00 per lamb for slaughter, and a further £6.00 for cutting and packing, but we are well aware that this is very cheap. Abattoirs that do take on private commissions tend to set aside a specific day of the week, and usually do them all first thing in the morning, so don't be surprised if you're expected to deliver your animals before 8am. Just as when carrying out slaughter at home, you should pen your lambs up overnight before taking them in. Make sure they're clean, or you might be turned away. Crutch them, and, if necessary, shear their bellies. (Some abattoirs may insist upon this). Remember that you'll need to take a completed movement license with you (form AML 1, see appendix), but you won't have had to observe a six day standstill, as the animals are moving directly to slaughter (page 134). You may have to sign a declaration stating that all veterinary medicine withdrawal periods have been properly observed.

The Slaughterhouse Staff

We've always found slaughterhouse staff to be very helpful, and extremely compassionate in their handling of stock. However, one does occasionally hear reports of smallholders who've not experienced such an understanding attitude. To avoid this happening to you, consider the following points:

• Remember, there are two sides to every coin. Your seemingly simple requests can cause significant inconvenience to the smooth running of the system, and you may be considered to be very unhelpful!
• Nothing that you say / do / request / write down etc. will be taken seriously by staff if you turn up with a bunch of striped / speckled / spotted lambs being cuddled by weeping children in the back of the family car. A similar attitude will be engendered by a brand new Landrover Discovery towing a brand new horsebox which you can't reverse!
• Attend your local livestock market regularly, and become absorbed into the local farming community. If you know, and are known by, other farmers queuing to unload at the abattoir, and can engage in meaningful agricultural conversation while waiting your turn to unload, everything will go more smoothly, your requests will be taken seriously, and someone will offer to reverse your trailer for you instead of cursing the delay while watching you get in a muddle!

If the abattoir has also agreed to cut the carcasses for you, then ensure that you provide a detailed written cutting list. If there's no cutting facility at the abattoir then you might be able to arrange for the carcasses to be transferred to your local butcher as part of a regular delivery run. Generally, the butcher will be able to supply bags or polytrays for the individual cuts (vacuum packing is not generally recommended for lamb, as the meat discolours and may develop an unpleasant taint), but it will be up to you to provide the outer packaging. We use really thick insulated boxes made from recycled cardboard, which we prefer to polystyrene. Delivery drivers prefer them too, as they're pretty well indestructible. It is ideal if you can be present when the butcher is doing the cutting, in order to pack each lamb in its delivery box straight away. Once the boxes are sealed, complete with ice packs, they must set off on their journey more or less immediately. If there'll be a bit of a delay, for example if the carcasses were cut up in the evening and your courier can't collect them 'til the next morning, leave the boxes open in the butcher's cold room overnight. Pop in the frozen ice packs and tape up the boxes first thing in the morning. What you can't do is take the meat home and store it in your own fridge or freezer prior to distribution, unless you are a registered food premises. On the whole, it is best to send out meat fresh and chilled, rather than frozen, and don't forget to include some promotional material and recipe leaflets in the box. It's tempting to put in a little "freebie" such as a jar of home made redcurrant jelly, but unfortunately you're not allowed to place other food items in the same package.

Fresh carcasses and cuts must be maintained at less than 7°C during transport, and offal at below 3°C. Frozen meat must remain below -12° C throughout the journey.

COSTING

Despite the high price you receive by selling direct to the end user, there's actually very little extra margin to be made by marketing conventionally produced lamb in this way. Don't be too disappointed if you find that you've put in a lot of hard work for not much gain!

EXAMPLE	
Whole Welsh Mountain lamb, 14kg carcass, sold to abattoir at £2.40/kg	£33.60
Whole Welsh Mountain lamb, 14kg carcass, sold direct to customer COST	
Transport to abattoir	£3.00
Cost of slaughter	£7.00
Cost of butchery	£6.00
Packaging & ice packs	£6.00
Delivery	£10.00
TOTAL COST	£32.00
Value of lamb	£33.60
Break even price	£65.60

Therefore, in order to generate any profit by direct marketing it is necessary to charge your customers more than double the deadweight value of the lamb in this case, even with the calculation based on a very low price for slaughter & butchery.

Later in the season, as the abattoir price falls, the figures look more attractive, so generally it's better to sell faster growing earlier born lambs through conventional routes, and later maturing animals to private customers.

BY-PRODUCTS

CURING SKINS

Having successfully removed the skin from the animal, lay it out and scrape off any remaining flesh, fat, and connective tissue. It may also be possible to remove some of the membrane at this stage. Most books recommend the use of a blunt blade for this job, but we have found that a very sharp knife and a steady hand are preferable. The skin can now be salted if it is not possible to continue with the process at this stage.

Place the skin in a container containing a 10% solution of formaldehyde and leave for 1 week, stirring occasionally. If the skins have been salted they should be thoroughly rinsed before placing in the formalin solution.

At the end of the week the skin is removed from the solution and thoroughly washed several times with clean cold water. Shake off as much water as possible and tack out the skin (wool down) on a board. Rub Lankroline into the skin with a nailbrush. (We have used ordinary leather dressing for this. Neatsfoot oil would probably do).

Allow to dry for a day or two, and before completely dry rub all over in a circular motion using coarse sandpaper on a block. Apply more dressing and leave until the next day. Repeat the sanding and dressing, and continue to repeat day after day until all the membrane is removed.

Remove the skin from the board and wash carefully in warm water with a little washing up liquid. Rinse in clean cold water and leave to dry out at room temperature. It may be desirable to re-fix the skin to the board (fur uppermost this time) whilst it dries, but we do not usually do this.

SAUSAGE SKINS

Unravel the large intestines of a sheep, separating them from all attached membranes. Place one end over a cold tap and turn on a steady flow to wash all the excrement from within. Beware of twists!

Cut the intestines into manageable lengths (5′ seems to be about right. One whole intestine is about 45′). Turn each length inside out by turning back one end and putting it over the tap (imagine that you're putting the tap into the turn-up of a pair of jeans). Run the water in slowly, and it'll all turn inside out in an instant. Lay out each length on the kitchen table and scrape off the gut lining with the back of a knife. What you're left with will look something like a semi transparent bootlace!

Store the skins until required in strong brine solution (4oz salt / pint of water).

See appendix for mutton sausage recipe.

SHEPHERDS' CROOKS

All sheep keepers will have a crook of one sort or another for everyday use, probably a utilitarian article made from aluminium or fibreglass, but there'll also be a special horn-topped stick, reserved for outings such as shows and sheepdog trials. It's almost a "badge of office". Stick makers can be seen exhibiting their craft at many country events, sometimes competitively, and often have sticks for sale. Really, though, what could be better than to make one yourself, from the horn of an old favourite tup who lived a long and useful life on your holding, and eventually died peacefully in old age? (I say "old age" for a reason – the horns of older sheep are far denser, and make a better finished article. A lamb's or wether's horns are soft centred, and aren't really any use).

"Professional" crook makers will use dies and

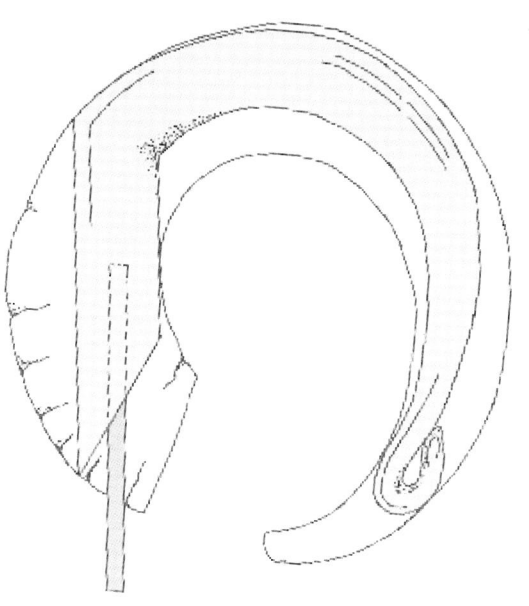

Progress through several grades of increasingly finer sandpaper before joining the crook to its shaft, which should be of well seasoned hazel. Many good crooks are spoiled by an untidy union between horn and stick, so it's worth taking the trouble to do a neat job. I always use a scarph joint, and insert a length of studding into holes carefully drilled down into the top of the stick and up into the base of the horn. The whole assembly is bonded with an epoxy resin. A final careful sanding ensures that the joint is more or less seamless. I then polish the horn with fine emery paper before applying several coats of yacht varnish to the whole thing. The last job is to fit a ferrule to the foot of the shaft – a little bit of copper pipe will do.

presses to help condense and shape the horn, but the casual craftsman can produce an attractive product without any specialist equipment:

Generally the first step is to take the twist out of the horn. It'll need to be boiled in water for half an hour or so to soften up. When it has cooked for long enough take it out and squash it between two boards in a vice (or use big G clamps). Just as when turning the horn on a live sheep (page 174), you'll need to over bend it a bit, so use small wooden wedges where appropriate. Leave it in the press until cool.

Lay the flattened horn on a sheet of thick paper or card and draw around it. Now you can fiddle about to decide the best crook shape you can make out of it. When you're happy with the drawing, transfer your design to the horn by cutting it out of the paper and using it as a template. Use an ordinary woodsaw to cut out the rough shape, then set to work with rasps and coarse sandpaper. Keep plenty of the powdered horn – this can be mixed with epoxy and used to fill any cracks or imperfections.

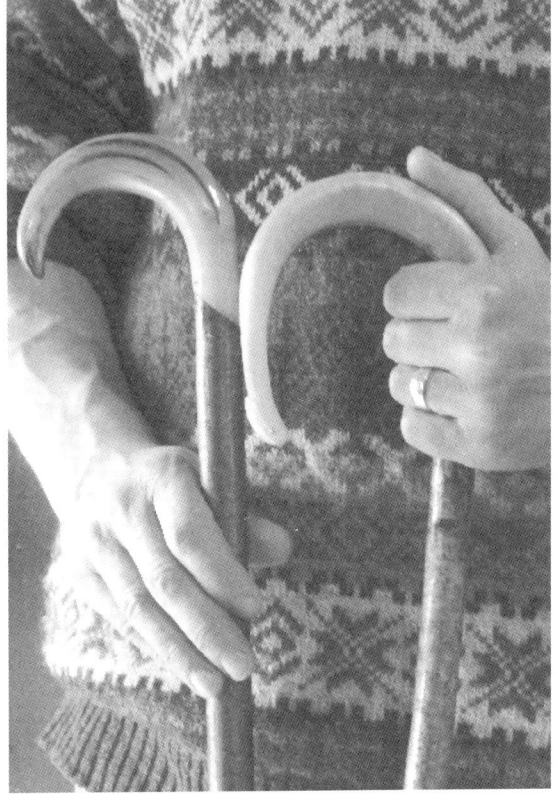

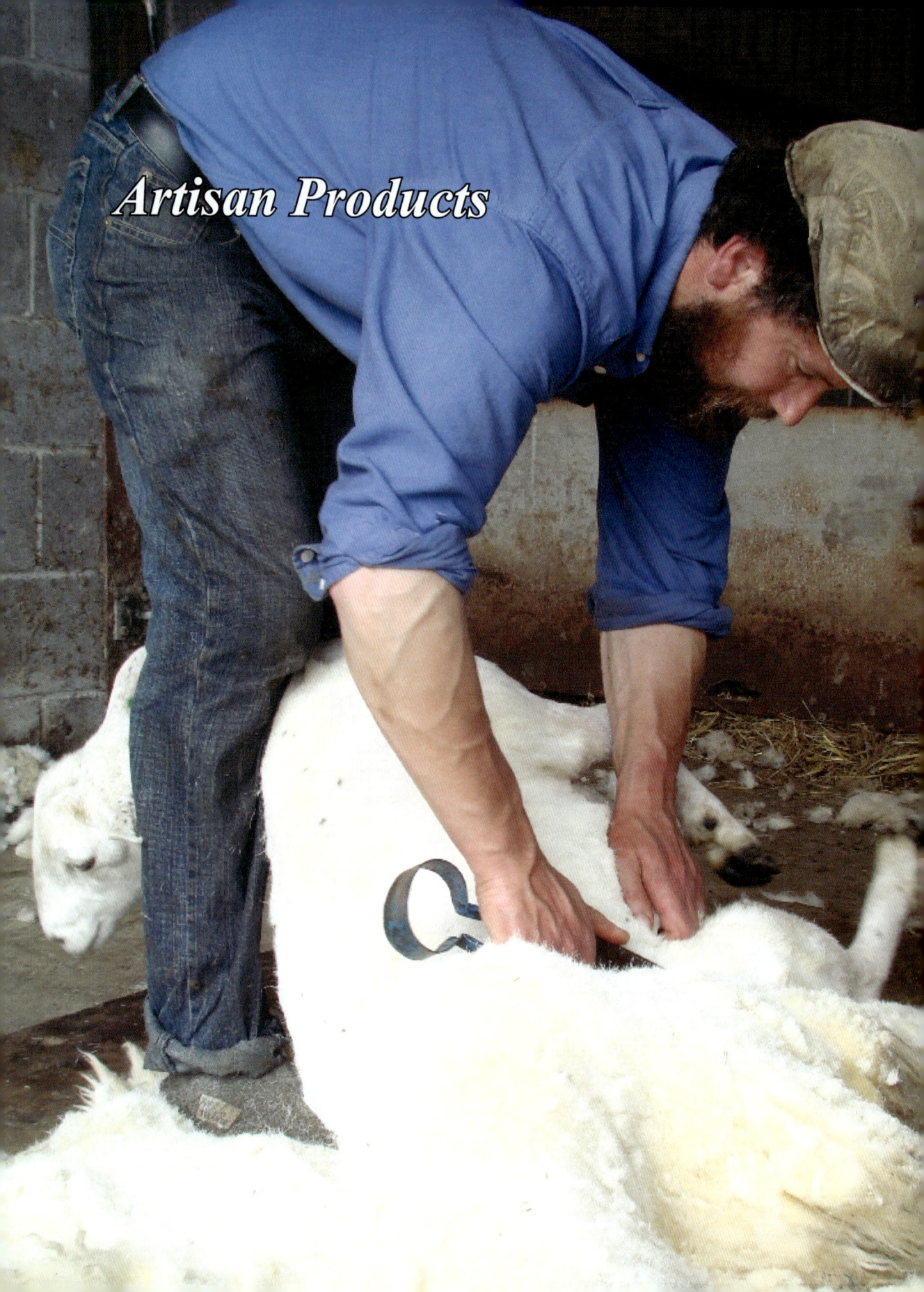

Artisan Products

USING WOOL: SPINNING, WEAVING, DYEING AND FELTING

Unfortunately, domestic processing of home produced fleeces into everyday items of family clothing cannot be considered to be a worthwhile activity, even for the most dyed-in-the-wool self supporter. In terms of the time and effort involved, you'd end up with very expensive garments! What's the point, when you can pick up a pure wool jumper in a charity shop for a couple of quid? Yet producing wool for sale to the British Wool Marketing Board (BWMB) isn't a financially justifiable activity either, with the price received scarcely covering the cost of shearing. In the case of coloured sheep, where the fleece may fetch less than 10p per kg, shearing can become a rather a costly business, unless you do it yourself. So what's the answer? Should we all keep wool-less breeds of sheep? I don't think so. As global oil reserves dwindle, man's reliance on petrochemical based synthetic fibres cannot continue, and wool must surely reassume its rightful place. We sheepkeepers just have to hang on in there!

Although I've said that small scale hand spinning of your own fleeces won't be justifiable as the sole source of the family's winter woollies, that shouldn't preclude the smallholder from having a go at producing a few things – it's great to have a jumper made from the fleece of a favourite animal, or, in my case, from the fleece of the very first sheep I ever sheared. (I can't recommend socks though – Dot made me a pair once, from hand spun Welsh wool, and they were like straight jackets for the feet!). As your skill develops, it may be possible to start adding value to those otherwise worthless coloured fleeces, by producing yarns for sale in a variety of natural shades. We sold quite a bit of wool like this when we lived on Bardsey Island, with Dot spending some of the long winter evenings spinning both white and coloured fleeces (and mixtures of the two), for sale via an "honesty box" system to summer visitors. She provided a pot of assorted knitting needles, and a simple pattern, and, after wet weather had kept the holiday makers indoors, people would emerge wearing Bardsey wool hats!

From this point, it is but a short step to producing a range of finished items in natural colours – artisan products for sale at the higher end of the market. Suddenly the idea of keeping sheep for their fleeces doesn't seem so bad after all! And there are, of course, a number of other applications for wool that can be explored, such as its use as an eco-friendly insulation material, both in the building trade and in food packaging.

PREPARING THE FLEECE

Fleeces stored in paper sacks in a suitable environment (cool and dry) will remain useable for a long time – years, even – and, provided that they were properly rolled at shearing time, it should be a fairly simple matter to unroll them and lay them out correctly for sorting. Alternatively, if it's convenient to do so, fleeces destined for home spinning can be sorted at shearing time, and the various different parts packed separately ready for use. Remember to label the bags!

Lay the fleece out skin side down on a clean surface, and begin by removing any contamination. Very heavily contaminated fleeces, for example from sheep that have regularly slept under the hayrack during the preceding winter, are of no use to the hand spinner, and, if sold to the BWMB, will be subject to price deductions. Matted fleeces should also be rejected. Any second cut wool (little short pieces, caused by the shearer going over the same part of the sheep twice, in order to tidy up a bit he's missed) should be picked or shaken off, and the belly wool can be put aside for felting (or for lining hanging baskets, if it's very poor). The remainder of the fleece is then divided up into about 7 different areas. There is a traditional pattern to this, but use your judgement a bit as individual fleeces vary. Basically, put wool that looks and feels similar, together. The best quality wool is found on the shoulder (known as Super Diamond), and then the back (Diamond). Other areas that should be separated out are the sides (Extra Diamond), neck (Shafty), the haunches, tail and hind legs (Prime Britch and Britch) and the forelegs (Brokes). In the case of coloured sheep you may wish to further sub-divide each category according to shade, but this is not essential – some nice random effects can be created by spinning the wool as it comes.

Carding

Before spinning, wool needs to be teased out and fluffed up. This could be done by simply pulling apart a piece of the fleece with your fingers, but a bit more care in preparation will result in a better yarn, so it is best to card the wool. This is also a good opportunity to remove any remaining debris, and to blend different colours.

A carding comb is a slightly curved rectangular wooden bat, about 8″ by 5″, with a handle on the long edge. The convex face of the bat is covered with "card cloth" which has closely spaced rows of bent wire teeth. Carders are always used in pairs.

Top and above: Carding combs and rolags prepared from different coloured wools

Method

With one card held face-up in the left hand (if you're right handed), lay staples of wool on the teeth, side by side, parallel with the handle. Cover the surface completely, but not too thickly. Commence carding by holding the other card face down in the right hand and brushing it gently across the wool on the first. Repeat this a few times. Most of the wool will transfer to the other card, and you'll see that the fibres are gradually being combed into line with one another. Next, transfer all the wool back onto the teeth of the first card. Repeat the process of combing and transferring several times, until you've created an even mat of wool with all the fibres more or less aligned. Remove the mat from the combs by stripping it gently from one card to the other, and back again. Lay the resultant light, fluffy rectangle of wool (batt) on the concave back of one of the cards, and roll it into a sort of long, airy, sausage shape, such that all the fibres are curled. This is called a rolag. Make quite a few rolags while you're at it, and stack them loosely in a box or basket, so that once you start spinning you won't have to keep stopping in order to prepare more wool.

If you intend to develop your wool processing activities into a cottage industry scale operation, then it would be worth buying a drum carder.

SPINNING

Spinning basically consists of drawing out the parallel fibres of a rolag, and twisting them into a continuous thread of the required thickness. This can be used directly for weaving, or 2 or more may be plied together to form a yarn suitable for knitting.

Drop Spindle

The drop spindle represents one of the earliest methods of converting raw fleece into yarn, preceded only by a simple weight, such as a stone with a hole in it, which would originally have been used in the process of drawing out and twisting the fibres. In Europe the spindle did not begin to be replaced by the wheel until some time in the fourteenth century; yet fine woollen cloth was being manufactured well before this, clearly demonstrating the quality of the yarn that can be produced using this primitive tool. Simple to make, or cheap to buy, the drop spindle remains an excellent

way for beginners to gain a "feel" for the wool, and develop the rhythm of co-ordinated hand movements required, before moving on to a more expensive spinning wheel.

To make a drop spindle, all you need is a slightly tapered stick, around 10″ long, and a wooden (or clay) disc about 4″ in diameter by ¾″ thick. The stick should have a notch in its thinner end (like a crochet hook), and the disc (or whorl) needs a hole through the centre, slightly smaller in diameter than the thicker end of the stick. When the stick is pushed through the hole, thin end first, the whorl should slide almost to the other end, leaving an inch or so of the thicker end of the stick protruding. Now you're ready to spin!

Method

You'll need a short length of yarn, already spun, as a leader. Attach this to the spindle just above the whorl, then pass it under the whorl and wrap it round the spindle a couple of times. From here, it goes back over the whorl and all the way up to the thinner end of the spindle, where it is held in place by a half hitch around the notch. The free end needs to be about 8″ long. Now, dangle the whole assembly in front of you by holding the end of the leader between the finger and thumb of your left hand. Together with the leader, also between the finger and thumb of the left hand, pinch a piece of the fluffy end of a rolag. With the other hand, give the spindle a good twirl in a clockwise direction, then pinch the leader and rolag just below the left finger and thumb. The two hands now move apart, drafting some of the wool from the end of the rolag between the finger and thumb of the left hand. When a section of suitable length and density has been drawn from the rolag, pinch more tightly with the left hand and release your grip slightly with

Above and right: Using the drop spindle

the right. The right hand now slides up to meet the left, allowing the turning spindle to twist the fibres into a thread. The end of the leader and the new thread should join seamlessly. Don't allow the twist to pass the left finger and thumb or it'll tangle up the rolag, making subsequent drafting very difficult. Continue repeating the sequence of hand movements until you've spun sufficient yarn for the spindle to touch the floor. At this point you'll have to stop, undo the half hitch and the turns under the whorl, and wind the yarn onto the spindle. Once again pass the yarn under the whorl before taking it back to the top of the spindle and securing it with a new half hitch. If you didn't give the spindle enough of a spin it may well stop rotating before it reaches the floor, then begin to unwind – if you don't spot this happening, your yarn will break. Experience

will soon teach you how much spin to give, and how finely you need to draft the fibres from the rolag.

Spinning Wheel

Spinning with a wheel uses exactly the same sequence of hand movements, although it's simplified by the fact you're working in a horizontal plane, and you get to sit down! There is, of course, the added complication of having to operate the treadle, but once you get the hang of it you'll find that a natural rhythm develops, and you'll carry on without even thinking about it. The action of the treadle

Using the wheel: Begin by attaching a leader of spun yarn

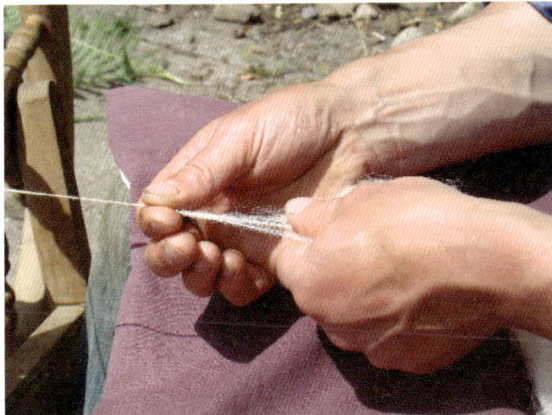

Drafting fibres from the rolag

turns the large wheel, which in turn drives the bobbin and flyer mechanism via a drive belt (just a piece of string, really). The bobbin rotates at a slightly different speed than the flyer, either faster by virtue of there being two drive belts on different sized pulleys, or more slowly by using a braking system, depending on the design of the wheel. In the first case, the spun yarn is drawn onto the bobbin, and in the second case the yarn is wound onto the bobbin by the flyer.

Just as when using a drop spindle, it's necessary to begin by attaching a leader of spun yarn. This is tied to the shaft of the bobbin, and wrapped around a few times. It is then carried

over the hooks of the flyer, before exiting via the spindle eye – you'll need a little wire hook to pull it through the hole. As before, spinning commences (in a clockwise "Z twist" direction,) by joining the leader to the end of a rolag, then just carry on! From time to time you'll need to shift the yarn onto the next hook of the flyer, to ensure that the bobbin is evenly filled.

If you're really serious about generating an income from your wool, you may consider sending fleeces away to be spun, leaving you free to concentrate on producing the finished articles for sale. Several companies that offer this service to small flock owners are listed in the resource section at the back of this book, including those that specialise in handling organic fibres. In many cases these businesses are run by smallholders who've diversified into wool processing.

Plying

A single strand of twisted yarn will, if given half a chance, begin to unravel itself. However, if two or more yarns are run together, whilst applying a twist in the opposite direction, a state of equilibrium is reached, resulting in a stable yarn. This is called plying. Single strand yarns are used for weaving, but for knitting you'll need 2- or 3-ply. Some interesting variations can be created by plying together bobbins containing different coloured wools.

Method

Put two or three filled bobbins in the Lazy Kate, making sure that the larger diameter ends are all at the same side. Place the Lazy Kate behind and slightly to the left of you, either on the floor or on another chair. Fit an empty bobbin on the wheel, and attach a new leader. Tie the ends of the yarns to be plied to the free

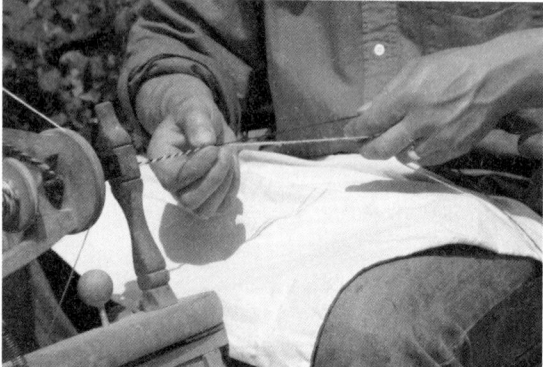

Top: Spinning wheel and Lazy Kate
Above: Plying together two different coloured yarns

length from the bobbins held on the Lazy Kate. Repeat these movements, over and over, until the bobbin on the wheel is full, i.e., when you've used up half the wool on the Lazy Kate. Now, to empty the bobbin of plied wool, you'll need to wind a skein.

Winding And Washing

Skeins are wound using a Niddy Noddy, which isn't half so bizarre as it sounds! The Niddy Noddy is made of wood, about 18″ long overall, and shaped like a capital letter I, except that the two cross pieces are perpendicular to one another. You could easily make one, by pushing pieces of dowel through holes drilled at each end of a length of broom handle. The wool is wound onto the Niddy Noddy in a continuous fashion, around one end of one peg, then one end of the other, then back to the other end of the first peg, then the other end of the second. As with everything we've seen so far, there's a rhythm to this – up and over, down and under, up and over, down and under with the right hand, winding the yarn, while the left hand, holding the centre of the shaft of the Niddy Noddy, twists it back and forth in time with the movement of the right. When you've wound a whole bobbin full of plied wool

end of the leader, then begin to treadle the wheel in an anti-clockwise ("S twist") direction. The action of the hands is very similar to when spinning, except that there is no drafting to be done so the left hand stays still. For making 2-ply, one yarn passes between the thumb and first finger of the left hand, and the other between the first and middle fingers. Pressure of the fingers controls the speed at which the yarn is pulled off the bobbins on the Lazy Kate. The right hand gently holds the two threads together and slides back towards the left, allowing the twist to develop in front of the fingers. When the two hands meet, the right pinches the yarn and moves forward again, allowing the plied yarn to be wound onto the bobbin on the wheel, whilst drawing another

Two Niddy Noddies, one purchased (front) and the other home-made (behind)

A skein wound on the Niddy Noddy

onto the Niddy Noddy, tie around the skein fairly loosely in four places using short lengths of yarn, before slipping it off. You could now simply roll this into balls and start knitting, but you'll end up with a greasy "fisherman's jersey". True enough, it'll turn the weather, but in respectable company it might just turn a few stomachs, as it'll pong pretty strongly of sheep! It's best, then, to wash the skeins at this stage, and, if you do want to make a weather proof sweater, re-oil it with purified lanolin during the final rinse. Or soak the skeins for a couple of hours and wash in cold water only (no soap) – this will remove most of the odour but leave the lanolin, which will be OK for outdoor garments. (Your sheep won't object to you smelling like one of them, whatever!)

For more thorough washing, soak the wool in warm water before washing in "hand hot" water with detergent – fairy liquid is fine. Very grubby wool will need washing more than once. Rinse several times in tepid water, before hanging the skeins up to dry on a broom handle. Another broom handle slipped through the bottom of the skeins will keep them taught and prevent kinking. Excess water can be gently squeezed out, but not wrung – handle the wool very tenderly throughout the whole washing

process, or you'll end up felting it!

Wool that has been thoroughly washed in this way can be dyed.

Some spinners prefer to wash (and dye) fleece wool before spinning. Certainly this keeps the spinning wheel and other equipment much cleaner, but washed wool has a "dry" feel to it, and is much harder to spin than its greasy counterpart.

DYEING

If you'd like to introduce a splash of colour to your creations, then it adds a nice touch if you use natural dyes from plants which can be grown in the kitchen garden or encouraged to proliferate naturally around the holding. I've included a pretty comprehensive list of recognised dye plants, together with a colour guide and suggested mordants, in an appendix at the back of the book, but, despite the information I've provided, you'll find each dye bath you make up takes you on an interesting voyage of discovery – the colours never come out quite like you expected, and very rarely do you get the same colour twice! Some plants will give a fairly predictable result, but others can be quite startling! The least likely looking plants often provide the stronger colours, whereas others – such as beetroot – are surprisingly disappointing. Possibly the most historically significant English dye plant is woad, with which the ancient Britons stained their bodies bright blue in an attempt to frighten the invading Roman army.

The basic method of obtaining dye remains pretty much the same whatever plant is used, in that the vegetable material is boiled to extract the pigment, although there are a few (lichens in particular) that need to be fermented – in urine, of all things! A mordant is used to ensure that the dye is colour fast, and may also be used to modify (bloom or sadden) the tone.

Mordants

Some dye plants contain their own mordant, such as the tannin in oak, or the oxalic acid in rhubarb, in which case no fixative will be required. However, in most cases a chemical (metal salt) mordant will need to be used, the commonest being alum (potassium aluminium sulphate), chrome (potassium dichromate), tin (stannous chloride) and iron (ferrous sulphate). A couple of less frequently used mordants are lime (calcium oxide – used with woad) and copper sulphate. Cream of tartar is often added to alum in order to brighten a colour, and in some cases washing soda or even common salt can be used for mordanting. Depending on the chemical being used, and the intended effect, mordanting can be carried out before, during, or after dyeing, or before and after.

Mordanting With Alum And Cream Of Tartar Before Dyeing

It is not possible here to give detailed instructions for the use of all the various mordants, but alum, being the commonest used (and the safest), seems like an appropriate example. The addition of cream of tartar will result in brighter colours. Too much alum makes the wool go all sticky, so make sure you weigh the ingredients carefully.

Method

For each 4oz (approx. 2 skeins) of washed yarn you'll need 1oz of alum, ¼oz of cream of tartar and 1 gallon of soft water (rain water is good). Heat the water to around 30°C in an old saucepan. Stainless steel is preferable, as it won't have any effect on the final colour, but aluminium pans are OK for use with alum. Don't use chipped enamel saucepans, or the iron will "sadden" the dye – if you want to produce subdued colours, then ferrous sulphate can be added later. Dissolve the alum and cream of tartar in a small quantity of boiling water, and stir the solution into the saucepan. Thoroughly wet the wool with clean, fresh water, and place it in the saucepan containing the mordant solution. Gradually, over a period of about an hour, bring the saucepan to the boil, turning the wool gently a few times. As soon as boiling point is reached reduce the heat and simmer gently for a further 45 minutes. Turn off the heat, and allow the yarn to cool in the liquor. After this, the wool can either be stored damp for a couple of days, after which it will be in an ideal state for dyeing, or it can be dried and stored for future use, in which case it will have to be thoroughly wetted again. If you're storing skeins of pre-mordanted yarn, remember to label each one according to the substance used.

A couple of simple dye recipes, using 4 ounces of wool pre-mordanted with alum and thoroughly wetted:

Yellow / green

Simmer 8-10 handfuls of onion skins in soft water for about 30 minutes. Strain out the skins and allow the liquor to cool to hand heat. Enter the wetted wool, making sure that it is completely submerged, and bring the pan slowly to the boil. Simmer until the wool reaches the desired colour. This method will result in varying shades of yellow. The addition of iron to the dyebath gives greens. (A bright orange can be obtained from onion skins by mordanting with tin, but this is not such an easy substance to use as alum).

WEAVING

Dyeing at home

My experience of weaving is limited to the use of a mediaeval style loom that my father built, on which we made some interesting wall hangings, but even where a more up-to-date piece of equipment is used the basic principle remains the same.

A loom, of any design, is basically a frame (either vertical or horizontally mounted) on which one set of threads (the warps) are held taught, while another thread (the weft) is woven in and out of them, from one side to the other, and back again, and so on, until an appropriate length of cloth has been formed. The weft is generally wound onto some sort of shuttle (we used a stick shuttle), or can be pulled through using a long flat piece of wood with a notch forming a hook at the end. Care must be taken not to pull the weft too tight, or the shape of the finished piece will be distorted. After every few passes with the shuttle the weft can be pressed firmly into place with a heavy wooden comb.

Our mediaeval vertical frame loom is a "weighted warp loom", in that the warps are held taught by weights tied to the ends (in our case, stones with holes in, gradually accumulated over many trips to a shingle beach!). The same design can be seen in Scandinavian and Icelandic illustrations. The two uprights of the vertical frame have holes drilled at intervals along their lengths to allow for re-positioning of other components as the length of cloth on the loom grows. The top cross member is the beam, over which the threads are hung when warping up. The warps are then tied in place, tight to the beam, with a "spacing chain". The lower cross member of the rectangle is also the shed rod. Alternate warp threads hang in front of or behind the rod, creating the "shed" – the gap through which

Purple / blue

Crush 2lb of elderberries, and put them in a pan of cold water. Heat slowly and simmer for 45 minutes or so. Strain off the berries and allow the liquor to cool to hand heat. Enter the wetted wool, making sure that it is completely submerged, and bring the pan slowly to the boil. Simmer for half an hour or more, until sufficient depth of colour is achieved. Leave the wool to soak in the dye overnight to give a lovely purple colour. This can be made bluer by the addition of a handful of common salt to the dye while the wool is soaking.

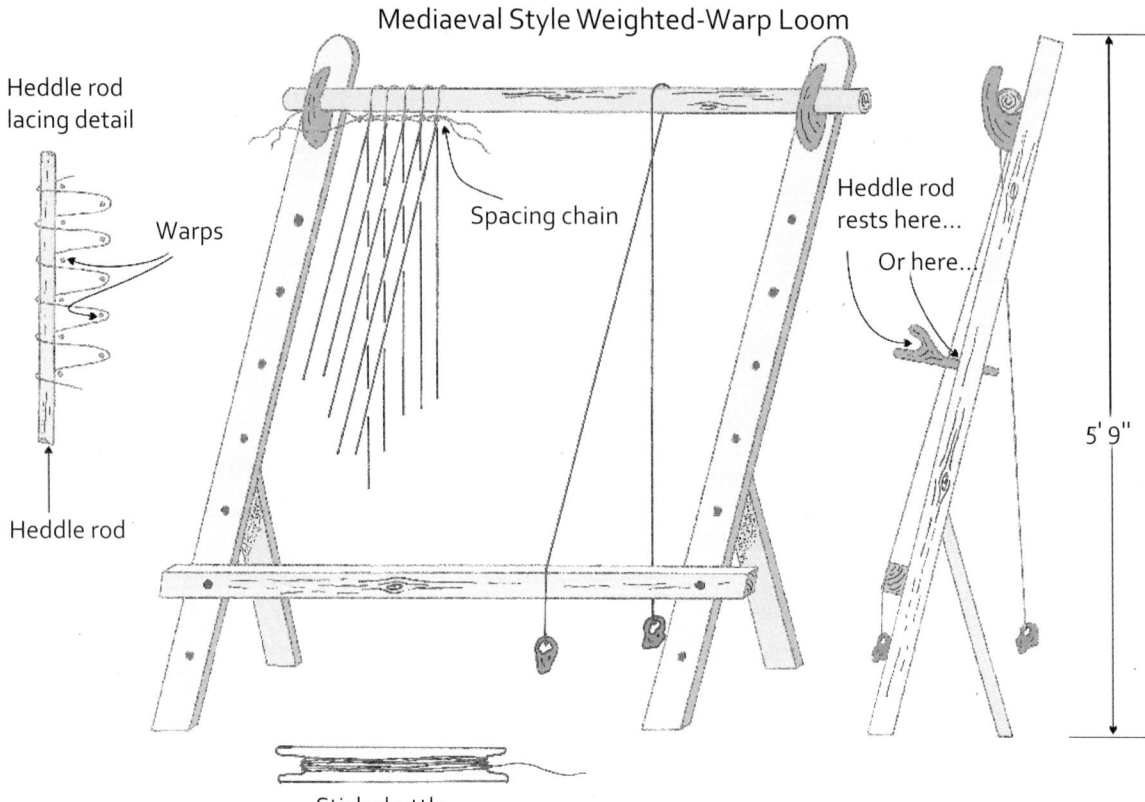

Mediaeval Style Weighted-Warp Loom

Heddle rod lacing detail

Warps

Spacing chain

Heddle rod rests here...

Or here...

Heddle rod

5' 9"

Stick shuttle

the shuttle containing the weft is passed. Of course, it is necessary to bring the hindermost warps to the front before sending the shuttle on its return journey, otherwise you'd simply undo what you'd just done. For this, another, movable, horizontal bar, called the heddle rod, is used. All the warp threads that hang behind the shed rod are loosely laced to the heddle rod. Pulling the heddle rod forward (where it is held by a couple of notched pegs located in the holes in the uprights) moves the warps into the alternate position.

FELTING

(Information kindly supplied by Val Grainger, the "Woolly Shepherd").

Everyone has a story about how their nice woollen jumper shrank and became felt in the wash, and they usually believe it's because they used too hot a wash. This is incorrect – you can actually boil wool! What actually makes felt is the combination of shocking the wool with hot water, and agitation. If aided by a little soap, the result is felt. Tribespeople of Mongolia make felt for their yurts by putting their wool on the floor, wetting it, and inviting as many people as possible to walk all over it for several hours, if not days!

So, how can you make felt at home, and are all sheep's fleeces equally suitable?

Well, sheep have been bred for a range of purposes, and therefore there are different types of wool with different properties. Downland sheep, such as the Dorset and the Suffolk were bred to have a naturally "superwash" type wool, which is soft yet hard wearing – it is difficult to felt without a lot of effort, as it is effectively shrink resistant. Merino wool, on the other hand, is so soft that it will felt very well, very quickly.

To make felt the wool must first be washed, a process known as scouring, and carded into batts (big webs of wool) or rovings (long strands of carded wool). It is then layered in opposite directions so as to produce a criss-cross of fibres. The wool can then be either needlefelted or wet felted.

Needlefelt does not depend on water, heat, soap and friction to make felt. Instead, long barbed needles are pushed into the wool causing the scales on the wool fibre to lock together into a thick mat. This mat can be very tight if a lot of needling is done over a small area, or loose and full of air if less needling is carried out. This is how needlefelt insulation is made on large commercial needlefelt machines.

Wet felting is achieved by pouring hot soapy water over the layered wool and applying some sort of friction. On a small scale, the wool is often layered on bubblewrap and, after pouring the soapy water on, the whole lot is rolled around a rolling pin and rolled, re- wrapped, and rolled again, several times. Mechanical wet felting can be achieved with the use of large felting machines that provide the friction needed to make large sheets of tight felt.

Other methods of making felt include knitted felt, where a garment or bag is knitted several times too big, then washed on a very hot programme in a washing machine with a pair of jeans or an old trainer to knock it around and provide the friction needed to shrink it into felt.

Placing sheep's milk cheeses in presses

DAIRY PRODUCTION

Given that sheep can provide the self supporter with milk, meat and wool, all from the same animal, why aren't more smallholders keeping multi-purpose flocks? I don't know the answer to this, I'm afraid. I do know that domestic milk production is one of the least common smallholding activities these days, due to the unavoidable tie of twice daily milking – many people are running their holdings on a part-time basis, with all enterprises having to be fairly flexible in nature in order to fit in around other, off farm, commitments. This, together with a marked reluctance on the part of British people to try anything "different", means that small scale (subsistence) sheep dairying has never really taken off in this country, which is a shame, given the historical importance of ewe's milk. And, in this age of allergies and intolerances, ewe's milk has a very modern significance too – most people (unless they are lactose intolerant) who have been told that they must cut out dairy based foods are fine with sheep's milk and sheep's milk products. Even in cases of severe lactose intolerance, sheep's milk is well worth trying, particularly if consumed as yoghurt or cheese.

Sheep's milk is easily digested, higher in

solids than goat's or cow's milk (making it ideal for cheese production), and contains a greater percentage of essential minerals (calcium, phosphorous, sodium, magnesium, zinc & iron) and B vitamins (see table below).

Comparative analysis of different milks.			
	Sheep	Goat	Cow
Water (%)	80.82	85.71	87.08
Fat (%)	6.86	4.78	3.78
Protein (%)	6.52	4.29	3.58
Lactose (%)	4.91	4.46	4.90
Casein (%)	5.00	2.50	3.20
Minerals (%)	0.89	0.76	0.71
B Vitamins (mg/l)	22.49	12.67	9.83

The Milking Flock

As I've already mentioned (page 26), individual animals of just about any breed of sheep can be milked, at a push, particularly if they've previously shown a capacity for a higher than average sustained yield by successfully rearing triplet lambs to satisfactory weights at weaning. If you don't fancy starting afresh, then make a selection of the milkier ewes in your existing flock, and cross them with a ram of a dairy breed. A couple of generations of this policy, combined with careful use of performance records, should result in the beginnings of a productive dairy flock. This will have been achieved with considerably less initial outlay than would be the case if a new flock were to be purchased, although it will take longer for yields to reach high enough levels to give a reasonable return. Having said that, you might be glad of a year or two at a lower level of production, while you get the system up and running.

As with cattle, the dairy sheep exhibit the classic "dairy wedge" conformation. Think of the difference in body shape between a Jersey cow (dairy) and a Hereford bull (beef) and you'll know what I mean. When selecting your foundation dairy ewes, look for the high tapered shoulder, light forequarter, narrow head and fine bone. These characteristics are totally contrary to what we're looking for when choosing breeding stock for other purposes! Additionally, in common with other types, the dairy ewe should have good length, a broad deep pelvis, and well spaced straight moving hind legs. Udder attachment must be good (i.e., no droopy tits!), with fairly large teats placed well underneath, like a goat's. (This teat shape and placement would not be so desirable in a ewe that suckles her offspring). During a lactation period of just over 200 days, a good

Milking parlour

ewe of one of the recognised dairy breeds may produce a total of 600 litres, although a more realistic flock average of 250 litres should give a justifiable return. Crossbred ewes and ewes of other (non-dairy) breeds used for domestic milk production are unlikely to exceed 150 litres / lactation, although individual animals may well top this figure. These are the ones whose daughters you should retain.

Management

Routine seasonal management of the dairy sheep flock does not vary greatly to that of flocks kept for other purposes. They'll need tupping, lambing, shearing, hoof trimming, etc. just the same. One significant difference lies in the fact that many of the veterinary products used regularly in conventional flock husbandry are not licensed for use in sheep producing milk for human consumption. This, perhaps, paves the way towards organic production methods, but it really isn't much help if you've got a sick ewe or a persistent flock health problem that needs to be tackled. A quick flick through the *Compendium of Data Sheets for Veterinary Products* has shown me that none of the products that we use routinely in our flock for internal and external parasite control, and none of the antibiotics that we would use after difficult lambings, or to treat infections, are licensed for use in dairy sheep. In fact, just about the only things I can find that are OK are the various clostridial vaccines, such as Heptavac P and Covexin 8.

It's not that there's anything actually wrong with the products, it's just there's insufficient demand for the pharmaceutical companies to justify the expensive licensing tests. Vets can prescribe unlicensed products via a process known as the prescription cascade, in which case standard milk withhold periods will be applied. These could be quite long – there's a higher perceived risk with milk than meat, as several months may elapse before a treated meat animal enters the food chain, whereas milk is produced daily. Treating dairy sheep with veterinary medicines on a conventional routine basis could result in only 85% of the milking portion of the flock producing a saleable product at any one time. This stresses the importance of tackling flock health issues outside the lactation period. It also suggests that there's considerable potential for organic methods and the use of homeopathy within the dairy sheep industry.

Incidentally, withheld milk can safely be used for rearing surplus lambs, so it needn't be wasted.

Lambing

You need to be aware that the specialist dairy breeds are extremely prolific, with litters of 3, 4 and even 5 lambs being reasonably common. In consequence, newborn lambs are small, lack vigour and are very vulnerable. Be prepared to accept mortality rates that wouldn't be tolerable in a conventional sheep flock. All lambs should be stomach tubed with about 120ml (or approx 40 – 50ml per kg bodyweight) of fresh colostrum as soon as possible after birth. In total, they'll need about a litre of colostrum during the first day, but once you've given an initial boost by stomach tube they'll hopefully suckle the remainder of their requirement from the ewe.

General practice is to remove the lambs from their mothers at 24 – 48 hours old, and rear them artificially, although some people advocate leaving the lambs with their dams for 30 days before weaning abruptly. Given that yield peaks at around 40 days of lactation,

this method would seriously compromise the overall milk output of the flock, but the loss of production may be offset against reduced lamb rearing costs. Either way, the most important thing is to get the lambs weaned and onto good quality pelleted feed as early as possible, say 4 – 6 weeks of age. In an early (December / January) lambing flock it would be worthwhile pushing the lambs on with an ad-lib diet in order to get them to saleable weights from about 14 weeks old – just in time to catch the Easter trade, when lamb prices peak. For later lambing flocks the economics of artificial rearing do not look so good. On a small scale I'd suggest turning the later born lambs out to grass once weaned, in a sheltered paddock, and keep them growing on with a minimal diet for slaughter at about one year old for home consumption. However, this will impact on the amount of grazing land available for the milking portion of the flock.

Some of the problems associated with lack of vigour and poor economic return can be minimised by the use of terminal sire tups on the bulk of the flock, with only the very best ewes being bred pure in order to provide replacements.

Feeding

As with any dairy enterprise, correct nutrition of the lactating animal is essential if optimum output is to be obtained, yet the self supporter may well be content with a low input / low output system, thus reducing the dependence on bought in feeds. Basically, the more you want to get out of your flock, the more you will need to put in. The science behind the correct formulation of rations for lactating ruminants is complex and fascinating, and it's not only relevant to the managers of high yielding Holstein herds chasing every last pennyworth of milk – anyone aiming to achieve that little bit of extra performance from their animals would be well advised to read around the subject, even if only to give a better understanding of the analysis given on bags of proprietary pre-prepared feedstuffs. For the majority of smallholders though, running fairly low-key enterprises, a bit of general knowledge and a few rules of thumb will usually suffice to ensure that animals are well fed, healthy and reasonably productive.

Rules Of Thumb For The Feeding Of Lactating Dairy Ewes

Record milk yields on a regular basis (daily, if possible), and, for a 70kg ewe, feed 0.5kg of concentrate feed per litre of milk produced. This type of feed has a dry matter (DM) content in the region of 86%. Non-productive ewes will typically consume around 2.5% of their own body weight (BW) in DM per day, but high yielding dairy sheep may manage as much as 4.5%. For the sake of our example here, let's assume a DM intake at peak lactation of 4% of BW.

Now, supposing that our 70kg ewe is, at present, producing 3 litres of milk daily. She'll be getting 1.5kg of concentrates, giving 1.29kg DM per day. Her total daily DM requirement is 2.8kg (70kg x 4%), so there's a shortfall of 1.51kg per day that needs to be provided in the form of forage. This can be obtained by feeding 1.7kg of hay (85% DM), 6.04kg silage (25% DM), or from 7.55kg of grazed grass (20% DM).

At times, this level of feeding may not provide sufficient energy and protein, in which case the ewe can be expected to mobilise body reserves. BW losses of up to 100g per day during early lactation are acceptable, but this does emphasise the need for ewes to be in

excellent body condition at lambing.

With a bit of commonsense and general stockmanship you'll be able to tweak these figures according to the nature of your enterprise, the size of your sheep, and your target level of production.

Milk Production For Home Use

A bucket and something to sit on. That's the basic essentials to get you started. Preferably the bucket will be of stainless steel (we've got a couple of splendid ones bearing the legends "Special Diet Kitchen" and "Rothschild Ward" – no prizes for guessing where they came from!) and ideally should have a lid, but any food grade receptacle can be pressed into service initially. You could easily end up spending a fortune on dairy paraphernalia and, whilst it would be jolly handy to have, much of what is available is not absolutely essential. It would be best to gradually build up a stock of equipment over a period of time as your experience and commitment develops – the requirements do become more complex as the scale of the operation increases. For home use, it's fine to milk a few ewes by hand in a clean outhouse, and carry the milk to the kitchen in a lidded bucket or churn, for further processing. Hot and cold water taps close to the milking area are refinements worth having, enabling you to keep everything reasonably clean. For udder washing I would use warm water with a splash of iodine, though if you haven't the luxury of a hot tap you can use anti-bacterial udder wipes. A small platform or milking stand is a good idea, in order to raise the animal to a convenient height. The ewe is encouraged to jump on by an offer of feed, and generally there's a yoke at the front of the stand through which she has to put her head to reach the food in the bucket. Under the relaxed regime of subsistence dairying, the ewe should be given time to finish her feed before you start milking. (You can be washing her udder in the meantime). Sheep (and other herbivores) in a natural environment are prey species and are at most risk from predators when they've got their heads down in the grass, so tend to be tense and wary when feeding. Countless generations of domesticity have not altered this, and in fact may have added to it – a ewe with her head in a bucket will not settle 'til the last crumb is gone, and will be on the defensive lest her flock mates try to steal from under her nose. A relaxed ewe has her head up and is probably chewing the cud. Her eyes may be half shut, but her ears are wide open. Now the milk can flow – you only have to observe sheep, goats or cows suckling their own offspring to see that this is so. It never ceases to surprise me that so many smallholders seem to need to use feed as a means to get their animal to stand still whilst being milked. If she won't stand to be milked without bribery then in all likelihood there is something amiss, either in the environment of the milking area or, more probably, in your relationship with your stock. In our experience, animals fed before milking give better results than those that are feeding whilst milking takes place.

By Hand Or Machine?

Hand milking is cheaper, quieter, makes less washing up and is undoubtedly more intimate, a very pleasant and relaxing occupation… when the weather's nice! When it's freezing cold, your hands are chapped and you've cut your thumb or sprained a wrist you may view things in a different light, and wish you'd opted for mechanisation. Don't be conned into the belief that hand milking is more "natural". More old fashioned it may be, but there is nothing natural about a human being squatting down next to an animal and squeezing away at her udder! Come to that, there is nothing natural about attaching a machine to her teats either, but at least the sucking and pulsating action of the milking machine does very closely resemble the suckling action of her offspring, an effect you cannot hope to achieve by hand. Machine milking will give you cleaner milk – no bits of wool or dust dropping in the bucket – and, what's more, it's consistent. Regardless of who places the cluster on the udder, the sensation for the ewe remains the same – an important consideration if you ever need to be away from home. You can very quickly teach your animal-sitter the correct way to use the machine, but finding a competent hand milker to take over in your absence can be a devil of a job.

TECHNIQUE

Hand Milking

Traditional practice is to milk sheep from behind, although I don't suppose it really matters a great deal. Sit down at a convenient height, either behind the animal or alongside her, whichever you feel most comfortable with. With a slight downward stroking motion, take a teat in each hand, and, using the thumb and forefinger, squeeze shut the top of the teat – you must be pretty firm here, as milk flowing back up the wrong way during milking is painful for the ewe and could cause damage to the udder. Now, without any pulling (if anything you should raise your hand slightly) bring the other fingers of that hand to bear on the teat, one by one, gently but firmly forcing all the milk trapped in the teat cistern (by your thumb and forefinger) out through the teat canal and into your bucket. Repeat with the other hand on the other teat, and in the meantime release the thumb and forefinger of the first hand to allow that teat cistern to refill. You'll soon fall into the rhythm of it....first one hand, then the other.......back and forth.... back and forth....relaxed and unhurried, until it seems there's no more milk to be had. Stripping is then carried out by gently drawing the thumb and forefinger down the teat to eject the last few drops, but I stress that it really must be only the last drops – should you commence the stripping action whilst any significant amount of milk remains then damage to the delicate membranes lining the teats is likely to occur. It is essential that stripping be carried out thoroughly, for if milk remains in the udder after each milking then yield will fall day by day, and eventually the ewe will dry off. Having said that, do not fully strip out for the first few days of a new lactation.

Machine Milking

Small, trolley mounted units, suitable for milking up to 2 ewes simultaneously (or a single cow), are readily available, but at close to £1000 they represent a considerable investment. However, if you'll be milking half-a-dozen sheep twice a day, every day, you may consider it a worthwhile one-off expense. We were lucky, and bought our machine in excellent condition second hand, but they don't crop up very often. Whether you have purchased a machine new or second hand it is just as well to get it checked over by a qualified dairy engineer before you start using it. On a second hand machine there could be parts that need replacing, and a new machine purchased for sheep may well have been pre-adjusted to the settings required for cows or goats. You'll need teat cups and liners specifically for sheep – goat ones, although they look similar, are actually too long. Once the machine is properly set up for the job, and you are familiar with its

use, you'll be able to carry out routine checks and adjustments yourself. Vacuum pressure should be between -40 and -45kpa. Pulsation rate needs to be set at 120 per minute (that's 3 times faster than for cows!), with a pulsation ratio of 50:50 (time liner open : time liner closed).

Trolley mounted milking machine suitable for small scale production

With the machine running at the correct vacuum level, and a ewe on the milking platform, move the cluster forward between her hind legs and under her udder. Push the small valve that allows suction to commence, and place the teat cups into position one by one. Just pop them onto the appropriate teats and they'll stay there. Keep your hand on the cluster for a moment or two, just to stop it swinging around, until the ewe settles down. Don't be tempted to support the cluster in any way – the weight is important to prevent the teat cups from gradually creeping up and injuring the udder. A clear section in the milk pipe is handy, enabling you to see when milking is almost over – avoid over milking which could damage the sphincter at the teat orifice. Machine stripping is carried out by pulling down gently on each teat cup with one hand while massaging the corresponding quarter with the other. To remove the cluster from the udder, simply stop the suction (same valve as before), whereupon it'll just slip off.

After emptying the churn, rinse everything out with cold water, before drawing a few gallons of really hot water through the whole system. Periodically dismantle everything and thoroughly clean with dairy hypochlorite.

Whether you milk by hand or machine, apply some udder cream to the teats when the job is done. There are a number of brands of medicated ointment sold for the purpose, but we find plain Vaseline to be better than anything else.

Processing

Sheep's milk has a mild, slightly sweet flavour, and is naturally homogenised. The high proportion of solids-not-fat (casein in particular) means that as little as 4 – 5 litres may be required to produce 1kg of cheese, as opposed to 9 – 10 litres of cow's or goat's milk.

The high fat content enables ice-cream to be made using the whole milk, without the need to add additional cream or egg yolk, resulting in a product that actually has a lower fat content than cow's milk ice-cream. Sheep's milk yoghurt is naturally thick, and can be used as a low fat substitute for cream in cooking. Frozen sheep's milk will store well for more than 4 months, and does not separate on defrosting.

Pasteurisation

Pasteurisation of sheep's milk is not compulsory, and, with good hygiene, it is easily possible to produce milk that is clean enough to satisfy all the regulations. When pasteurisation is carried out it can either be by holding the milk at 73ºC for 15 seconds, or 65ºC for 30 minutes. Small scale domestic pasteurisers are available from smallholding supplies companies.

COMMERCIAL PRODUCTION

Be aware that producing milk for sale is a whole different ball game, involving the Food Standards Agency, Environmental Health and your Local Planning Department, amongst others. If contemplating an expansion into commercial production, you must engage in discussions with the relevant authorities before committing yourself. Work with these people, and you'll be guided through the maze of legislation. If you decide to go ahead without consultation you may well find yourself in trouble.

You also need to be aware that, if you process your milk before sale, your processing facility may be liable for business rates. Don't forget to factor this into your costings.

Small Scale Commercial Production

Here, I'm talking about flocks in the region of 30 – 50 ewes. An enterprise of this size wouldn't be expensive to establish, but would rely on direct marketing of all produce to generate a reasonable return. Sales of whole milk, either fresh or frozen, are limited by lack of demand, although a ready market may be found amongst allergy sufferers. The same can probably be said for yoghurt. Principal outlets are likely to be cheeses or ice-cream sold through farmers' markets and other similar events. Of the two, ice-cream is perhaps the easiest to make on a small scale, but possibly has only seasonal appeal. Cheese making is a complex art, and training will be required, yet this remains the best way to add value to your raw product.

Simple milking parlour installations that can be erected inside existing buildings are available. Usually a single module will milk up to 6 ewes at a time, although multiples of modules can be put up together to increase capacity. These installations do crop up for sale second hand from time to time, being suitable for either sheep or goats – a neighbour of ours bought a six yoke module second hand, and I think it cost around £1,700. Re-assembling the whole thing was not difficult at all. Basically it consisted of a framework of pipes that clamped together, giving a platform (thick plywood and rubber matting) at about waist height, which the animals accessed via a ramp. The six yokes, with integral feed bucket holders, were constructed as a single unit that was mounted along one side of the platform. Each yoke closed automatically as an animal put its head through to reach the feed, but all were released simultaneously by operating a lever at one end. Along the other side of the platform, behind the animals, ran the vacuum pipeline. This actually formed part of the framework of the whole thing. Six taps enabled a separate milking unit to be attached to the vacuum line for each animal. These could either be complete bucket units, as seen on trolley mounted machines, or clusters only, with milk being transferred via a second pipeline to a refrigerated bulk tank. The operator stood on the floor behind the platform, so all udders were presented at a convenient working height! The vacuum motor

itself was situated in an adjoining outhouse, in order to reduce noise levels in the parlour.

Obviously there were considerable establishment costs incurred over and above the purchase price of the second hand equipment, including additional and replacement parts, conversion and upgrading of the existing outbuildings, and labour.

Large Scale Commercial Production

The British Sheep Dairying Association (BSDA) suggests that the minimum size of flock required for financially justifiable commercial production of sheep's milk is in the region of 250 – 300 ewes, with up to 500 ewes being preferable. That probably puts it outside the scope of this book!

As far as I can see, most of the large scale sheep's milk producers in the UK operate in groups or co-operatives, or supply raw milk to larger dairies for processing into cheeses and ice-cream. Branded products are on sale in various supermarkets, as well as specialist stores. Many of the producers are large farms that have diversified, or that have established a milking flock as a second enterprise in order for a younger family member to join the business. All have made substantial financial investments in equipment and facilities, and are fully committed to dairying.

See appendix for dairy recipes.

Land Management

Grassland management on the smallholding is not an easy topic to cover! Although high densities of smaller, traditional holdings are found in the predominantly grassland / livestock rearing areas of the UK (e.g., western regions), many new smallholdings have been created in other parts of the country following the amalgamation of large farms and the subsequent selling off of one of the houses, together with a few acres. As a result, we have to consider management of a diverse range of grazing land, from re-seeded former arable fields to upland heather moors!

SWARD TYPES

What Is Grass?

To the farmer or smallholder, grass is a crop. It's an easy crop to grow, but a very difficult crop to grow well! In many cases, grassland is rather taken for granted, but when you consider that well managed grazing provides food for livestock at less than 20% of the cost of purchased concentrates, you'll realise it's worth putting a bit of effort into! Really, when we say "grass", what we're referring to is the whole sward, consisting of a variety of grass species, clovers and broad leaved weeds, some of which are useful, many others less so, and a few of which are a downright nuisance. More than 50 plant species may combine in varying proportions to form the diversity of habitats that we collectively refer to as pasture. Let's take a look at a few of the commoner species, and their place within the sward:

• Ryegrass

There are two basic types of ryegrass – Perennial and Italian. Italian ryegrass probably has no place on the average smallholding – it is fast growing and high yielding, but requires high inputs of artificial fertilisers in order to achieve its potential. It has a low level of persistency (1 – 2 years) so is used in seed mixtures for heavy cropping short term leys. Perennial ryegrass, on the other hand, forms the mainstay of most long term leys, and, in conjunction with white clover, should be the dominant species found in improved permanent pasture.

• Annual Meadow Grass

Annual meadow grass thrives in areas damaged by poaching, such as in gateways or where feed troughs have been sited. After resting a field for a while, this species will be seen to be growing thick and strong on the previously damaged area, giving the misleading impression that there's plenty of fresh grazing! Annual meadow grass is, in fact, a low yielding, un-palatable weed species, and of no value to livestock at all.

• Cocksfoot

This large, coarse grass is a component of more traditional seed mixtures. It thrives under lax grazing or conservation regimes. It has very good drought resistance, so is useful on thin soils.

• Timothy

This is another component of traditional leys, but is used less often in seed mixtures nowadays as it cannot compete with the more productive ryegrass. It is winter hardy, has a moderately high yield potential, and is very palatable to grazing livestock. Timothy thrives on heavy soils or in high rainfall areas.

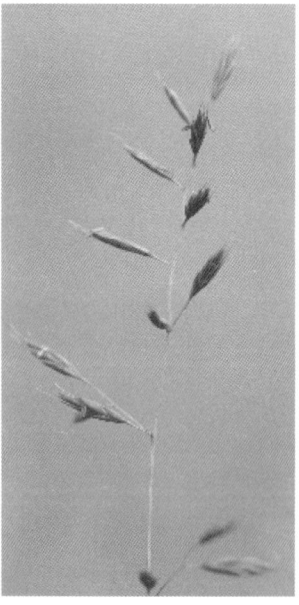

• Fescues

These are hill and upland grasses of moderate / low productivity. Red fescue is a valuable species under poor conditions, and is often included in seed mixtures on hill farms. Sheep's fescue is less useful. Both species would be considered undesirable in a lowland sward. Meadow fescue combines well with timothy and white clover to give a seed mixture suitable for non- intensive livestock holdings.

• Mat Grass

This is another upland grass, commonly found together with fescues on peaty soils. It is low yielding and not particularly palatable to stock, but its presence indicates good potential for improvement.

• Purple Moor Grass (Or Flying Bent)

Yet another upland species, common on deep, wet peat. It has a low level of digestibility, and is palatable only when very young. It is really only grazeable for 2 – 3 months of the year (June, July, August).

• Tufted Hair Grass

A weed species of poor pastures in both lowland and upland situations. Tufted hair grass thrives on heavy or badly drained soils, and is unpalatable to grazing livestock.

• Yorkshire Fog

This is a common component of underutilised lowland swards. It will tolerate low fertility and low pH, and is typically found in pastures cut late for hay. Yorkshire fog produces a relatively high yield, but has low palatability.

• Creeping / Common Bent

These are low yielding grasses of both upland and lowland environments. Bents tolerate low fertility, and are quite valuable in hill swards in conjunction with fescue. However, a high proportion of bent is a common cause of sward degeneration.

- ### *Sweet Vernal*

This is a common species found in poorer swards. It is drought resistant and tolerates low fertility. Sweet vernal is what gives traditional hay from old fashioned meadows its beautiful aroma, however it has a high proportion of stem to leaf and is unpalatable to stock. Clearly, the sweetest smelling hay is not necessarily the best thing to feed to your animals!

- ### *Meadow Foxtail*

An early growing grass of lowland pastures, meadow foxtail is palatable, nutritious and fairly high yielding, making it a valuable part of a traditional hay meadow sward. Unfortunately it is a difficult species to establish.

- ### *Crested Dogstail*

This is a valuable species of less intensively managed swards, and is commonly found in sheep pastures. It is fairly hardy, but not particularly competitive, so will grow well late in the season when the growth rates of more dominant species slow down.

- ### *Rough Stalked Meadow Grass*

This palatable species is commonly found in fertile lowland swards in conjunction with perennial ryegrass (with which it may easily be confused). It produces moderate yields early in the season, but may suffer from moisture stress later on in the year.

- ### *Smooth Stalked Meadow Grass*

This meadow grass is not so productive as the rough stalked variety, but has good drought tolerance, so is suited to light land. It provides early spring growth, and produces a leafy aftermath following topping or mowing. It is quite a useful species for sowing in poultry runs.

- ### *Clover*

Both red and white clover, in common with other leguminous plants, have the ability to "fix" nitrogen from the atmosphere – a dense clover sward may contribute the equivalent of up to 250kg N / ha. This is why ryegrass and clover work so well together. Red clover is not very persistent (up to 3 years) so is best suited to short or medium term leys. Its high protein content makes it ideal for fattening lambs, or for conserving as hay or silage. Don't graze with ewes for 6 weeks before or after tupping, due to negative affects on fertility. White clover is of more interest to the smallholder – it will occur naturally in many swards, and should be encouraged to become well established.

Notice the grass that the cattle have left around this thistle. It may not look like much but, when multiplied by the number of thistles in the field, it represents a considerable loss of grazing

Broad Leaved "Weeds"

Deep rooted herbs such as plantain and yarrow provide essential trace elements for grazing stock. Chickweed can be a nuisance during the establishment of new leys, as it quickly colonises areas of bare soil – control is difficult without also damaging clover. Pineapple weed (a member of the chamomile family, I think) will appear, together with annual meadow grass, in gateways and other damaged areas – if you allow your house cow, goats, or dairy sheep to eat it, their milk will taste dreadful for a few days! Dandelions, daisies and buttercups have no agronomic value, but together with many other wildflower species such as knapweed and cat's ear they present an uplifting splash of colour in any traditional meadow, as do vetches and tares.

Thistles, nettles and docks are a real pest, and will considerably decrease the productivity of grassland – as few as 5 thistles per 20m² can reduce the output of a sheep flock by more than £150 per hectare per year. Ragwort is a poisonous weed that you are legally obliged to control. It is rife on many smallholdings due to poor land management and overgrazing with horses. It is also worth mentioning bracken (an invasive fern) and soft rush (common on overgrazed, poorly drained, acid soils). Both of these species are capable of rapidly dominating large tracts of grassland, almost to the exclusion of all other species, rendering the area more or less non-productive. Their only saving grace is that they can be cut and carried for use as animal bedding.

SWARD TYPES

It is perhaps worthwhile to note at this point that grassland is not the natural ecosystem of any part of the British Isles. At one time, woodland of varying types would have covered the lower lying areas, becoming scrub at higher altitudes, and moorland above 2000 feet. This means that the grasslands of today are a rather unstable habitat, and any change of management will quickly be reflected by a change in sward composition.

Upland: Moor & Mountain

These are the most natural of our grassland habitats, but at best, can only be described as semi-natural. Generations of management, such as the burning of heather on the grouse moors, and extensive grazing by sheep, have suppressed many indigenous species and encouraged others to dominate. The soil will be acidic – many of the species present, including the heather, rely upon a type of root fungus for their nutrition, which cannot thrive under alkaline conditions. The predominant grass species are likely to be bent and fescue on the better areas, with flying bent, mat grass, mountain heath grass and wavy hair grass elsewhere. Particularly poorly drained areas, and bogs, will be colonised by cotton sedge. Bracken will often be found encroaching into the better, well drained areas. This type of grassland is only likely to be significant on

smallholdings that enjoy common or mountain grazing rights, and, although the stock carrying capacity is very low, I cannot really advocate carrying out major improvements on such pasture, as moorland has become something of an endangered habitat. Whereas previously, government funding was available to increase the productivity of these areas, nowadays grants can be obtained to allow them to remain in their existing state, and I believe that there may also be funding available to encourage regression of previously improved areas. However, there may be small, localised areas on the smallholding where the types of grasses described above are found to be dominant due to factors such as acidity and poor drainage, and where some reasonably gentle improvement can be carried out.

Firstly, open out some ditches or land drains. If bracken is a problem, then turn in your pigs, without rings in their snouts – whilst engaged in rootling for buried delicacies they'll bring the rhizomes of the bracken plants to the surface, where they'll die (the rhizome requires anaerobic conditions in order to survive). Electric fencing should be used to concentrate the pig's attention onto a small area at a time, and to prevent them gaining access to your newly dug ditches. Use your poultry to scratch out all the dead grass and moss. Run over the land a couple of times with a spiked harrow, spread a bit of lime and then broadcast your new seed mixture (the sweepings off the hayshed floor will probably do here). A few passes with the chain harrows (or an old gorse bush) should suffice to ensure that the seed is buried. Squash it all down a bit with the roller, and hope for the best. The result will be what is known as a partial re-seed, i.e., you haven't completely killed off the existing species before sowing the new. By good grazing management you can ensure that the more desirable species will thrive. At best, the sward composition will now be something like this: bent, fescue, perennial ryegrass, cocksfoot and the meadow grasses - rough stalked, smooth stalked and annual. If you're lucky, some white clover will also be present.

Low Quality Permanent Pastures

This is the type of sward that will probably be encountered when taking over a grassland smallholding – unimproved (or semi-improved) permanent grassland, either currently or previously employed for the continuous grazing of cattle and sheep, with perhaps an occasional cut of poor quality hay being taken. Fields may have been ploughed only once or twice within living memory, or not at all. These swards are likely to be dominated by bent grasses and fescues, together with both rough and smooth stalked meadow grasses. Other species of low productivity such as Yorkshire fog, creeping soft grass, downy oat, meadow barley, soft brome and wavy hair grass will be seen in varying proportions. Grasses of higher agronomic value (perennial ryegrass, cocksfoot, timothy, meadow fescue and meadow foxtail) are likely to be present in small quantities – it is the density of these species that determines the overall quality of the sward. White clover is unlikely to be found in any significant quantity. Poorly drained areas will be indicated by clumps of rushes.

Don't be too disheartened by the sight of a mass of thistles. Remember the story of the blind farmer who went out one day in his pony and trap to "view" a piece of land that was for sale. Upon arrival, he asked his boy, who had accompanied him, to tie the pony to a thistle. "I'm sorry, sir" replied the boy, "I can't do that. There's no thistle big enough!" "Then we'd best go home again!" said the blind man. A strong crop of weeds will be indicative of good fertile soil, albeit under rather poor management.

The smallholder has two courses of action open to him when faced with this type of sward – cultivate and begin again (with either a partial or full re-seed), or improve (and subsequently maintain) the sward composition through grazing management.

High Quality Permanent Pastures

Consisting almost entirely of perennial ryegrass and wild white clover, these traditional fattening pastures are maintained in a highly productive state through careful grazing management, and are capable of carrying high densities of grazing livestock without resorting to artificial fertilisers. Historically, probably the most famous example of this type of grassland was found in the Romney Marsh area of Kent, from whence comes the Romney Marsh breed of sheep. At one time, this was said to be the most densely stocked sheep pasture in the world, with set stocking rates of up to 12 ewes (+ lambs) per acre. Heavy grazing during the summer months ensures that the clover is kept free of competition from the ryegrass, with occasional rest periods in spring, autumn and winter, as appropriate to each species, allowing both the clover and the ryegrass to make up growth. It should be possible to take a good crop of hay off such land from time to time, but not too often from the same area, or the sward composition will alter. Most of these traditional pastures were lost during the ploughing campaign of the Second World War, and many more have subsequently been ploughed and re-seeded with fast growing, heavy cropping short term leys for silage making.

It must be noted that the types of high quality and low quality permanent pastures outlined here are extremes within their categories. Many intermediates will be found, and management should be varied accordingly. In addition, there are other very specific types of pasture that I've not mentioned, such as chalk downland, wetlands and watermeadows, each forming its own distinctive habitat.

Leys

It is likely that, with the exception of any out-lying inaccessible fields, steep slopes, or very poor pieces of land, most of the grassland on the smallholding will be managed as either a long or short term ley. The dividing line between the two seems somewhat blurred, so we also have "one year" and "medium term" leys. At what point does a long term ley cease to be classified as such, and become permanent pasture? Your guess is as good as mine! Basically, a ley is a temporary pasture – a field is put under grass for a number of years, grazed and mown accordingly, then ploughed under and the land used for another type of crop (or re-seeded with grass). A ley forms an integral part of many traditional rotational systems of land management.

Short Term Leys

The short term ley of 1 or 2 years is a particularly useful part of traditional arable rotations, such as the famous Norfolk Four Course (wheat – roots – barley – grass). As such, it'll be of interest to smallholders who wish to grow small areas of cereals and root crops in addition to grass, in order to be self-sufficient in livestock feeds. The barley in the rotation can be undersown with the grass mixture, thus acting as a nurse crop, protecting the grass seed during germination. Once the barley is harvested the grass will soon grow up through the stubble. For a 1 year ley – primarily for hay or silage – commercially available seed mixtures usually consist of Italian rye grass and red clover, however as we've already seen, Italian ryegrass may not

be an ideal species under more traditional management systems. Having said that, it has the potential to outyield all other varieties when cut for hay or silage in its first year, so perhaps it should be given consideration in situations where only a limited area of the smallholding is suitable for mowing.

Medium Term Leys

A medium term ley is likely to be intended for 3 – 4 years of use. Seed mixtures may contain both perennial and Italian ryegrasses. Management of this type of sward will be as for the short term ley in the first 2 years. By the 3rd year the Italian ryegrass will have come to the end of its useful span, making way for the more persistent perennial. During years 3 and 4 a more "intensive" grazing regime can be employed – probably mixed rotational grazing with cattle and sheep – with definite grazing and resting periods. At the end of year 4 it is time to plough in the grass to make way for the following crop. By this stage quite a few weed species will have crept into the sward. Perhaps the livestock used for the last grazing period of the season should be your pigs – they'll grub up the weeds, and carry out a fair proportion of the cultivation required.

Long Term Leys

The long term ley is land put under grass for a period of 5 – 10 years, but don't be surprised if, after 10 years, you start calling it permanent pasture! The simplest form of long term ley is a basic perennial ryegrass / white clover mixture, with several different strains of the grass and clover included, to give as wide a range of heading dates as possible. For a longer term ley that may well end up as "semi- permanent" it is important to include a few of the more persistent grass species such as timothy, cocksfoot and crested dogstail, which will enable the sward to withstand the various regimes of mowing and hard grazing to which it will be subjected. Seed mixtures can be tailored to suit specific conditions, and traditional herb species, old fashioned grasses and wild flowers can be incorporated.

Grassland Re-creation

So far we've worked on the assumption that the smallholding will consist of existing grassland, of one type or another, that will be managed for mowing and grazing in a traditional rotational fashion. Established swards will contain a diversity of grass species that can be augmented by sown varieties. Occasional ploughing of worn out leys may turn up a long forgotten seed bank of wildflowers. Plants found in field margins, hedge bottoms and rough corners will contribute to the general bio-diversity of the holding, and the overall result will be an attractive, ever changing, patchwork of habitats. But what if there is no grass on the holding? As I mentioned earlier, many "new" smallholdings are formed from small pockets of former arable land, and, after many years of intensive monoculture, this type of land will be more or less devoid of naturally occurring grassland vegetation. Simply leaving it alone and expecting re-generation to occur is not really an option, since the only colonisers (other than arable weed species) will be wind blown – witness the ragwort infested mess known as "set-aside". Don't despair though! Grass seed mixtures are available specifically for arable reversion (see supplier's details in the appendix), containing all the traditional hay meadow grass species, together with wild flowers, clovers and vetches. Independent advice (i.e., not sponsored by one of the agro-chemical giants) is readily available, and mixes can be tailored to ensure that varieties suitable to the locality, soil type and proposed management regime are included. This type of re-seed, used in conjunction with long and short term leys on other areas of the holding, can, with correct management, transform a former arable desert into a wildlife rich, diverse ecosystem that is also capable of putting food on your plate.

PASTURE RESTORATION

As we've seen, there is a broad diversity of sward types and grassland habitats, each requiring a different system of management. It is not practical or possible to discuss all options here, so let's consider the following all-too-common scenario: More often than not, smallholdings change hands in a run-down state, and the newcomer will be faced with the task of rejuvenating worn-out pasture. In most cases this will consist of previously improved (but now sadly neglected) permanent pasture, which has become dominated by weed species of little agricultural value. Overgrazing by unsuitable classes of livestock (equines, mostly) will have reduced the environmental value of the habitat too. Following a period without livestock (i.e., after the previous owners have removed their stock, but before you've introduced your own), this type of sward will be characterised by clumps of couch and other coarse grasses, growing through a dense mat of flattened and decaying vegetation. Thistles, docks and ragwort will be widespread, and patches of nettles abound. It's enough to make your heart sink, particularly if the word "organic" appears in the strategy for future management of your holding!

There are a number of different options that can be considered when carrying out grassland improvements, but in the long run it generally pays to be pretty heavy handed early on, even if this means resorting to the use of some chemical treatments and artificial fertilisers in the first year or so. This shouldn't jeopardise any long term ambition to manage the land organically. Whatever your principles, there's no point starting off with your hands tied – get the land and pasture in good shape first, and then concentrate on the long term view. Remember, you don't become a martyr until you're dead!

Some Options For Pasture Improvement

The appropriate course of action will depend upon a number of physical factors (e.g., accessibility, aspect, altitude, drainage, etc.), soil type (structure and analysis), and the degree of degradation of the existing sward. The intended use of the land (e.g., grazing / hay making / amenity / wildlife conservation) and any long term management ideals must also be considered, particularly when choosing seed mixtures.

Physical factors are largely fixed. Although drainage can be installed, and access improved, at the end of the day a field that's too steep to plough will always be too steep! An indication of soil type / structure / analysis can, to a certain extent, be gained by studying the plant species present, although it may not be a good idea to rely too heavily on this method of evaluation, as this brief extract from *"Hovel in the Hills"* by Elizabeth West, shows –

"We have learned to keep an open mind on the theory that plants can be used as soil indicators. I can show you a patch of ground here where bird's-foot trefoil, clover, moss, sow-thistle, chickweed, groundsel, bracken, goosefoot, buttercup, foxglove,

Top: Sheep's sorrel
Above: Knapweed

Top: Self heal Above: Bird's foot trefoil

self-heal, rush, daisy, sheep's sorrel, yellow rocket and knapweed all grow together. According to our various reference books, this indicates that we have rich, very acid, gravely, calcareous, well drained, poor, wet, sterile, loamy, undrained, fertile, dry, sour, markedly alkaline, marshy clay soil."

The simplest option is probably to buy a basic soil testing kit from your local garden centre, and, if you do nothing else, at least check the pH, which should ideally be about 6.0

The quality of the existing sward is also determined by examining the species present, and in what proportions. Remember to take into account the amount of bare ground. In some cases a reasonable density (more than 40%) of desirable grasses and clovers will be found – these can be encouraged to proliferate through careful grazing management. At the other end of the scale there may be very few species of agricultural or environmental importance in the existing sward, in which case it's best to plough up the whole lot and begin again.

Total Re-seed

(Usually spring or early autumn)

The idea of ploughing up an established piece of grassland can seem a rather drastic step to take, particularly if you're new to this way of life! Really, though, it's the best way to get your smallholding livestock venture off to a flying start.

As the name implies, a total re-seed means you start from scratch, usually by first spraying off the existing sward with glyphosate ("Round-up"). Glyphosate is a broad spectrum systemic herbicide that is transported throughout the plant, ensuring that no part survives. It is de-activated in contact with soil. There is no persistent or residual effect, so a new crop can be sown almost immediately. Of course, it is possible to re-seed without killing off the old sward, but establishment of the new grassland will be hampered by competition from the existing weed species, resulting in a rather short-lived improvement. Probably, the decision as to whether or not to use a spray will depend upon the density of species such as docks and couch in the old pasture – carrying out cultivations without first killing them off will simply cause these species to multiply, as each chopped up piece of root will form a new plant! Personally, I would view this type of treatment as a one-off occurrence, in order to tackle a badly neglected field, returning to more traditional management methods thereafter.

Method

1. Graze down the old sward really hard. You'll need high stocking rates for this to be effective.

A Word About Fertiliser

Unfortunately it is not possible to devise a fully sustainable self-contained management system for the smallholding: No matter how careful you are to spread manure, compost your waste and rotate your crops, you'll find there's an overall deficit of nutrients, particularly phosphates. From time to time you'll need to import fertility in one form or another. The alternative is to witness a progressive decline in the productivity of your holding and the health of your stock. You might get away with it for 20 years or so, lulled into a false sense of security by an initial increase in output following generous use of FYM (Farm Yard Manure) on land previously not managed in that way.

Now, before you all start jumping about and yelling at me that farmers managed perfectly well before the advent of artificial fertilisers, so why can't we do it now?, let me remind you of the millions of tonnes of guano imported from the islands of the South Pacific during the 19th century, and the contents of the "night soil" cart, and the Thames sailing barges carrying stable litter from the hundreds of working horses in the capital to farms on the east coast, and the dung from city dairy cows, many thousands of which were housed in cellars and basements up and down the land. (During the 1850s, one dairy in Glasgow housed 1,700 cattle, hand milked three times a day by a regular army of dairymaids!). Also waste fish from the docks, and blood-and-bone meal from the abattoirs and knacker's yards, and seaweed spread on the land, and manure from pigs fed on waste food from the cities (waste food that now goes to land-fill I suppose, now swill feeding is banned), and industrial waste such as basic slag. These were all bought in fertilisers, most of which are no longer available, for a range of political, environmental, human health and economical reasons. And if we turn the clock back even further, to before the enclosures of the 18th century, to a time when a large portion of the British countryside was open common and deciduous woodlands, peasants had rights of venville and pannage. Livestock grazing the moors and commons (venville) and pigs foraging for acorns and beech mast in the woods (pannage) during the day time would be penned up at night on the smallholding, in order that their dung could be collected for use on the land – a very convenient way of importing fertiliser from the surrounding countryside! The modern smallholder will also need to import some fertility. This could be in a variety of forms, such as by buying in hay rather than making it on the holding (see page 249), but if you have to resort to a bit of artificial fertiliser from time to time, then so be it – as the late John Seymour said, on the same subject, "A man can't really hoist himself up by his own boot-laces, nor can a farm".

When re-seeding very small areas, your pigs can be usefully employed carrying out steps 1, 2, 3 & 4 in their own fashion.

Ground wheel drive fertiliser spinner, suitable for smaller acreages

A lot of animals for a few days will do a better job than a few animals over a longer period. Cattle are useful if there's a dense mat of vegetation that needs breaking up. You could ask a neighbouring larger farmer if he'd like to turn in a lot of bullocks for a week or so. If you've only got a small number of animals to work with, then divide the field up into more manageable sections. Failing this, get a contractor in to marmalize the whole lot with a flail mower.

2. Kill off the redundant sward by spraying with glyphosate as soon as it shows signs that it is beginning to recover from step 1. (Systemic herbicides are more effective on rapidly growing vegetation).

3. Spread manure if required. If you are new to smallholding, you probably won't have accumulated much muck, in which case some artificial fertiliser can be applied later (see point 10).

4. Cultivation: if you've sprayed off the old pasture, then heavy discing, or repeated harrowing, may be sufficient to break up the surface. However, if you've decided not to use chemicals, you'll need to plough the field in order to bury the existing vegetation, where hopefully it'll die and rot down. You'll also need to rotovate after ploughing, in order to produce some sort of seed bed.

5. Broadcast the new seed mixture. A fertiliser spreader can be used for this (but not on a windy day!). A little Lister ground wheel drive spinner is ideal, as it can be towed behind a car or quad-bike.

6. Spread lime if required.

7. Chain harrow.

8. Roll.

9. The new sward should be *very lightly* grazed by sheep as soon as ground conditions and growth stage allow. This may be within a few weeks in the case of a spring re-seed, or perhaps not until the following spring if the re-seeding was carried out in the autumn. Nipping off the young plants encourages them to "tiller" out from the base, resulting in a denser sward.

10. Fertiliser should be applied after this initial light grazing. The amount required

Fattening store lambs on forage rape.

will be dependant on the intended use – if a high yielding ley has been sown, it is quite reasonable to expect to be able to take a crop of hay or silage in the first year, in which case fairly high fertiliser inputs will be needed. If, on the other hand, you've used a more traditional seed mixture, use less fertiliser for a rotational grazing regime. Hay can be cut in the second year, when the sward has become well established.

Using A "Nurse" Crop

A nurse crop is sown at the same time as the grass, in order to protect the developing sward. It is, in fact, doubtful if any "nursing" actually takes place (there is more likely to be competition between species!), but it is a useful way of getting two crops off one piece of land. A spring re-seed may be sown together with a cereal crop (usually barley, sometimes oats, but never wheat), with the grass growing up

TIP! Grass seed is fairly fine and free flowing. If you're not careful, you'll find you've spread the whole lot in one pass, leaving nothing for the rest of the field! The trick is to thoroughly mix the seed with sand before you start. This gives you a greater volume to work with, making it a simple matter to ensure even distribution across the whole area.

through the stubble for autumn grazing as soon as the corn is harvested. In the case of a late summer re-seed, for example after hay making, forage rape can be included in the seed mixture. This is probably most appropriate for the smallholder. Rape can be grazed from about 10 weeks after sowing, and is useful for fattening store lambs through the autumn and winter. Its rapid growth will protect the recently ploughed field against soil erosion, and the dung from the folded sheep will be thoroughly trodden in, providing nutrients for the emerging grass.

HABITAT CREATION

If you're struggling to balance the need for a productive holding with a desire for wildlife conservation, consider the following:

Hay cut from the "headlands" (round the edges) of a field is typically harder to make and of poorer quality. When choosing your new seed mixtures, why not sow a traditional, old fashioned wildflower mix for the first pass right round the headland, with conventional higher yielding species being sown on the remainder of the field? This will give you the best of both worlds – an attractive wildlife habitat on the poorer area, and a good crop on the better land. (Remember not to spread manure or fertiliser right up to the field margin). At haymaking time, use the rough headland bales to form a sacrifice layer at the bottom of the stack, and the following year they can be broken up and spread on the land, putting the wildflower seeds back where they belong!

Wildflower field margin

Once the rape has been eaten off the new sward will grow on, to give a useful bite in the spring.

Partial Re-seed

(Usually spring, after grazing with ewes and lambs, or late summer, after mowing for hay or silage).

Partially re-seeding basically consists of giving a "face lift" to a tired pasture. It is a less drastic procedure than completely re-seeding, and so may be more attractive to the smallholder, particularly if the existing sward is reasonably bio-diverse (but not too weedy). Considerable increases in yield can be achieved, although the improvement will be short lived unless management is amended in accordance with the new composition of the sward.

On a commercial scale, partial re-seeding is carried out by "slot seeding." This involves the use of a pretty complex seed drill that cuts narrow slots in the ground at approximately nine inch intervals. The machine deposits grass seeds, fertiliser granules and slug pellets into each of the slots, before pressing them shut. It then applies a small amount of herbicide along the top of each slot, in order to reduce competition from existing species. On smaller holdings the process can be carried out more simply (albeit with more passes), as follows:

Method (for spring sowing)

1. Check pH during the preceding autumn, and apply lime if required.
2. Graze down the existing sward. Assuming that the old pasture is in reasonable condition (if it isn't, you should be considering a total re-seed) then normal hard grazing by ewes and lambs in late March / early April will suffice.
3. After removing the stock (say mid-April) make passes in two directions with spiked

harrows, in order to open up a criss-cross pattern of scratches.

4. Broadcast grass seed mixture.

5. Chain harrow.

6. Roll.

7. Re-introduce sheep to the field, and continue to graze until the new plants are seen to be emerging.

8. Remove the sheep (probably at the beginning of May), and apply fertiliser according to intended usage. Subsequent management should be tailored to favour the newly introduced species, otherwise you'll soon be back to square one!

Partial re-seeding can also be carried out in late summer, on fields recently cut for hay or silage. In this case, forage rape is often included in the seed mixture for autumn grazing.

ROUTINE MANAGEMENT

In the previous section I looked at a "worst case scenario", taking rather drastic steps to restore a run-down and neglected pasture. Now let's consider a few of the routine procedures involved in grassland management, which should help maintain the pasture in a healthy, productive state, and prevent sward degeneration. Remember that good grassland management and good livestock husbandry go hand in hand. There are a surprising number of parallels, too: just as we need to ensure that our animals are kept free of pests and diseases, and receive an adequate level of nutrition for good health and growth, so must we apply these same principles to our grassland – it's a living thing, after all.

Intensive v Extensive

Although the phrase "intensive farming" conjures up negative connotations, it is a fact that (like it or not) smallholdings are, on the whole, very intensive small farms. Stocking rates tend to be much higher than on larger farms, for the simple reason that smallholders do not have the luxury of the acreage required for extensive systems. Extensive farming is perceived to be a cleaner, greener method, yet it only works by virtue of high acreages and low stocking rates. The same principles cannot be applied to a productive smallholding. However, as John Seymour said, *"it is far better to have a small acreage of land and really do it well, than have a large acreage and scratch over it. It takes very little land to grow the vegetables for a family, if that little land is farmed to the utmost"* and *"everyone who owns a piece of land should husband that land as wisely, knowledgeably, and intensively as possible. The so called self-supporter sitting among a riot of docks and thistles and talking philosophy ought to go back to town. He is not doing any good at all, and is occupying land which should be occupied by someone who can really use it"*. Clearly, then, there's a problem of nomenclature, and in our minds we're burdening the word "intensive" with images of factory farming and industrial scale agriculture, when in reality we should be proud of the high standards of husbandry, stockmanship, and welfare required in order to produce good yields of healthy crops and livestock from very small holdings. So, on that happy note, we'll look at some of the tasks you can carry out in order to manage your grassland "intensively", and not let it degenerate into "a riot of docks and thistles"!

Re-seeding

I've already talked about re-seeding as a method for creating new grassland, or for replacing neglected pastures, but there may well be areas of the holding where regular re-seeding is part of the routine husbandry. A temporary grassland or "ley" forms an integral part of most traditional rotational systems of land management, with fields being under grass for only a few years, before being ploughed up to make way for the next crop in the sequence. The burden of disease and the drain of nutrients are considerably less under a rotational system than where land is continuously cropped or grazed in the same fashion year on year. And I'm not just talking about whole fields here. It's quite a good idea to put a spare plot in the vegetable garden down to grass from time to time, and fold over it with poultry for a year or two. We've even grazed a cow on a short term ley in our vegetable garden, with only a single strand of electric fence between her and our crops!

You'll also need to periodically re-seed any poached areas in fields, for example around gateways or where feed troughs have been sited.

Fertilising

When we talk about fertilisers we include farm yard manure (FYM). Generally speaking, I don't think that smallholders make enough use of manure. The first muck heap is usually a source of great enthusiasm, particularly where previous experience has been limited to the constrained environment of a suburban back garden, but, as the amount of livestock on the holding increases, so too does the amount of manure produced. Enthusiasm wanes as muck handling becomes a chore, which always seems to get put off for another day! I've seen vast accumulated muck heaps on smallholdings, perhaps ⅓ of an acre in extent (yet scarcely more than a wheelbarrow high), just sitting there in the rain as all the essential nutrients leach away down the nearest land drain.

Spreading manure in late autumn

What a waste! True enough, you'll need some sort of smallish muck heap to supply well rotted manure for the vegetable garden (we carry about 5 tonnes of rotted muck into our garden each year, putting the whole lot on one plot. Each of the five plots gets mucked once in five years), but the remainder should be spread straight on the land, whenever weather conditions allow, where it'll do a lot of good. Perhaps it's not so much a lack of enthusiasm that causes the backlog, but a lack of suitable equipment to take the backache out of the job – I believe that a muckspreader is one of the essential bits of kit that every livestock keeping smallholder should have. And if you don't own a tractor, don't worry – there are several small muckspreaders available that can be towed behind a 4x4 or ATV.

It is fairly common practice, on mixed farms nowadays, to have samples of manure and slurry analysed in order to determine its fertilising value. This allows it to be applied accurately in accordance with the requirements of the growing crop, and if there's a shortfall of nutrients, then (and only then) the deficit is made up with artificial fertilisers. It strikes me that this is a very sensible compromise.

Manure is usually spread on grazing land

Preparing to apply artificial fertiliser to hay fields at the end of April, after grazing with ewes and lambs

during the spring and autumn, at rates of up to 15 tonnes per acre. Be aware, when applying manure to fields that are soon to be grazed, that the dung from pigs fed on proprietary feeds will contain high levels of copper, which is toxic to sheep. Some breeds are more susceptible than others (see page 243). It's also a good idea to identify areas of the holding where ground conditions will allow some spreading to be carried out during the winter, in order to limit the amount of muck that needs to be stored (an inappropriately sited muck heap can lead to prosecution and hefty fines). Fields that are due to be ploughed up in the spring are good candidates for winter spreading, as it doesn't matter about churning up the surface. If hay or silage fields are to be mucked this must be carried out early in the year (or in the preceding autumn), in order that all the manure has had time to break down and disappear from the surface before mowing, otherwise you'll end up incorporating lumps of dung in the crop, with possible health consequences for the stock that eat it. Muck can be spread fairly liberally onto the stubble of a mown field, as soon as the bales are shifted, but not if you're aiming to take a second cut – the 6-8 week interval between 1st and 2nd cut isn't long enough for

manure to break down. Slurry would be OK in this situation, or an artificial fertiliser (probably straight nitrogen).

Artificial fertilisers can be used "little and often" on grazing land, from spring (for early bite) through to late summer (for the autumn flush). There is nothing to be gained by very early applications – wait until soil temperature starts to rise. On mowing land, apply an appropriate amount when closing the fields off in April. The main thing is not to use too much – artificial fertilisers are just basic plant food (nitrogen, phosphate and potash) and, as with animals, overfeeding is wasteful! A bit of light drizzle after spreading is ideal as it will gently dissolve the prills and put the nutrients where they're needed. Heavy rain will simply wash it all away, and in very dry weather it'll just sit on the surface and scorch the plants.

Grasslands that are rich in clover will not require applications of nitrogen, which is why clover is such an important part of organic farming systems – the clover itself may contribute the equivalent of up to 250kg N per ha. However, the sward will need to be managed in a way that encourages the clover to proliferate, and it might be necessary to supply other essential nutrients – phosphates, in particular.

From time to time a top dressing of lime may be required, in order to keep soil pH between 6.0 – 6.5. We've found that applying calcified seaweed to fields that have been mucked helps to maintain the right balance.

Harrowing

It's a good thing to harrow grassland in May or thereabouts. Harrowing will disperse mole hills, rake out dead grass and moss, and let some air and light into the base of the sward, to the benefit of the finer grasses and clovers.

Chain harrowing

Also, in response to the bruising effect, the higher value grasses will grow more vigorously, producing a denser, heavier yielding crop. Harrowing should be carried out again after grazing with cattle or horses, as this will spread out dung pats and prevent the formation of sour areas. The ubiquitous chain harrow is ideal in most situations, but where there's a lot of dead material, or where a bit more aeration is required, use a spiked harrow or a spring tined grass harrow. Chain harrows can also be used to good advantage on fields recently spread with manure, in order to break up lumps, distribute the muck more evenly, and generally work it down into the surface. This is particularly useful where muck has been spread by hand, for example by forking it off a slow moving trailer. In fact, you can attach the harrow to the back of the trailer, and chuck the muck off in its path.

Rolling

The rolling of hay or silage fields, using a heavy ballast roller, is usually carried out towards the middle of the growing period. It may seem rather harsh, this business of squashing down all that lovely long grass, but don't worry – it'll soon spring back up, thicker and stronger than before! In addition to its stimulating effect, rolling will squash mole hills, level wheel ruts, and bury any small stones that may inadvertently have been distributed when muck spreading, removing these obstructions from the path of the mower. Quite apart from preventing damage to machinery, this reduces the risk of soil contamination to the crop, with the accompanying risk of listeriosis (see page 250). Listeria is generally associated with big bale silage, but can also occur in small bale hay. Affected animals usually die, despite treatment, so it is definitely a case of prevention being better than cure! The disease is also transmissible to humans.

Make your own roller using a 45 gallon oil drum filled with concrete. Remember to put a piece of pipe through the middle of the drum before pouring the concrete in!

Topping

Topping is one of the most satisfying aspects of grassland management – it keeps weeds in check, reduces the incidence of lameness and orf in sheep, and gives straggly summer pastures a real boost.

Basically, plants have two types of growth – vegetative and reproductive. Grasses, however, do not need reproductive growth (seed heads) in order to reproduce – they are able to multiply by vegetative means. It is this dense vegetative growth that we, as farmers, need to encourage on our grazing land. Reproductive growth is wasteful, unpalatable to stock, and results in sward degeneration. For the most part of the grazing season the livestock themselves keep grass growth within reasonable bounds, by constantly nipping off excessive vegetation, which encourages further dense growth. But there comes a time in the summer when

Topping stemmy pasture in mid-summer

the grass just shoots away, throwing up seed heads all over the place, and this is when it needs topping to bring it back under control, otherwise both the sward, and the livestock that depend upon it, will suffer. Topping is a particularly important part of organic grassland management regimes.

Topping simply consists of cutting off all the unnecessary seed heads (thereby stimulating vigorous vegetative growth), and, by timing it right, it also plays an important part in weed control, particularly thistles which, unlike grasses, don't like being cut down – by allowing them to produce flowers (but not set seed) before cutting, the plants are eventually exhausted and die. But timing is crucial. Remember the old adage:

"Cut a thistle in May and it's back next day, cut a thistle in June and expect more soon, cut a thistle in July and then it'll surely die!"

Having said that, thistles seem to have been setting seed much earlier in the last few years, so it's also worth remembering that "one year's seeding makes seven years weeding". Top a bit sooner than is customary, if you think it's

necessary.

One type of weed that should never be topped is ragwort. This poisonous plant becomes more palatable to stock (and more toxic) when cut and dried. Nor should ragwort be allowed to set seed. Mature ragwort plants should be pulled up and burnt, as soon as the clusters of yellow flowers betray their presence. Better still, nip it in the bud, quite literally, by grazing. (See overleaf).

The topper itself is like a rather crude, heavy duty mower. In fact, an old mower set to cut high (8″ or so) will do just as well – for years we used an old finger bar mower on the back of a grey Fergie for cutting thistles. The machine I use now is not really a topper at all, but a jungle buster. This is something like having a giant strimmer with serious attitude attached to the back of the tractor, with 3′ lengths of heavy chain in place of thin nylon cords! It doesn't cut as neatly as a topper, but takes pretty well everything in its stride, including bracken and brambles. The only thing it doesn't like is dense rushes, which are best tackled using a mower.

Gorse should be burnt off in a seven year rotation

Burning

Where gorse is encroaching on pasture it should be burnt off in a seven year rotation, so divide the area into seven blocks and burn one section each year. The fresh young growth can provide an important winter feedstuff, particularly when other herbage is covered in snow. Some animals are particularly well adapted to this diet – New Forest ponies, for example, grow thick bristly moustaches in winter to protect their sensitive lips from the prickles! In some areas young gorse used to be harvested and put through a chaff cutter, for inclusion in livestock rations.

Gorse can be burnt between October and March without the need to apply for a permit, although the local fire department should be advised of your activities.

Heather habitats will also benefit from a similar regime.

Grazing As A Management Tool

Grazing management is very much a two way thing – we manage the grassland in order to maintain and improve our livestock, and we manage our livestock in order to maintain and improve our grassland!

• Weed Control

Ragwort

Many smallholdings, particularly those that have been overgrazed with horses, are badly infested by ragwort. That was the situation here at Ty'n y Mynydd when we moved in, yet within a very few years we'd totally eliminated it from our land, without resorting to any chemical treatments. Although we pulled up quite a lot, the demise was largely achieved through grazing management.

Mature ragwort plants are not palatable

to stock unless cut and dried, for example by accidental inclusion in hay. However, in spring, the tender young shoots (which are not nearly so toxic) will be consumed readily by sheep, and, if they are repeatedly nipped off, the root reserves will be exhausted. The secret is to graze hard with sheep during the whole of April, and when I say hard, I mean really hard – the poor things need to be more or less licking bare earth! But it's worth it in the long run. Provided that you remain vigilant about pulling up any ragwort that occurs in inaccessible places, you'll find that just two seasons of this regime will be the end of it, and you'll be able to bale future crops of hay with confidence.

Thistles

Thistles thrive under low nitrogen input, extensive systems. As we've already seen thistles can be controlled by cutting, but they can also be discouraged by good grazing management. Cattle are more effective at reducing thistle numbers than sheep. (An Irishman once told me that all you had to do was to sprinkle a bit of salt on each thistle, whereupon the cows would eat the lot…). Avoid over grazing in winter and early spring, and avoid under grazing in summer. But above all, don't let them set seed! And the same goes for docks, too – if organic options are preferred, it's even worthwhile hand weeding standing crops of hay.

• Encouraging Clover

Contrary to popular belief, clover is not directly killed off by applications of artificial fertiliser, but, as it is not very competitive, it is rapidly ousted by the more dominant nitrogen loving grass species, if the management regime favours them. In order to avoid having to apply chemical nitrogen we need clover, which

Dense white clover sward

fixes nitrogen from the atmosphere…which encourages the high yielding grasses…which shade out the clover…and so we're back to square one again! This is where good grazing management comes into the equation.

A suitable grazing pattern to encourage the spread of white clover in a sward would be as follows:

• Soil pH should be in the range of 6.0 – 6.5

• Soil potash and phosphate indexes should be around 2.

• Eat the pasture down with sheep during November.

• Graze heavily in March and April. (Clover is very tolerant of heavy stocking rates).

• Heavy grazing after April will have a negative effect on clover, as it favours the more dominant ryegrasses.

• Graze rotationally throughout the summer, with well defined rest periods of 3 – 4 weeks, followed by fairly rapid defoliation.

Strip grazing. The cattle graze under the wire, so cannot trample or dung on any of the crop

(Interestingly, this regime seems completely at odds with that required for thistle control. It seems that you just can't have it both ways!)

We've found that the greatest natural proliferation of white clover has occurred in the field where we strip graze our cows. The field is grazed by sheep from early March to Mid April. We then apply a light top dressing of artificial fertiliser (100kg per acre of 20:10:10) and rest the field for a few weeks before putting the cows in at the beginning of May. In some years we omit the fertiliser and apply calcified seaweed instead – this depends upon the amount of manure the field has received, which in turn depended on ground conditions in the preceding autumn. The cows are strip grazed behind a single strand of electric fence, which is moved forward by about a yard a day. Ideally they graze under the wire, so don't trample or dung on any of the crop. After a while we introduce a back fence behind the cows, to allow re-growth on the grazed area. Any thistles occurring behind the back fence are topped, and dung pats spread. By the time the cows reach the bottom of the field there's sufficient new growth back where they started from, so the sequence begins again. In good grass growing years we've moved the cows back to the top of the field before reaching

the end, and made hay from the surplus. The cows are housed at the beginning of November, whereupon we put sheep in the field to tidy up, before spreading manure. The field is then rested over the winter.

In this way, our one-and-a-half acre field provides all the grazing required for 3 milking cows, and sometimes a portion of their winter fodder, too. We have all the milk and other dairy products we require, including cream, butter, ice-cream and yoghurt. The remainder of the milk is used to rear around 12 calves each year, the sale of which gives us some cash income. Any surplus milk helps to fatten our pigs. The dung from the calf pens and pigsties is spread on the field at the end of the grazing period, where it contributes to the next year's crop. Grazing for a small number of sheep is available at two key times of the year – lambing and tupping. This, I think, is fairly close to the type of intensive land use that old John Seymour was talking about.

GRAZING REQUIREMENTS

So far I've considered grassland management in fairly general terms, as appropriate to a mixed smallholding enterprise. Much of what I've discussed is applicable to all grassland holdings, regardless of the type of stock kept, and can be described as "routine and remedial" management. Sheep can be (and often are) used as a tool within a pasture maintenance program, for example by running them after cattle or horses to tidy up what's left over, and to graze around the dung pats, but sheep do, of course, also have their own specific requirements.

Stocking Rates / Density

Stocking rates are generally calculated using

Livestock Units (LSUs), with different classes of stock being given standard values, according to their feed requirements. The baseline figure of 1 LSU is based on the requirements of a single dairy cow (see table below). This is a useful system, as it enables you to calculate requirements for grazed and conserved forage for each of the different categories of animal on the farm. Having said that, you'll often see stocking rates and densities given as so many ewes per acre or per hectare, which is fine. You should be reasonably conversant with both methods, probably using ewes per acre for your own sums, and LSU per hectare on official forms from Defra! The overall stocking rate refers to the whole of the area of land used for grazing livestock, including land closed off in the summer for the production of hay or silage for use on the farm, whereas stocking density refers to the number of ewes and /or lambs

(or LSUs) per acre on a specific area during the grazing period. Thus you'll have a summer stocking density, when grass is growing fast but available acreage is reduced (due to a portion having been kept for mowing), and a winter stocking density, when more fields may be available for grazing, but the grass isn't growing.

The total number of ewes (and lambs) that can be stocked on a given farm has been shown to be directly proportional to the amount of artificial fertiliser applied, up to a maximum of around 20 ewes per ha at fertiliser application rates in the region of 350kgN/ha/year. The total weight of lamb sold per ha also increases, although it should be noted that, at higher stocking rates, the proportion of the lamb crop sold directly off grass (i.e., without supplementary feeding) falls. These figures, however, are based on research carried out in the 1980s (Rosemaund EHF), and times have

Livestock units allocated to different classes of grazing animal		
Species	Category	LSUs
Cattle	Dairy Cows	1.00
	Suckler Cows (excluding Calves)	0.75
	Bulls	0.65
	Calves up to One Year	0.34
	Weaned calves & stores 1 - 2 years old	0.65
	Stores & replacement heifers over 2 years old	0.80
Sheep	Ewes (excluding suckling lambs) and shearling flock replacements	
	Light breeds (hill and primitive)	0.06
	Medium (most lowland breeds and commerical crossbreds)	0.08
	Heavy (traditional large down breeds & longwools)	0.11
	Rams	0.08
	Lambs, birth to fat or store	0.04
	Store lambs	0.04
	Lambs, birth to hogget	0.08
	Breeding ewes, including lambs (for general calculations)	0.15
Goats		0.15
Horses	All categories	0.80

Potential Stock Carrying Capacity of Various Sward Types			
Pasture Type (predominant species)	Growing Season	Yield	Stock Carrying capacity
Perennial Ryegrass & White Clover	6-9 months	8 - 11 tonnes DM / ha/ year	10-15 ewes / ha (4 -6 ewes / acre)
Cocksfoot, Timothy & Yorkshire fog	5 - 7 months	5 - 7 tonnes DM / ha / year	6 - 8 ewes / ha (2.4 - 3.2 ewes/ acre)
Bents, Fescues & Molinia	3 - 4 months	2.5 - 5 tonnes DM / ha / year	2 -4 ewes / ha (0.8 - 1.6 ewes / acre)

changed. The era of cheap fertiliser is over, feed costs have rocketed, and there's no longer a government incentive (in the form of the Annual Ewe Premium) to increase stocking rates. And, what's more, we're now well aware of our environmental responsibilities. Put simply, high stocking rates on a big scale don't make sense these days, either ethically or economically. The smallholder, however, faces something of a dilemma – serious small scale operations need to be fairly intensive in order to survive, and others become intensive through mis-management (e.g., retaining too many animals on the available area, for "sentimental" reasons), yet morally and financially the high level of inputs required may be difficult to justify. The net result, unfortunately, tends to be a steady decline in animal health and welfare. The saying "stock-and-a-half is half-a-stock" is very, very true indeed. Overstocking and non-productive livestock units are two of the commonest factors that lead to the downfall of many budding small scale livestock enterprises.

Non-productive Livestock Units

Ask yourself whether you really need to keep a horse, and the answer will probably be no. Ask the same question about pot-bellied pigs, or llamas, and you'll get the same answer, unless of course it is your intention to breed these animals for sale to other people who don't really need them. Non-productive livestock are one of the biggest drains on the resources and finances of potentially productive smallholdings. The problem doesn't end with species / breed selection, either: recently I visited a smallholding to look over the sheep flock. "200% lambing this year" I was told, and I have to say that when I saw 20 ewes in a field with 40 lambs between them I was pretty impressed. After a while I noticed that an adjoining field also held about 20 ewes, without lambs this time. "What about that lot?" I asked. "Oh," replied the owner, "they're the old girls, we don't breed from them any more!" So, in fact, for their crop of 40 lambs they were feeding a flock of 40 ewes, and their lambing percentage, in truth, was far less impressive.

Now, we all have one or two old favourites amongst our flocks that are kept on well beyond their useful sell-by date, but there are limits! A productive and profitable smallholding really cannot carry so many passengers; neither can a geriatric population be considered "balanced". There may be difficult decisions to make, but if you really are aiming for an economically and environmentally sustainable holding, you'll have to make those decisions.

Our additional mountain grazing actually has very little impact on our overall level of production

Overstocking

Consider the following:

"A smallholder decides to keep a few sheep and, finding that he does rather well out of them, decides to keep a few more. Once again he has a good year, his freezer packs of lamb finding a ready market amongst friends and family, so again he increases his flock. This pattern of behaviour continues for a year or two more, until one season our smallholder finds that the profits from his sheep enterprise are way below the expected level. Non-existent to be precise. He blames the weather, the sheep, market prices, everything, in fact, except himself, and, on doing a few sums, comes to the conclusion that if he'd had just a few more lambs to sell he would at least have broken even. Once again the flock is increased, and the following year is a total disaster…"

If you see reflections of your own situation in this scenario don't worry, I can wryly assure you that you are not alone! It is not difficult to identify the point at which things went wrong – your bank balance can tell you that; the difficulty lies in admitting to yourself you got it wrong, and taking the necessary steps to cut the flock back down to its optimum size. And of course this does not only apply to sheep.

It would be better, I think, to be slightly understocked, and make any surplus grass into hay or rent it out to a neighbour, than to be faced with the embarrassment of having to sell off a load of painfully thin animals simply because you went over the top. Fewer animals of higher quality will generally give better returns, particularly now that there are no headage based subsidies to be claimed.

It is clear, then, that having some idea of the potential stock carrying capacity of your grassland (see table opposite), together with the requirements of the stock you intend to keep, and having a preferred management plan in place before embarking on a potentially expensive project, will greatly increase your chances of success!

Quite considerable variations in stock carrying capacity can be seen between different sward types, and, on larger holdings, it is unlikely that all the available land will fall within one category. This is best illustrated by outlining our own position at Ty'n y Mynydd. Here, our overall stocking rate is in the region of 0.5 LSU per ha, yet within that total we have areas capable of supporting summer stocking densities in excess of 2 LSU per ha (6 ewes + twin lambs per acre) and other areas that struggle to cope with 0.2 LSU per ha (0.8 ewes + single lambs per acre). Prior to the reform of the CAP our overall stocking rate was something like 0.9 LSU per ha. This was only achievable by the use of a fair amount of artificial fertiliser on the better land, and the taking of two silage cuts each year. The stock carrying capacity of the poorer land was as now. When the incentive of the ewe premium was removed we reduced stocking levels, and consequently cut our fertilizer use right down (with none at all used in some years). Although we're now selling fewer lambs per year we're finding that a higher proportion of the crop are finishing at reasonable weights, and the bottom line shows us that less really is more. We do, however, depend heavily upon that portion of our total acreage that is capable of carrying fairly high densities during the main grazing period, as would the majority of small scale producers. Our additional mountain grazing, despite the acreage involved, actually has very little impact on our overall level of production.

In the light of all this, I would suggest that a reasonable starting point for the average smallholding would be an overall stocking rate of 3 – 4 ewes per acre (or the equivalent in LSU, if other classes of stock are to be kept

Mixed grazing with cattle and sheep can increase the output of both species

alongside the sheep flock), without fertiliser. Fertiliser applications totalling around 120kg N per ha (or the equivalent contribution from white clover in well managed swards, and FYM) could see the figure rise to 6 ewes per acre (15 ewes per ha), and / or enable some fields to be closed off for the production of hay or silage.

If the preferred option is to keep fertiliser applications to a bare minimum, or not to use any at all, there are a few grazing strategies that can help optimise output:

Mixed Grazing

Stocking density can be increased by as much as 10% by grazing sheep and cattle together, without additional inputs, and without any detrimental effect on the performance of either species. In fact, individual animal performance (e.g., daily liveweight gain) is seen to increase by up to 6% in the case of cattle, and 10% in sheep, with an increase of up to 40% in the proportion of the lamb crop that are sold fat off their dams. (Researched by An Foras Talúntais, 1977 – 1981). The optimum ratio for efficient land use is probably in the region of 1 suckler cow and calf to each 6 ewes with lambs, although, in our experience, having fewer sheep in the mix (e.g., 3 sheep per cow) results in much heavier lambs at weaning. We tend to graze the few ewes that are suckling ram lambs with breeding potential together with some of our cattle, for this very reason.

Strict Rotational Grazing

While rotational (or "paddock") grazing systems are generally impractical on large scale and extensive sheep farms, they do lend themselves to the more intensive land use requirements of the smallholding, as a means of maintaining higher stocking rates. Docile breeds are particularly well suited, especially those that evolved under similar conditions, such as the Hampshire Down.

Generally, the area available should be divided into 6 roughly equal paddocks, around which the flock are moved in a fairly strict rotation. The process of rapid defoliation followed by periods of rest is inclined to encourage the spread of white clover, so rotational grazing ought to result in an overall improvement in sward composition. You should aim to have the sheep grazing on sward heights in the range 4 – 8cm, with 6cm being the optimum. As soon as sward height falls below 4cm then it's time to move the flock into the next paddock. When grass growth is exceeding the rate of consumption, take a couple of fields out of the rotation for the making of hay or silage, or introduce a few cattle to eat off the stemmy stuff.

The establishment of a permanent paddock grazing system can be expensive, due to the amount of fencing involved, and the fact that each enclosure will need a water supply. However, on a small scale, the whole thing can be set up fairly cheaply using electric fencing.

Forward Creep Grazing

Forward creep grazing fits in very well with a rotational system of management. Basically, the lambs are allowed access to the next paddock in the cycle, via a pop-hole or "creep". By this means, the lambs always get first choice of what's on offer. In early lambing flocks this also provides an opportunity to push the lambs on a bit with some trough feed. Forward creep grazing is more difficult to manage later in the season, after shearing – by this time the lambs should be well grown, thus requiring a fair sized pop-hole, and newly shorn ewes are adept at squeezing through small spaces, particularly when the grass really is greener on the other side!

Smaller Ewes

Changing to a smaller, thriftier breed of ewe will enable much higher stocking rates to be maintained. Draft hill ewes are readily available, and cheap, so annual flock replacement costs are kept to a minimum. On better grazing, hill ewes can be stocked far tighter than would have been the case on their native heath, and far tighter than their lowland counterparts. Look at the table of LSUs, and you'll see that a heavy breed ewe has almost twice the forage requirement of the small hill sheep. Most of the hill breed ewes are maternal and milky, and, on good grazing, will be quite capable of rearing quality prime lambs by a terminal sire. Winter feed costs will be lower, too. An old boy once told me that the secret of profitable sheep farming was to have little ewes rearing big lambs, and he was probably right!

Extended Rest Period

Where grazing is limited, there's a lot to be said for housing the flock for an extended period during the winter – there's not much nutrition to be had from grazed grass during the winter, anyway. Fields that are badly poached seldom recover in time to give an early bite, and sacrifice areas may need to be re-seeded. Where there is sufficient acreage for the winter stocking density to be kept low, or where annual spring re-seeding of a portion of the available land is part of the usual pattern of management, then it would be normal practice to house the flock for only a few weeks around lambing time. On smaller holdings, however, it may be beneficial to house a March lambing flock as early as December. This fits in well with organic systems of grassland improvement, as white clover needs to be eaten off in November and March, and rested in between.

GENERAL REQUIREMENTS

Water

Don't listen to anyone who tells you that grazing sheep don't drink. That's utter nonsense. All fields should have an accessible water supply, either piped or natural (or bucket and watering-can for small flocks!). It's true that there are certain times of the year when their needs may be satisfied by the moisture that is found in and on the herbage that they're eating, but at other times (e.g., when suckling lambs, or when receiving supplementary feed) their water requirements are quite high. Any restriction in water supply will have a knock on effect on feed intake. At best, this will affect the growth rates of suckled lambs. At worst, reduced feed intake in late gestation (for example, if the available water is frozen) can cause pregnancy toxaemia and death.

Don't site water troughs under trees or hedges if this can be avoided

Insufficient water intake in fattening wethers and growing rams may lead to the formation of urinary stones – the provision of rock salt can help alleviate this risk, by stimulating thirst.

Care should be taken over the siting of water troughs: they should not be so high that the sheep can't reach them (provide a little step if necessary), nor so low that they're easily fouled. In the event of a lamb inadvertently jumping in, it needs to be able to get out again – a few bricks placed in the trough will help. Don't site water troughs under trees or hedges if this can be avoided, or they'll get full of leaves in the autumn, which, when they begin to decompose, will make the water smell (and taste) horrible! The area around the trough should be easy for you to get to (for maintenance) and free-draining.

To keep costs at a reasonable level, water troughs are often sited in a fence line, in order that stock in two fields have access to one installation.

Somewhere To Scratch

Sheep generally get "cast" whilst rolling over in an attempt to scratch their backs, particularly if they are overweight, in full fleece, or heavily pregnant. It only takes a small rut or depression

This ewe with mastitis has hidden herself away under a gorse bush. It is important that places like this are checked regularly

in the field for them to get stuck. Warm, still days after heavy rain are the worst – they appear to get very itchy as the fleece dries out. It doesn't seem to be so bad if it's dry and windy after rain. Sheep seldom get cast in fields where there are low, overhanging branches, or some similar place for them to scratch. Nice neatly fenced paddocks, or former arable land sub-divided by electric fencing, are particularly bad! Provide somewhere for your sheep to scratch by knocking in a couple of fence posts 4 foot apart, and nailing a piece of timber between them at an angle – 2 foot above the ground at one end, rising to 3 foot at the other. Years ago, shepherds on arable farms would have put a similar device – called a "roller wattle" – in every field.

Sheep that are stuck on their backs quickly become bloated, as they are unable to belch up rumenal gases, and will die if help is not forthcoming. They're also at risk from predators, and many a cast ewe has lost an eye to crows or magpies by the time she's found. I've also known some to be pretty badly damaged by foxes. After turning the animal back over, it can take some time for the bloating to go down, so she should be held for a while to recover. Sheep should always be righted end over end, not by rolling to the side – rolling them over when bloated or pregnant can lead to torsion, which is bad news! Bear in mind that a sheep that has been down for some time might be suffering from "pins & needles" just as you or I would, so she may simply keep on falling over until her circulation has got going again.

Shade From The Sun

Although sheep can be seen grazing quite contentedly in the worst of weather conditions, they do appreciate some shelter from the heat of the sun. Rams, in particular, need to be provided with shade – overheating in late summer, during the period that sperm are being produced in readiness for the forthcoming mating season, may have an impact on fertility at tupping time. The semen production cycle is about 7 weeks, so a hot spell in mid-August can adversely affect the performance of a ram expected to work in early October.

In fields that have naturally shaded spots, such as under hedges or in overgrown dry ditches, it is important that you know where these areas are, and check them regularly. They are just the sort of places where poorly sheep will hide away, particularly ewes with mastitis or lambs with maggots.

No Entanglements

Long bramble shoots encroaching into the field from surrounding hedgerows should be cut back, and any stray lengths of barbed wire should be removed. Not only does it look untidy, having tufts of wool snagged all over the place, but it's easy for a sheep in full fleece to be completely ensnared. The more she goes around and around trying to free herself, the more firmly she's held, as her wool twists into a rope that's inextricably bound up with the offending item. First winter lambs are the worst affected, as they have plenty of wool but not enough body weight to break free. For this reason, many hill farmers will shear their ewe lambs in the autumn, before sending them to the lowlands for the winter. Quite apart from the risk to the animal if she's not discovered, lengths of wire hidden in the fleece will result in expensive breakages at shearing time.

Avoid storing trailers and out-of-use machinery in fields where sheep are grazing – it's amazing what they'll get caught up in given half a chance, particularly male lambs of horned breeds. If necessary, cordon off an area for the purpose, using hurdles or flexinet.

Dry Lying

Although sheep don't seem to mind paddling about in fairly wet places when they're feeding, they do like to have a dry spot to lie down. If the flock is grazing a particularly boggy field it may be necessary to allow access to part of another, drier, field, as a lie-back area.

Environmental Enrichment

We've all watched young lambs playing, and marvelled at their seemingly boundless energy and sheer joy of living. A hillock or boulder becomes the hotly contested site of endless rounds of "King o' the Castle", and races are run along earth banks and fallen boughs. In small, well maintained paddocks, lambs are often deprived of these natural obstacles, and I think it's only fair to provide them with some source of amusement, such as a big log or a couple of straw bales to jump on. If nothing else is available they'll play their games on their mother's backs. This can't be much fun for the ewes, and will make a frightful mess of their fleeces.

Freedom From Poisonous Plants & Toxins

Sheep are pretty astute when it comes to knowing what is safe to eat and what isn't, and cases of plant poisoning are rare. Generally, poisoning will only occur when sheep are forced to eat plants that they'd not normally touch, for example during drought conditions, or when other herbage is covered with snow, or where mis-management has lead to extreme overgrazing and potential starvation. Occasionally an individual animal will develop a depraved appetite for a particular poisonous plant, even when there is an abundance of more suitable forage, and plants such as ragwort may be inadvertently incorporated into hay or silage. Quite a number of ornamental shrubs are toxic, and can become more palatable when wilted or dry, so livestock mustn't be given access to garden prunings.

Non-plant poisoning is also fairly unusual, although the sites of old bonfires, where large amounts of domestic rubbish and old timber have been burnt, are a possible source of toxins such as heavy metals and arsenic. For some inexplicable reason sheep seem to be drawn to these places!

Due to the high levels of copper in proprietary pig feeds, sheep grazing land on which pig manure has been spread are at risk from copper poisoning, with some breeds being more susceptible than others. The North Ronaldsay sheep in particular are extremely susceptible when removed from their native Orkney environment – in 1832 a dry stone wall was built around the island (the longest wall of its kind in the world), in order to restrict the sheep to the foreshore, where they have adapted to exist on an unusually low copper diet, consisting almost exclusively of seaweed! Anecdotal evidence suggests that a similar problem may exist in the Portland breed, but to a lesser degree. Among the commoner breeds, the Texel is known to be susceptible to copper poisoning, whereas the Scottish Blackface and some strains of Welsh Mountain sheep are quite copper tolerant.

Some Common Poisonous Plants

Rhododendron

This is the only type of severe plant poisoning I've seen in sheep. A bought-in ram made a beeline for a few branches of rhododendron poking through the fence, and scoffed the lot! Goodness knows why, as we'd kept sheep in that field for 10 years at least, and none had shown the slightest inclination to sample the shrubbery.

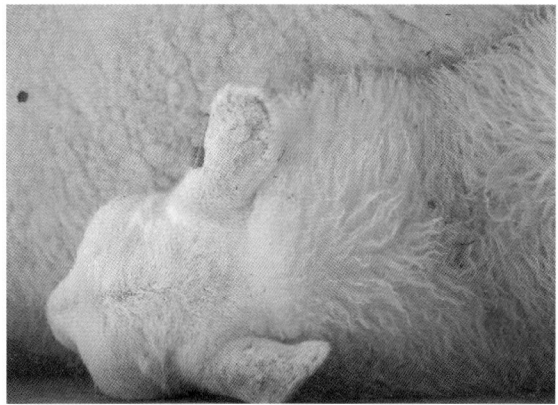

Signs of photosensitivity can be seen on the ears of this lamb

Rhododendron poisoning is just about the only thing that will cause sheep to vomit, and the presence of undigested leaves in the regurgitated rumen contents makes diagnosis easy. Other symptoms include signs of severe gastric discomfort and salivation. The outlook is poor, but affected animals should be given cold tea (black, two sugars).

St. John's Wort

A useful dye plant (see appendix), but if consumed will cause photosensitivity leading to sloughing of the skin on non-woolly parts of white faced sheep. The ears in particular are affected, and may completely drop off. A similar, but less severe, effect can be seen in Texel X store lambs fattened on rape. Animals showing signs of photosensitivity should be kept indoors until healed, perhaps being let out at night to graze.

Water Dropwort & Hemlock

Two fairly similar looking poisonous plants likely to be found in unkempt wet meadows and ditches. Sheep are at risk during dry summers, particularly when vegetation in overgrown drainage channels is cleaned out and spoil heaps left on the bank. *Water Horsetail*

Bracken covers huge areas of grazing land

Ragwort

is another poisonous plant of wetland areas, which sheep may be exposed to in the same way.

Bracken

The fact that bracken is poisonous is indisputable, yet it covers a huge area of grazing land, and is increasing its range each year. Traditionally it has been used as a livestock bedding material, and remains popular in some areas – I have first hand experience of bedding sheep, cattle, pigs, goats and ponies on bracken, with no ill effects whatsoever. Bracken has also historically been used in dressing leather, for fuel, for thatching and in the process of cleaning linen. I've even come across an old reference to bracken silage being fed to sheep and cattle, where it was found that the heat generated in the silo appeared to have destroyed its toxic properties. I don't recommend that you try this, though! The poisonous effects of bracken – mostly carcinogenic in nature – are seen when the plant is consumed in large quantities over long periods.

Ragwort

Ragwort is most toxic when adult plants are dried, for example by accidental inclusion in hay. Sheep will nibble off emerging shoots of ragwort in the spring without harm, so can be used to help eliminate the weed. If mature plants are controlled by spraying, cutting or pulling they should be removed from the field and burnt. Under the Weeds Act of 1959 you are legally obliged to control ragwort on your land.

Yew

This traditional tree of graveyards is worth a mention here, as a surprising number of smallholdings are old vicarages! The smallholding on which I grew up was called "The Old Rectory", and we had mature yew trees lining our driveway, and pasture alongside the churchyard. Sheep are most at risk from fallen branches or prunings. Death occurs very soon after ingestion.

HAY OR SILAGE?

Often on smallholdings, the decision as to whether a particular crop of grass intended for winter forage production ends up as hay or silage is made under great pressure, on the spur of the moment. As rain clouds gather to threaten an almost perfect crop of hay, frantic phone calls are made to secure the services of a contractor, to get it baled, wrapped, and above all safe before the heavens open.

That is pretty much what happened to us during our second summer at Ty'n y Mynydd. The weather clouded over and a light drizzle set in almost as the mower left the field, so our crop sat untouched in its rows, safe enough for a few days, but certain to deteriorate thereafter. On the fifth day we got some sunshine, but, with heavy rain forecast, all hopes of getting it in as hay were dashed. Eventually we got hold of a contractor, who of course had been inundated with calls from farmers in the same predicament as our own. "I'll do my best", he told us. True to his word, the machinery rolled on to our farm at 7.30pm and, about an hour and a half later, as the last bale came off the wrapper, the rain began to fall. It could rain as much as it liked for all I cared! It was a simple matter the next day to carry the bales off the field using a spike mounted on the three point linkage of our small tractor.

It was with some trepidation that we opened our first bale that winter, our previous experience having been almost entirely based on the use of hay. We were pleasantly surprised – the quality of our high dry matter silage far surpassed that of any hay we had ever made, and this was reflected in increased production from our livestock. Since that year most of our preserved forage has been in the form of silage, although we do make some small bale hay, weather permitting. My aim here,

however, is not to come down strongly in favour of either one method or the other, but to discuss the pros and cons of each, enabling forage conservation strategies to be planned in advance rather than on the spot, thereby avoiding much tearing out of hair!

In basic terms, the principal difference between hay and silage is that in the former, grass is preserved by drying, whereas for silage, preservation is by fermentation followed, in effect, by pickling, under anaerobic conditions. The natural juices and sugars within the crop are required to ensure the process is a success. It can be seen, then, that a crop which has been fully dried for hay, and subsequently spoiled by rain, cannot later be baled and wrapped as silage.

The belief that hay making is somehow more "old fashioned" and therefore better suited to smaller, more traditionally run holdings, together with romantic images of haytime and harvest, seems to form a stumbling block preventing many smallholders from considering silage as an alternative. Well, maybe summers were hotter and drier years ago – certainly there is nothing romantic about haymaking on this (western) side of the UK. Besides, the process of ensiling crops for winter fodder may not be so modern as you think - the ancient Egyptians were apparently very good at it, and even preserved whole cereal crops by this method, something considered very up to date these days! However this old-fangled technology did not catch on in the UK until MAFF's "make silage, make sure" campaign during the years of the Second World War, and even then it was really only the farms of the wetter western regions that took to it.

As the science behind the processes (and the problems) of silage making were researched, and people's knowledge of the nutritional requirements of ruminant animals increased, so more stock farmers dropped hay making from their annual routine until today the majority of conserved forage in the UK is ensiled. Initially this would always be in clamps, silos or pits, but the introduction of baling and bagging (and more recently wrapping) has made silage an extremely convenient process for smaller acreages. Indeed, I know of a number of smallholders, with 10 acres or less, who wouldn't dream of returning to the "good old days" of hay making!

MACHINERY

Theoretically, you can make hay without tractor mounted machinery. In fact, we have made perfectly reasonable hay off our rather unkempt lawn, using a brushcutter for mowing and a pitch fork for turning, then carried it in and stacked it loose. On any larger area this is hardly a realistic option. In the predominantly grassland areas of the country there is no shortage of contractors – both large concerns, and those who specialise in smaller acreages – so do not despair if you are not fully kitted out with your own machinery. If, however, you live in an arable area, it may be harder to find a local contractor with grassland machinery. If you do decide to employ a contractor then contact him early in the year to discuss your requirements, rather than leaving everything to the last minute. Certainly you must speak to him before you cut your crop. We once made the mistake of mowing our grass a few days before the Royal Welsh Show, when most of the population of rural Wales (including our contractor) de-camp to Builth Wells. Had we spoken to our man beforehand we could have avoided this blunder.

It is becoming increasingly more difficult to find contractors with conventional balers, so if hay is your aim you will probably have to

Our vintage Jones Mk. IV baler

own this piece of equipment yourself, baling being the most time sensitive part of the whole process. You may pick one up at a farm sale for a reasonable price. Our own, a vintage Jones Mk. IV, cost us £80, and we've spent about the same again on parts. You should get something far more up to date for under £500.

Believe it or not, the large machinery required for silage making will generally fit through a smaller gateway than an old conventional baler, so if your access is tricky, or your field entrances narrow, this is something to bear in mind.

Mowing and tedding requirements are about the same for both hay and silage, although for silage it is preferable (but not essential) to use a mower with a conditioner as this results in better wilting of the crop. Even under ideal weather conditions hay will need to be turned a few times to get it perfectly dry, though silage can be baled straight from the swath left by the mower and will only need tedding if it is a very heavy crop, or if it gets wetted by rain.

You'll need a trailer – preferably a large flatbed – to get your hay in off the field, and some willing helpers for loading and stacking. To move big bales a simple spike on the back of your tractor will suffice, though if the tractor is a small one put some weights on the front to

Wrapped bales can be stored outdoors – ideal where space is at a premium

keep the wheels in contact with the ground - a bale of silage may weigh ⅓ of a tonne or more!

STORAGE

If you are lucky enough to have plenty of outbuildings, then the storage of small bale hay should present no problems. A designated hay barn is best, but failing that, the small size and compact nature of conventional bales enables good use to be made of any available space. A traditional hayloft over animal housing is far from ideal. This reduces essential ventilation for the livestock below who, furthermore, will be living in an environment of dust and potentially harmful mould spores filtering down from above. An outdoor stack covered with tarpaulins is also unsatisfactory as high losses through weathering and condensation will occur. Where storage space is at a premium then wrapped silage may be the answer. The layers of stretch film so essential to the ensilage process also provide a perfect weatherproof coating to each bale. Store outdoors, either as a single layer, or stacked, depending on the capabilities of your tractor.

Any damage to the wrap must be repaired promptly to prevent deterioration of the crop which, if well looked after, can be carried over from one season to the next, although that is

about the limit. Hay by contrast, can be kept in store for a number of years, and remain perfectly usable.

Probably the single most off-putting aspect of silage making is the wrap itself, or rather, the disposal of used wrap. The UK agricultural industry (and if you have a holding number that includes you) produces in the region of 500,000 tonnes of non-natural waste per annum, and most of this is silage wrap. Traditionally these waste products have been disposed of by burning or burying on farm, or, on smaller holdings, by putting in the dustbin along with domestic rubbish. Amidst a number of concerns, environmental and otherwise, these practices have now been banned – Defra's farm waste regulations that came into force in 2006 require disposal only at a licensed site, for a fee of course. There are several companies that recycle agricultural waste plastics (which, incidentally, includes the polypropylene twine used on small bales, which also can't be burned, buried or chucked in the bin) into a number of useful products, and hopefully the idea will catch on.

Locate your nearest facilities via the online waste recycling directory www.wasterecycling.org.uk or contact the environment agency on 0845 6033113.

Whilst it is right and proper that we should take these environmental concerns seriously, and aim to keep our holdings as "green" as we practically can, I feel that the issue should be given some perspective; the 500,000 tonnes per year of agricultural waste pales into insignificance when compared to the fact that the great British public produces some three million tonnes of refuse over the Christmas period alone!

FEEDING AND HANDLING

With regard to the rationing of preserved forages, there is one thing to be clear about from the beginning – as a feedstuff, silage is far superior to hay, not only nutritionally (see table on page 250) but is on the whole more palatable to stock (try eating a few dry cream crackers and you'll soon understand why moist foodstuffs are more palatable!). However there are of course a number of other factors to be taken into account when deciding which type of feed is most appropriate for specific circumstances.

There is no doubt that the portability of individual small bales is a valid consideration – the ability to lift a single bale onto your back and carry it to the hay racks is very handy. However, as numbers of stock increase to say, more than 40 sheep, or 2 cows with calves, the enthusiasm for small bales begins to pall – there seems to be a constant trudging to and fro, often across muddy fields, to keep up with the seemingly insatiable demands of hungry animals. Big bale silage suddenly becomes more attractive, with several days worth of fodder being shifted in a few minutes. This may be of importance to part time farmers / smallholders who perhaps are not at home much during the week. What time they do have is best spent in observation of their stock, and routine tasks should be streamlined to allow this.

Where stock are fed indoors there is little to choose between hay and silage, and a decision may be based purely on the number of mouths to feed. Probably it is not worth opening a big bale for less than 30 sheep, 2 cows or 6 - 8 goats. If properly handled, a bale of high DM silage, once started, will keep for a surprisingly long time – we have had a bale open for as long as a month, feeding a bit off the outside

The portability of small bales is a valid consideration

layer each day, with very little evidence of deterioration.

When stock are allowed free access to feed or are in any other way fed on an ad-lib basis, then big bales are undoubtedly more convenient, but where animals are individually housed, numbers are small, or access is restricted then conventional hay bales are probably more practical.

The perceived high risk of listeriosis infection via contaminated forage is probably a factor which prevents many small scale stock keepers from having a go with silage. The fear, however, is far worse than the reality – it must be, for most ruminant animals in the UK consume silage during the winter, and the number of cases of listeria is relatively small. Of far greater concern, in my opinion, is the very real risk of introducing toxoplasmosis to sheep flocks (and to some extent goats) by feeding hay contaminated with cat faeces.

In order for either of these infections to be a problem, they must of course be present in the first place, so if you are feeding your own hay or silage then it is up to you to ensure that the crop is made and stored in such a way as to prevent contamination occurring.

A further concern when using silage, most usually expressed by goat keepers, is that of "milk taint" i.e., the flavour of the feed having an adverse affect on the taste of the milk. This is not an issue with the type of high dry matter crop that I am talking about, although problems may occur if very wet, poorer quality silage is fed. Goats, being fussy eaters, would probably turn their noses up at that sort of food anyway! Wherever animals are kept for their milk, a change from hay feeding to good palatable silage is likely to be followed by an increase in yield.

Whichever method you decide to use to preserve your winter forage, the process remains a stressful one. Once the mower has entered the field you are fully committed and at the mercy of the weather until the crop is safely gathered in. Perhaps you'll decide that it's not worth the trouble, and buy in your requirements. Well, that is nothing to be ashamed of, for it is said that "to buy hay is to buy land." There is more than a bit of truth in this – you will be buying in fertility. Up to 50% of the nitrogen used to grow a crop of grass can be returned to the pasture in the form of manure. Someone else's fertiliser applications will greatly benefit your soil, after the bought hay has been processed through your stock. The converse also applies – think twice before selling off any of your own crop, if you have a surplus.

Additional Considerations

Cost
- Based on an estimated 12 small bales to each big bale, contractors' charges per unit of feed will be about the same for each, although purchase of wrap will add an additional cost to silage making. Given that the wrap also provides winter storage this may be considered a worthwhile expense.
- If you buy in forage, or if it is your intention to sell any surplus, then be aware that the price (per unit of feed) of small bale hay is almost double that of big bale silage!

Other Types Of Bale
I have not discussed the following:-
- Big round bale hay - Makes poor use of storage space.
- Small bale silage - Expensive to make and awkward to handle.
- Heston bales - Have not generally found favour on smallholdings, although may be worth considering if you buy in hay or straw as price per tonne is a lot less than conventional bales.

Sisal Twine
- Can be used with some older conventional balers. Being a natural product, it does not present any disposal problems.

Weather
- The shorter weather window required for silage making allows greater flexibility in the timing of harvest. This may be an important consideration if you work part time on your holding, or if you intend to take two crops in one season.
- Be guided by what has become the normal practice in your area. If you live in the drier South East, hay making should present no

problems, however in the wetter regions silage may be the most practical option.

Lie of the land

- Consider not only the aspect of each field but other physical factors which may affect your activities. For example, it can be very frustrating trying to make hay in small fields bordered by large trees, as much of the crop will be in shadow for part of the day.

Relative Nutrititional Values Of Hay And Silage		
	Metabolisable Energy (MJ/Kg DM)	Metabolisable Protein (g/Kg DM)
Hay - High Quality	10.3	77
Hay - Medium Quality	8.7	70
Silage - High Quality	11.8	125
Silage - Medium Quality	10.6	105
Source: ADAS		

TOXOPLASMOSIS

- A protozoan parasite.
- Generally a problem encountered in in-lamb ewes, causing abortion.
- Pregnant women are at high risk of infection and should avoid all contact with sheep during the lambing period.
- Usually introduced to sheep flocks via hay contaminated with cat faeces, but can be contracted from contaminated pasture.
- Ewes infected at tupping time will repeat, and, if still running with the ram, may conceive and lamb late. Ewes infected in early pregnancy will appear barren. Ewes infected in mid-pregnancy will abort. Ewes infected in late pregnancy will produce stillborn lambs, or very weak live lambs with little chance of survival.
- Level of infection in a flock can be such that more than 25% of lambs may be lost.
- Reduce risks by keeping cats out of hay sheds and feedstores, neuter all farm cats and discourage feral cats from taking up residence. Isolate ewes that have aborted, and dispose of placentae and foetuses carefully.
- Infected sheep develop an immunity, so shouldn't be culled.
- Future outbreaks can be prevented by vaccination.

LISTERIOSIS

- A bacterial infection.
- Usually associated with big bale silage, although clamp silage, haylage, big bale hay and conventional hay bales have also been identified as the source of outbreaks.
- Generally introduced into silage via contaminated soil.
- Avoid soil contamination by chain harrowing pastures to disperse molehills, rolling to flatten ruts and by not setting the mower too low.
- Repair damage to wrapped bales immediately to ensure good fermentation and preservation.
- Symptoms include circling, apparent blindness, drooping of one ear and one side of the face, cud spilling, salivation and abortion.
- Affected ewes require massive doses of antibiotics (80 – 100ml per day for several days) and careful nursing. Most will die despite treatment.

FIELD BOUNDARIES

There is a saying that "good fences make good neighbours" but, having said that, I don't think I'd have met half so many helpful people, or made so many useful contacts, if I hadn't had to go knocking on doors and asking for my sheep back! Amazingly, our neighbours have been remarkably tolerant of our practice of grazing the whole parish, but you may not be so lucky, particularly if you're surrounded by crops and gardens. There may be heartwarming occasions, such as the return of an old favourite you'd long since given up for dead, or amusing occurrences, as in the case of the ewe we sold twice, but, generally speaking, stray sheep at best represent a considerable waste of time and energy. At worst, they could cost you a fair bit in damages, particularly if they get onto the highway and cause an accident – do make sure

that your farm insurance policy covers you in the event of a breakout. Either way, you can rest assured that if sheep can get out they will!

Traditional Boundaries

Dry Stone Walls

Undoubtedly the most effective (and most picturesque) traditional field boundary is the dry stone wall, but then, living in North Wales, I would say that! Clearly, for stone walls to be an option you need to be in an area that has plenty of stones, which generally means the poorer upland regions, where this type of boundary can be seen as a dominant feature of the landscape. Local walling styles will vary according to the nature of the raw material available. A properly constructed dry stone

Repairing the mountain boundary

Fencing alongside an old wall can save a lot of bother

wall will stand untended for centuries, and, provided that it's steep enough sided and has well projecting "brisket stones" just below the copings, it'll remain stock proof, too. However, once the integrity of the wall is disturbed, whole lengths may collapse. People are the worst culprits – no matter how many gates and stiles you provide, hikers seem to insist upon climbing over any wall that crosses their desired route. I've even seen groups of people walking along the tops of walls to get from A to B. Once the copings have fallen, sheep will launch themselves at the little blip in the skyline, until eventually, amidst a shower of falling stone, they scramble over. It's all downhill from there on. Provided that it's spotted in time this type of damage can be repaired relatively simply, but where larger sections have subsided you'll have to pull the whole piece down and build it afresh. Often the fault will be found to lie in the very base of the wall, where one of the foundation stones has settled over the years until it slopes away from, rather than into, the wall. In this case, everything above it, and to

each side, will have become unstable and at greater risk from damage by trespass. As a basic guide, when rebuilding stretches of wall, every stone you lay must feel like it would rather roll into the wall than out of it, and every uneven surface must slope into the wall, otherwise the next course of stone will simply slide off! Pinners or chock stones can be inserted from the back (i.e., from the middle of the wall, it being built as two faces with a rubble infill) to level things up. In this way, as the wall settles, it

Laid hedge

simply set on end in the ground, with their tops held together by a twisted wire.

Laid Hedges

A laid hedge is the lowland equivalent of the dry stone wall, in that a naturally occurring local material is used to create a stock proof enclosure. At the interface between the two styles we have the clawdd, which is a stone faced earth bank (or an earth filled stone wall, depending on your viewpoint!) topped by a hedge, but more commonly the hedge is planted on a low bank formed by heaping up the spoil from an adjacent ditch. The presence of the ditch serves to increase the overall height of the barrier, and provides an additional obstacle to inquisitive animals. Where this type of boundary marks the line between two properties, the official boundary is the ditch, with the hedge belonging to the landowner on whose side of the ditch it stands. The basic process of laying consists of partially cutting through the main stem ("pleacher") of each thorn bush, and carefully bending them over to lie horizontally along the line of the hedge, overlapping one another. Each laid pleacher should lie uphill, so if working on a slope you'll have to start at the top and work your way downwards. Lots of material will have to be cut out of the hedge to leave the main stems fairly evenly spaced, and side branches will need to be removed in order that the hedge doesn't become too wide, and to enable each pleacher to lie comfortably on its neighbour. Where possible, the stems can be roughly woven together. The result may look rather stark, initially, as it often appears that you've cut more out than you've left in! The laid pleachers can be held in place by stakes driven in at right angles to the direction of lay. Ideally these stakes will be cut from the brushwood that's been trimmed out.

simply becomes tighter and stronger. However, rebuilding stone walls is time consuming and costly and, unless you've really got the knack, you may have to do it all again in a few years time! Where a large proportion of the boundary is no longer stock proof I suggest you save yourself a lot of bother by simply fencing alongside the old wall. If it isn't too badly collapsed you might get away with 3 strands of barbed wire, strained really tight, with the bottom wire sufficiently close to the wall that sheep can't scramble under it, and the top strand high enough that they won't try to jump it. If, on the other hand, there really isn't much left of the original boundary, a permanent netting stock fence provides the best solution. Again, this should be alongside the old wall.

What you mustn't do is to knock fence posts into the top of the wall. The only time a fence should run along the top of a wall is when the posts were incorporated as the wall was built, but even then it puts considerable strain on the structure and, more often than not, the demise of the wall begins where there's a post.

One interesting variation of stone boundary is the slate fence, seen in the quarrying areas of North Wales. Here, large slabs of slate are

The final artisan touch is to twist long lengths of briar or withy willow between and around the protruding tops of the stakes, binding the whole thing together.

New growth will shoot up vertically from all along the length of each horizontal pleacher, resulting in a dense network of crossed stems that is more or less impenetrable. Clearly the hedge will need some protection from grazing stock until the new growth is well established. A properly laid hedge alone will be sufficient for the containment of docile cattle or heavy lowland sheep, with perhaps just a strand or two of barbed wire alongside for added peace of mind. For some of our smaller native sheep breeds it's a different matter entirely, for, as one farmer put it, "those little ewes would crawl up their own backsides to escape", implying that, no matter how small the hole, they'll find a way through! The best thing would be to erect a permanent stock fence along one side of the hedge, with the other side temporarily protected by electric wire if need be. This would still allow access to the hedge for maintenance. Re-laying may need to be carried out at 10 – 15 year intervals.

Post & Rail

Post and rail fences may look pretty, but really are of no practical use where sheep are concerned, as the spacing of the rails is generally sufficient to allow most ewes (and all lambs) to simply hop through the gaps or walk underneath. In order to be sheep proof a post and rail fence will need wire netting added to it, so you might as well save yourself a lot of expense and bother by simply putting up a proper stock fence to begin with. The same can be said of traditional wrought iron park railings, although these may be an existing historical feature that you'd like to retain.

Wattle Or Cleft Hurdles

Cleft wooden hurdles were the method by which the big arable flocks of yesteryear were folded over roots and the like, and wattle hurdles provided ideal shelters at lambing time. Old photographs show what appears to be a peaceful and romantic scene – certainly it was illustrations of this kind that drew me into the world of shepherding – but one gains a wholly different opinion by reading the accounts of the men who tended these sheep! In short, containing a flock by this means involved hours of backbreaking slog, carrying dozens (or maybe hundreds) of sodden heavy hurdles across muddy ground to set a fold for the next day's grazing. No wonder everyone uses flexinets these days! Wattle lambing enclosures have also been superseded by the fact that the majority of flocks are housed for lambing now. However, there's no getting away from the fact that these old hurdles do look nice, so it's good to have a few around the place, particularly if you host open days or demonstrations on your farm. They're prohibitively expensive to buy, but if you can make your own from material grown on the holding then that's good. They have the added advantage of being fully biodegradable, or when they start to look a bit tatty you can chop them up for kindling!

An equally attractive alternative is to make hurdles of sawn timber as I've described on page 95 (and following), but, useful as they are, they're hardly a practical solution for the subdivision of large areas of grazing.

Above: Post holes should be dug as neatly and
squarely as possible
Above right: Generally struts are set in line with the
fence, but where deviations are greater than 90° this
may not be the case

*The head of the strut should fit snugly into a notch
cut in the strainer*

Practical Solutions

Permanent Stock Wire

No matter how aesthetically pleasing the
traditionally made boundaries may be, at the
end of the day a permanent wire netting stock
fence remains the most practical and secure
solution. The trouble is that anyone can put
up a wire fence but not many people can
do it really well, and unless the job is done
properly you might as well not bother. I see
far too many new stock fences that have lost
their integrity – and hence their usefulness –
in only a few months, let alone years! It pays
to get it right in the first place. A good fence
has a life expectancy in excess of 15 years, and
the addition of a single strand of electrified
wire will probably add another decade to that!
The key to successful fence erection lies in the
straining posts. The minimum number you'll
require is one more than the number of coils

(or part coils) of wire you're intending to put
up, though clearly if the fence line includes
gateways, or major deviations from a straight
line, you'll need a few extras. Strainers should
be at least 6″ in diameter and 7′ long, 3′ of
which must be buried in the ground. Post
holes should be dug as neatly and squarely as
possible. Use a spirit level to ensure that the
post is set vertically, and then – this is most
important – put back into the hole *every
single bit* of earth that came out. Shovel it
in a little at a time, and tamp each shovelful
down hard. Never be tempted to set straining

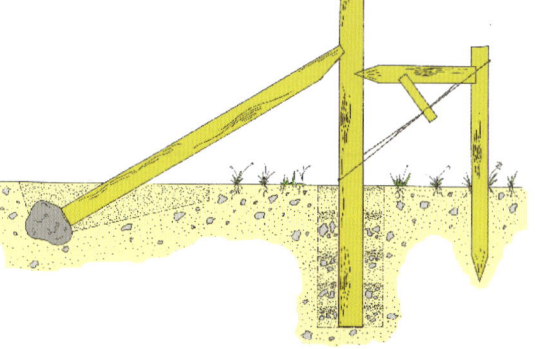

Above and diagram: An ordinary strut and a box or "H" strut used on either side of the same straining post

posts in concrete. Concrete won't bond with the surrounding earth, so, in time, the whole mass will simply rock back and forth in an enlarged hole. And, when the post eventually rots as a result of water pooling on the concrete, and snaps off, some poor devil has to dig the blasted thing out! Each straining post, unless flanking a gateway, will require two struts. Generally struts are set in line with the fence, however where corners or deviations to the line are greater than 90º this may not be the case. Minor changes in the line can be supported by less substantial intermediate strainers (provided that they do not coincide with a join in the wire), which may only need one small strut or stay to counteract the pull of the wire. Struts also need to be about 7´ long, though of a lesser diameter than the straining posts that they abut. The foot of the strut should be set against a large boulder buried in a trench, and the head should fit snugly into a notch cut in the strainer to receive it. Nails, if required, shouldn't be inserted until the fence is complete. If the angle at which the strut leans against the strainer is incorrect, then, as the wire is tightened, the straining post is simply levered out of the ground! This is probably one of the commonest mistakes made by amateur

Using a post knocker. It's much easier with two people!

fencers. Really, it shouldn't be a problem except on uneven ground, in which case you could install a box (or "H") strut. The important point to note in the construction of a box strut is that the twisted wire holding it all together runs from the base of the strainer to the top of the adjacent post, not the other way around. If you get this wrong then the whole assembly will serve no useful purpose whatsoever. A roll of netting is 50m long, so strainers should be set at intervals of no more than 48 – 49m, in order to allow sufficient surplus wire at each end of the run for straining and attachment.

All wire that's fixed to strainers should be carried right around the post, and joined back on itself, otherwise the tension of the wire will gradually cause the post to rotate, resulting in the whole fence going slack. Once the strainers are in you can run out and tension the bottom strand of barbed wire, which should be no more than a couple of inches off the ground. This then serves as a guide line for knocking in all the intermediate stakes, at intervals of about 8´. Use an alloy headed mallet, or a two handed post knocker. If you've got a lot to do, consider hiring a tractor mounted machine. All stakes and strainers should be tanalised – it's well worth paying a bit extra for this. When rolling out the netting, make sure that you get it the right way up, with the smaller holes at the bottom. This may sound like stating the obvious, but it's surprising how many people get it wrong! Attach one end of the roll to the appropriate strainer, 2″ above the barbed wire (remembering to take it right around the post), then pull the whole lot up tight from the other end, using a pair of monkey strainers (one on the top strand and one on the bottom). A parked tractor makes a suitable mobile anchoring point. When fixing wire to intermediate posts the staples should never be hammered fully home – this allows for a certain amount of "give" in the fence, and makes it easy to replace an individual post at a later stage if need be. You'll notice that fencing staples have one point slightly longer than the other – the longest point should always be at the top, so the staples penetrate at an angle, preventing rain water from gaining access to the timber at that point. Also, each staple should be skewed slightly, so both points don't enter the same grain. This reduces the likelihood of splitting. The final task is to fasten a top strand of barbed wire, some 4″ above the stock netting. Beware of over-tightening this – being

Flexinet electric fencing

attached near the top of the posts it'll exert considerable leverage, and, if you overdo it, the rest of the fence will slacken off. If you need to trim any posts to length make sure that you creosote the tops, and also splash plenty of creosote into notches cut in the strainers.

Electric Fencing

Electrified wire provides the ideal solution to most of the smallholder's fencing needs. It's relatively cheap, and quick and simple to erect. It can be used for both temporary subdivision of paddocks and for more permanent installations.

Flexinets, made from polywire, are ideal for splitting up larger areas of grazing on a short-term basis, and have replaced traditional wooden hurdles where flocks are folded. It's useful for creating a series of smaller enclosures for rotational grazing, and, by leaving a small gap where two lengths join (supported by a couple of fence posts) you can allow your lambs to forward creep graze. Be aware, though, that horned sheep grazing close to the netting can easily get caught, and also that this type of fence may pose a risk to very young lambs, as they're inclined to get tangled up in it. Once they're used to it it's OK, but you'll need to check on them fairly frequently to begin with.

Temporary electric fences consisting of a number of single wires are a cheaper option, but, on open ground, I've found that all but the most docile sheep tend to go through them in the end. Their efficacy is much improved if they can be erected in front of a visual barrier such as a hedge or bank.

If you've got existing traditional boundaries that are in a reasonable state of repair then the best option is to install permanent high tensile electric wires close alongside. This makes a very unobtrusive fence, as so few intermediate posts are required, and will preserve the aesthetic appearance of the existing boundary whilst providing full security.

With the exception of the high tensile wire, most electric fence installations can be powered by a battery or solar energizer, giving total versatility – ideal if you're renting grazing away from home. However, if your land is in a relatively compact block I'd strongly recommend that you invest in a mains powered unit, with insulated cables run out to various parts of the farm.

Galvanised Metal Hurdles

On a very small scale, such as where only a few sheep are kept as pets or "woolly lawnmowers" I've seen areas of grazing successfully subdivided using metal hurdles. In general, though, this wouldn't be a cost effective solution. Having said that, you'll need a few hurdles anyway for handling purposes, so they could be pressed into alternative service occasionally, perhaps to make a small grass enclosure for some orphan lambs or to contain a poorly ewe that needs to be kept close at hand but is not sufficiently ill to warrant being brought indoors, or for a convalescent patient to be given the chance to stretch her legs and benefit from a bite of "Doctor Green".

Drill a hole through the top hinge of each pair to take a linchpin

Gates & Stiles

Gates

Traditional wooden five-bar gates look lovely when they're new, but when they're old and starting to sag, or if they warp so the latches won't catch, you'll curse them! Some grant funded countryside management schemes will insist upon the installation of timber gates, but, on the whole, galvanized metal gates are a better bet. They're remarkably good value for money, and are extremely versatile – you can easily lift them off their hinges and use them temporarily elsewhere, perhaps for setting up handling pens or for sub-dividing a building when housing the flock at lambing. Half-meshed gates are handy, as young lambs can't hop through between the lower bars, but you may find that sheep catch and pull out their eartags, and male lambs of horned breeds tend to break their horns on them.

Field gates should, wherever possible, be hung such that they can swing both ways, and when moving the flock between fields you should always open the gate away from the animals. If you don't do this you can rest assured that the last lamb in the group will end up stuck behind the open gate as his mother runs through. Panic ensues, and often the lamb will bolt down the fence line, ever further from the open gate. It's not too bad with a single lamb, as eventually he'll jam his head through the wire, allowing you to pick him up and pop him over, but with cows & calves, or, worse still, mares & foals, the results can be quite horrific. If the gate will only open the wrong way then it's best to lift it off its hinges. Incidentally, if you do end up in a situation where several lambs have become separated from their dams in this way, you should always bring the ewes back through to join them, and start all over again, rather than try rounding up the errant offspring. I've known lambs to be lost for good as a result of people's failure to observe this simple guideline.

When hanging a new gate, always drill a ¼″ diameter hole through the top hinge of each pair to take a linchpin. Forget bent nails or reversing the top hinge (the former is shoddy, the latter too permanent) – this method is perfectly stockproof, yet still gives you the flexibility to re-hang gates as required. In fact I suggest you go round and do this with all your your existing field gates, and any gates in buildings used for animal housing, too. All stock, from sheep to Shires, seem adept at lifting gates off hinges, and there is nothing quite like the sound of a gate crashing to the ground to have you bolt upright at 3.00am. You wouldn't be the first person spotted rounding up errant livestock clad in nothing but wellington boots and a head torch…

Stiles

It is a sad fact that walkers seem unable to deviate by more than 25 yards from their chosen route, even if crossing points are clearly visible, preferring to climb over any obstacle that bars their way. Telephone and electricity linesmen are equally bad. If you suffer a situation where the public are allowed to wander freely over areas of your farm then, in order to prolong the life of your fence, you'll need to install at least one (preferably two) stiles for every roll of wire erected. Popular crossing places should have ladder stiles. At these points, barbed wire should be protected by encasing in a short length of alkathene pipe. The upside to this is that it also makes it a lot easier for you to get around your own property. Official footpaths may require kissing gates, in which case I suggest you fit a piece of bungee cord to ensure that they close properly – sheep are expert at wriggling through when they're partially open.

Ladder stile

Sheepdog
Training
Explained

INTRODUCTION

While nobody would argue with the fact that a well trained sheepdog is an asset on any farm or smallholding where sheep are kept, it is perhaps not fully realised that an untrained – or worse still, part trained or poorly trained – dog is more than just a nuisance. It can be a liability. There is a saying that "a little bit of knowledge is a dangerous thing", and this can be aptly applied to sheepdogs! The inherent knowledge with which they are born, which is nothing more than a modified hunting instinct, is enough to do a lot of damage, unless it's channelled in the appropriate directions, or completely suppressed. If you choose to keep a dog of a herding breed then you owe it to your sheep, and, perhaps more importantly, your neighbour's, to ensure that it's properly trained, and what could be better than to train it for the purpose for which it was bred? But this needn't be as daunting as you might think, and, contrary to popular belief, you don't need an already trained dog to help you, nor are there any mysterious trade secrets. You should be able to develop even the most unpromising puppy into a reasonable working dog, once you understand the processes involved. OK, so you might not win any prizes, but you'll have plenty of fun, and, what's more, you'll learn to understand your sheep a lot better, too!

If you're a newcomer, it can help you get a "foot in the door" of the rural community in which you find yourself – there's an active social scene surrounding sheepdog training and trialling, with events taking place almost every weekend during the summer in some areas, and many of the folk involved will be smallholders or hobbyists, like yourself. And, once the word gets around that you've got good dogs, you'll not be short of work either, if you're

Border Collie

in a predominantly sheep farming region – an important consideration, when you need to earn a bit of cash to help make ends meet, as you struggle to get your small scale farming venture up and running. During our first seven years at Ty'n y Mynydd, almost all of our supplementary income was from shepherding, and I could not have achieved this without my dogs.

And if you don't fancy going out to work, then let the work come to you! Once you've got the knack (and provided that you've got the time), taking in dogs for training can form a profitable sideline to running a smallholding.

BREEDS & TYPES

Border Collie

Here in the UK, when we refer to sheepdogs we're generally talking about the Border collie. In no other breed, anywhere in the world, is the herding instinct so highly developed as in the Border collie, yet even within the one breed there are considerable variations, and a number of sub-types. Choosing the right type can make a big difference to the likelihood of the animal ultimately making a good sheepdog. For the greatest chance of success, buy a puppy or young dog that is registered with the International Sheepdog Society (ISDS). Although there's no absolute guarantee that any dog will work sheep, the odds are considerably improved by choosing an ISDS registered dog, and you can also be sure that the parents and the offspring will have been tested for the presence of certain congenital eye defects. There is no standard of appearance in the ISDS registered dogs, so you'll find animals of all shapes and sizes and colour variations, the principal criteria being the ability to work sheep. The ISDS stud book, which was started in 1906 or thereabouts, is not a "closed" stud book, which means that it is still possible for an un-registered dog of exceptional quality to be accepted on merit, provided that certain criteria are met. Although one doesn't see the letters ROM (registered on merit) cropping up very often in the pedigrees of the modern working dogs, it does at least mean that the breed has retained a degree of "hybrid vigour". This, together with the absence of a breed standard for appearance, ensures that the essential strength and stamina is not lost.

There is another type of registered Border collie, and in this case the Kennel Club (KC) is the organisation responsible. Here, we see what happens when appearance becomes the all important factor in a breeding program. The KC registered dogs tend to be a bit larger (and clumsier!) than the working strains, and all have the classic Border collie markings. Problems such as hip dysplacia may be more common than in the ISDS registered dogs, and there's a

far greater chance of coming across an animal with no herding instincts whatsoever – it's been "bred out of them" the old shepherds say. Despite this, there are many good dogs that are registered with both societies, and the pedigree of a show bench winner may well include the names of famous sheepdog trial champions (although the converse is less likely!).

Generally it is best to avoid unregistered "farm" dogs, although you could be lucky and pick a real cracker! The incidental litters of collie type dogs that occur on farms are often the result of indiscriminate matings between very close relatives, or possibly the father wasn't a sheepdog at all – any black and white puppy out of a working bitch could be passed off as a collie, but if the sire of the litter happened to be a visiting retriever or spaniel then you'll end up with a sheepdog that wants to fetch the sheep in its mouth, which really isn't the done thing…

Welsh Sheepdog

The Welsh sheepdog is an old strain of collie type dog, found only in Wales, that had become very rare. Recently it has enjoyed something of a revival, and there is a society devoted to the registration and promotion of the breed. In appearance, the Welsh sheepdog is similar to a rather heavily built Border collie, with a much broader head and muzzle. Colours vary, although red appears to be the most popular, with an outer guard coat of black hairs often giving a slightly brindled appearance. At work, the Welsh sheepdog displays tremendous stamina, will bark on command, and readily takes to water if sheep are to be fetched from the other side of a stream. It'll work cattle as well as sheep, "heeling" them to keep them on the move – a characteristic which, together with the popular coat colour, makes me think that there's probably a bit of the corgi in its distant ancestry. (The Pembrokeshire corgi was originally a farmyard cattle dog – Wales' equivalent to the Lancashire heeler). The Welsh sheepdog has great strength of character, and is happiest behind big flocks of sheep, so may not be the best choice for a novice handler. The sale of registered dogs is restricted to working stock farms only.

Bearded Collie

Although roughly similar in appearance to the Old English sheepdog made famous by Dulux, the beardie has (in some strains) retained the working instinct. Originally a droving dog, the beardie is less inclined to head sheep than the Border collie, though can eventually be persuaded to do so. Given that the basic principles of sheepdog training are based around a dog's instinct to head the sheep, you'll understand that training a beardie can be a rather laborious affair! I've seen very few (if any) bearded collies working really well, but I've seen some smashing dogs that were bearded collie / Border collie crosses.

There is some overlap between the bearded and the Border collie, as can be seen by the occasional appearance of a beardie in the pedigree of an ISDS registered dog, but there is also a separate society dedicated to the promotion of the working bearded collie.

Kelpie

The kelpie is undoubtedly descended from the early types of Border collie exported to Australia in the early 1900s. These medium sized black-and-tan dogs are brash and boisterous, and, although they lack the finesse of the Border collie they can be trained to a fairly high standard, over a long period. The kelpie wouldn't be ideally suited to a smallholding, as handling them can be something of a challenge. They're very active, and need plenty of work on big flocks to keep them fully occupied.

Huntaway

There's plenty of people who swear by the New Zealand huntaway, and plenty of people who swear at them, too! Big and clumsy, the huntaway's sole purpose appears to be to make as much noise as possible – all very well for scaring mobs of sheep out of patches of scrub in the outback, but hardly necessary in the sedate environment of a smallholder's back paddock! Having said that, a dog that is ¾ Border collie and ¼ huntaway is a very handy dog indeed.

Already Got A Dog?

As we've seen, the most sensible course of action is to purchase a youngster from an ISDS registered litter, but what if you've got a collie type dog already? Until now he's just been a much loved family pet, but, in your new role as a sheep keeping smallholder, you feel that everyone should earn their keep. Can an old dog be taught new tricks? In most cases yes, it can. Up to about 4 years old there's a fairly reasonable chance that a pet collie can be brought into work, provided that the necessary instinct can be roused, but it's unlikely that it'll ever develop the high level of skill of a dog that was started younger, and you will find that training takes longer, with frequent "revision" sessions being required. There'll be a lot of previously formed bad habits and sloppy behaviour to correct too, in you and the dog!

Buying A Puppy Or Young Dog

Whether you decide to start with a young puppy or a slightly older dog is largely a matter of personal preference, but the decision should also be influenced by the nature of the environment in which your new "team member" will be living. Have you got suitable accommodation for a very young pup? Have you got the time and patience required? Are the rest of the family as keen as you are? Can you provide a safe area for a puppy to play, not only away from risk of injury, but where bad habits such as chasing vehicles, constant barking, and wandering off, can't develop? Mostly, these habits are due to boredom, which brings us straight back to the issue of time – will the puppy be left unsupervised and unsocialised for long periods during the day?

Personally, I don't find puppies particularly appealing – they're noisy, smelly, demanding creatures – and would rather start with a young dog of 8 – 10 months. You could, of course, leave the rearing of a puppy to other family members and simply take over responsibility when it's time to start formal training, but really, if it's to be your dog you should look after it. I did once suggest to my own children that they might like to play with a litter of puppies that I had at the time (in order to socialise them a bit), but they just looked at me scornfully and carried on with what they were doing. Clearly they share my sentiments entirely!

Whatever age you're buying your youngster, up to about 8 months, the same basic guidelines will apply. You can't really tell what a pup will be like when it's grown up, so you just have to go by general health and wellbeing and hope for the best! The signs of good health should be self-evident in glossy coats and clear

Choose the pup that gazes up at you thoughtfully, as if he's taking in everything you say

eyes, but always ask the breeder if they've been wormed, and check that the bitch was vaccinated. The puppies may already have been vaccinated too. Both the dog and the bitch should have been eye tested (if old enough) to ensure that they don't have Progressive Retinal Atrophy (PRA), and the litter should have been screened for Collie Eye Anomaly (CEA). Both of these conditions are hereditary, and, prior to the eye testing program, had become quite common. Since eye testing has been a condition of registration with the ISDS the incidence of PRA and CEA has fallen to negligible levels. When it comes to actually making a choice, everyone has their own ideas: some people prefer tri-colours, others look for classic black and white collie markings, and some choose dogs with as little white on as possible. Some people like blue (wall) eyed dogs, and others can't stand them! Some handlers swear that a dog must have a black roof to its mouth, and others will make complicated calculations based on the width between the eyes and the length of the head (big head = big brains, I suppose!). Each to their own. I always pick the ugliest, in the vague hope that a deficiency in one area (appearance) may be compensated for by a bonus in another (intelligence)! It may be better for the novice to avoid the strong willed pup that's always at the front, and don't buy the little shy fellow at the back because you feel sorry for him. Look for the one in the middle of the pack that gazes up at you thoughtfully, as if he's taking in everything you say. Some indication of future character can be gained by observing the individuals within a litter at play, but not until a pup gets to around 8 months can you make a really valid assessment of its potential "trainability". At this age I'd expect to see some semblance of basic obedience and willingness to learn, and be able to get an idea of how the dog responds to discipline and handling.

Registration

The puppy may already have been registered with the ISDS, in which case the registration certificate will need to be returned to them, together with notification of transfer of ownership. The certificate will be updated with the new details and sent back to you. If the pup hasn't already been registered, then, before you take it away, the breeder will need to make a sketch of its markings in the litter registration folder, together with details of colour, variety, size and sex. Filling in the relevant part of the form with your name and address ensures that the pup is registered directly to you, and no transfer of ownership fee will be payable. You'll also need a name for your puppy. One syllable names are best (easier to shout!), and it's quite common for pups to be named after a famous ancestor. However, there can be no duplicate names in a litter, so it's not unusual for a dog to have a kennel name in addition to its registered name.

Cost

When compared to the prices that people pay for other breeds of dog (most of which serve no useful purpose whatsoever) a registered Border collie is ridiculously cheap! A well bred puppy may set you back somewhere between one and two hundred pounds (depending on the breeder and the bloodlines), and for just a little more outlay you'll be able to buy a youngster that has already started basic training. £600 would probably get you a rough-and-ready trained (but not polished) farm dog, but for something with a bit more flair expect to pay twice that. Even well trained dogs with a touch of class rarely make £2000.

Housing

Should a working dog be kept outdoors, or allowed into the house? It's an age old question. Many people cite health as a reason for keeping sheepdogs outside and, in the case of a full-time working farm dog they're probably right – a dog can't remove a layer of clothing each time he comes in, as we do, nor can he replace his wet coat with a dry one. If, after doing the morning rounds, in winter, in the pouring rain, a wet dog spends half an hour in front of the fire while his master has lunch, before plunging out into the cold again, his health will begin to suffer, particularly as this routine is repeated day after day. However, the smallholder's dog has an easier time of things, with only sporadic periods of work (preferably on dry days!), so in this case it's largely a matter of preference. Personally, I wouldn't have a dog indoors.

Once training begins there are a whole new set of problems associated with keeping a dog in the house. Training sessions can be fairly intense, and the dog needs his own refuge to relax and think things over afterwards, just as you do, particularly if things aren't going particularly well. If you've been in conflict on the training field, then you'll be surly with the dog in the house, which is unfair. But if you're not he'll be confused, as he knows he's done wrong. Either way, he'll turn to other members of the household looking for sympathy, which will undermine your authority as trainer. In its worst extreme this can result in the dog bolting for home during any training session where he feels under pressure, looking for fuss from the rest of the family. Another confusing situation arises as a result of different people in the household using different commands, and each applying a different interpretation of the ground rules. I remember watching a part-time smallholder competing in a sheepdog trial – he sent his dog on a beautiful outrun, then whistled it to stop at the top of the field behind the sheep, whereupon the dog ran straight back to him! Three times he repeated the attempt, and on each occasion the dog returned! In the meantime his wife, in the audience, was having hysterics – it transpired that every day, while her husband was at work, she let the family dogs out for a run, and the whistle with which she recalled them was identical to the one that her husband was using as a "lie down" command! No prizes for guessing who wore the trousers in that household! I suspect that had she not been present on the trial field then the dog would have stopped.

On balance, it's probably best for the dog to be kept in its own kennel, outdoors, away from disturbing influences such as wives and children!

Kennels for my sheepdogs at Ty'n y Mynydd. I built these nine years ago, at a total cost of £370

Individual kennels with runs are best, and it wouldn't be beyond the wit of most smallholders to build something suitable, or to convert an existing outbuilding. Complete units can be purchased, but they're awfully expensive. Alternatively a dog can be kept on a chain when not working, provided that this is done properly, and that there is access to snug sleeping quarters. You'd also need to be able to provide alternative accommodation during critical periods, for example if a bitch is expecting pups, or if a dog is unwell. I admit that this situation is far from ideal, but it is infinitely better than having a young dog running loose and getting into mischief. Even if you wouldn't wish to keep a dog in this way, a youngster should be taught to accept a chain from an early age. It's a simple matter then, when you're carrying out some task in pens with the sheep (such as hoof trimming), to tether the trainee where he can see what's going on and feel involved, without the risk of conflict. And, later on, if you decide to compete in sheepdog trials, you'll be able to tie up your dog in some shady place while waiting for your run, rather than letting him stew in the back of a hot car all day.

Feeding

This needn't be expensive. Clearly there are times when you'll need to spend a bit more for a higher specification diet, for example for breeding bitches, growing puppies, and during periods of very intensive work or trialling, but on the whole a basic diet will suffice. There are a number of complete rations specifically for sheepdogs that are available in either maintenance or working formulas, and are VAT free. The one I use costs £12.80 for a 20kg bag, which works out at under £2.00 per dog per week, during periods of low activity. If you slaughter your own stock, you can supplement the dog's diet with offal at no cost. There are some prejudices against feeding offal, but in reality it is very close to a dog's natural diet, and they do very well on it. At one time I fed my dogs almost exclusively on whole rabbit carcasses, and they've never looked so fit, before or since!

A healthy adult sheepdog should be fed once a day, last thing in the evening, 6 days a week.

GETTING STARTED

One of the questions I'm most frequently asked is "what is the correct age to start training a sheepdog?" and, in truth, there is no simple answer to this query. Training begins on the very day that you choose a pup from a litter to be your future work mate. You'll take him home with you, chatting to him in the car as you travel, which teaches him to recognise your voice. Back at the farm you'll praise him when he's good and scold him when he isn't. You'll teach him to bumble along behind you as you go about your chores, and curse him when he gets under your feet. You'll house train him if he's to be an indoor dog, teach him where his bed is, and generally lay down the ground

rules. He'll learn his name, and to come when he's called, and you'll introduce a few basic commands, such as "lie down" and "stay there" without even realising you're doing it.

Really, though, when people ask that question, they mean "what age should we start formal training?" but, alas, there's no simple answer here, either! Some dogs never have any formal training, as such, and become fairly useful farm dogs just through learning on the job. Ideally, though, this type of education should be supplemented by some periods of more structured tuition. For smallholders and hobby handlers, who don't have access to large and varied flocks on a regular basis, most training will be carried out in formal sessions. In most cases, dogs will be started on sheep at between 10 and 15 months of age – I wouldn't recommend starting any younger than this (for reasons that I'll explain in a minute), and if it's left a bit later that's no big deal.

Seeing Sheep

A collie puppy's first introduction to sheep generally takes place in a very natural way – as he grows, he simply becomes aware of the livestock around him, while he's going about with you on the farm. Gradually he becomes more interested, and bolder, until the exciting day comes when he realises that things run when he chases them! Look out! From this moment, it's important that the puppy isn't allowed access to sheep (or other animals) until such time as he's big enough, strong enough and has enough stamina to outpace them. This doesn't mean he should be kept away from stock altogether – the more time spent looking at sheep the better – but he mustn't be allowed to run free. Young dogs that are started too soon get frustrated by the fact that the sheep can travel faster than they can, and quickly develop bad habits such as randomly chasing individual sheep, running too tight, crossing over in front of the handler, biting the sheep's back ends and incessant yelping. This is why I suggest waiting until the dog is at least ten months old before starting. In the meantime, continue to take the dog around with you, ensuring that he's kept off the sheep (though if he does inadvertently get loose and start chasing sheep, for goodness sake don't scold him for it! Try to turn the situation into a positive experience), and keep his mind fully occupied by teaching some basic commands, such as "stand", "lie down" and "stay there". Bear in mind that 120% obedience at home will only give you 20% (if that) on the field, once those thrilling things called sheep are introduced into the equation!

The Commands

In general, the 3 or 4 basic commands used by sheepdog handlers are universally recognised, but there does tend to be regional and personal differences between the commands used at a more advanced level. Some handlers prefer to train their dogs in Welsh. Whistled commands are a different matter altogether, with each shepherd having his own repertoire. The only whistle that's used more or less unanimously is a long, drawn out blast, to stop the dog in his tracks. (The better the dog, the less drawn out the whistle needs to be!).

The most commonly used, universally understood cry, typically uttered by shepherds and sheepdog handlers, sounds something like "damandbloodyblastyoudog!" although even here there are some interesting vernacular variations to be heard.

Basic

- *That'll do:* Calls the dog off the sheep and back to the handler.
- *Come bye:* Sends the dog in a clockwise direction around the sheep. Shortened to "come" for smaller flanking movements.
- *Away here:* Sends the dog in an anti-clockwise direction around the sheep. Shortened to "'way" for smaller flanking movements.
- *Lie down:* Fairly self explanatory, except that it's not absolutely essential for the dog to lie down, provided that it stops. Most beginners feel more in control of the situation if the dog actually lies down, whereas more experienced handlers will never lie the dog down at all.
- *Stand:* As for "lie down", really. Having a dog that stops on its feet is handy when working in heather or long grass!
- *Stay there:* This really means "wait", in that a dog that's been told to "stay there" must wait until the same person gives another command to release it. So, if you put the dog in its kennel at night and say "stay there", it should be you that lets it out in the morning, not another member of the household.
- *Get up / Walk on:* The dog should walk straight towards the sheep from wherever he is. A few "steady" commands will probably be needed here, too!

In addition to these basic commands, each handler will have a way of keeping the dog at heel. In most instances this will be by using a command, such as "stay close", but in some cases the dog will be expected to be at heel at all times, unless its been instructed otherwise.

Note that there is no "sit" command – sheepdogs either stand up, lie down, or run about!

Advanced

• *There:* A slight pause, or half-stop, used to turn the dog on to the sheep at any point. This results in a much more fluid movement than is obtained by constantly giving "lie down" commands immediately followed by "get up".

• *Look back:* The dog leaves the group of sheep he's working, and goes to look for a second batch. These may be out of sight, in which case flanking commands are used to give direction.

• *Go out*: Keeps the dog further from the sheep, giving a wider outrun.

• *In:* Brings the dog closer to the sheep, and can be used to bring the dog through the sheep to shed off a section of the flock.

• *This:* Identifies a single sheep that needs to be split from the rest. Once the dog knows which sheep is wanted he won't forget it, so the two of you can work together to single it out, using very few additional commands.

• *Hold it:* Some handlers will train their dogs to catch an individual sheep that's been singled out as above, by gripping the wool on the shoulder. Essential for outdoor lambing in extensive flocks.

• *Speak up:* Although collies are quiet workers, they can be taught to bark on command, which is useful when running large flocks through yards and handling systems.

I've described just 14 fairly common commands here, but there are of course many, many more. Also remember that, for each dog, there may be two versions (long and short) of each spoken command, and perhaps two whistled versions, taking our initial fourteen to a total of 56 individually recognised instructions.

Teaching The Basics At Home

Here's a little routine to keep you and your pupil occupied, while you're waiting for his legs to grow long enough for the real fun to start!

A walled lane or farm track is an ideal site to begin, as it gives limited scope for the pup to dodge past you, and keeps him focused on the task in hand, as he can't see what's going on in the next field! Start with him on a fairly long, light lead – a piece of baler twine is just right, and is what most trainers use. Initially concentrate on keeping the pup at heel. I walk steadily up and down the lane, carrying a short piece of alkathene pipe which I swish from side to side in front of me to discourage the pup from passing, and all the while I'm keeping him close and repeating the "stay close" command.

If he hangs back (or moves to the side) I

Teaching Jan to "stay close"

pause and give a little tug on the string, saying "that'll do" to bring him back to my side. As soon as he's where he should be we set off walking again. After a day or two it should be possible to dispense with the alkathene pipe.

Next, introduce a "stand" and / or a "lie down" command:

Lie Down

Allow the lead to go slack enough for you to put your foot on it. Now, if you pull upwards on the end of the string you're holding, the dog feels a pressure pulling him downwards. It's important, though, that you don't yank the dog down – a steady pressure is enough,

Jan lying down, with my foot on the string

while giving the command. Once the dog is down, ease the pressure and continue giving the command, using the appropriate tone of voice (see page 272). Re-apply the pressure if required, and change voice tone. Don't overdo it though – a few seconds is enough, before resuming walking. You'll only need to put the string under your foot a few times – these dogs learn remarkably quickly.

Jan soon takes the "lie down" command without me having to step on the string

Holding out your hand, palm down, over the dog, can help to reinforce the command, but in general it's best to avoid using any hand signals, or later on the dog will be looking at you instead of the sheep.

Stand

Here, I give the dog a little more scope on the lead, and, when he looks like he might try passing me, I pull him up short while giving the command "stand". At the same time I step in front and turn to face him. I keep repeating the command, using my hand held out if it's

Moss learns to "stand" and "stay there", as I back away

necessary give some visual form of control. As before, keep any hand signals to a minimum – I often tell people that it's best to train a dog with your hands in your pockets!

For both the "lie down" and "stand" commands, I like to introduce a whistled version, alongside its spoken equivalent, right from the start.

Once the dog is taking these instructions fairly well, try adding "stay there":

Stay There

With the dog stopped, i.e., either lying down or standing, and me standing in front facing him, I begin to back slowly away while quietly repeating the stop command. When I reach the extremity of the string, I begin to alternate the stop command with "stay there".

Gently, I let go of the string and continue to back off, gradually reducing my use of the stop command, until I drop it altogether. By now I'll be about 5 yards from the dog, repeating "stay there" over and over. If the dog attempts to move I'll immediately take a step towards him, giving the stop command in a stern voice, immediately repeated in a soft voice (assuming he obeys!), before resuming "stay there". Again, don't overdo it – if you feel any tenseness developing, recall the dog and call it a day.

Before long I'll be able to move further away and reduce the frequency of the commands. If there's room to do so, I'll walk in a circle around the stopped dog, or even momentarily duck out of sight.

Now, using these few basic commands you can work to a sequence, as follows:

Walk along for a bit with the dog at heel, then ask him to stop (using either "stand" or "lie down"). Continue walking for 50 yards or so, having told the dog to "stay there". Turn to face him, and, after a short pause, give a quiet, "that'll do". He'll move towards you slightly hesitantly. When he's come about half way, whistle him to stop (again, either "stand" or "lie down", whichever one you didn't use last time). Wait a moment, then recall him with a cheerful "that'll do", and give him some fuss before repeating the whole exercise. It's important that you vary things a bit, so don't always stop in the same place, don't always use "lie down" in preference to "stand" (even though you might feel more confident with the dog on its belly), and try to alternate the use of voice and whistle commands.

Using Your Voice As A Reward

It's not necessary (or sensible) to reward a working dog with tit bits, nor should you use excessive words of praise – the dog has enough vocabulary to learn without you adding a lot of sweet nothings! The tone of your voice, if used correctly, gives sufficient reward or reinforcement, and body language helps, too. For example, if the dog fails to lie down when asked to do so in a normal voice, you repeat the command in a tone that shows you mean business. I don't necessarily mean louder – there are far too many handlers that simply stand and bawl at their disobedient dogs – but perhaps in a slightly menacing way, whilst taking a step towards the dog and making yourself tall. Immediately the dog responds to the command you must ease the pressure – back off slightly and reduce your stature, while repeating the command several times in a very soft, friendly and encouraging voice. This is sufficient reward. Generally, you'll see the dog raise his head, look straight at you and twitch the tip of his tail to show that he's understood. As soon as you see

A bit of praise for a job well done

these signs you can move onto something else, or, in the case of a very young trainee, call him to you with a joyous "that'll do!", and give him a bit of fuss.

STARTING ON SHEEP

Hopefully, by the time your youngster is sufficiently well grown to move on to the real thing, you'll have him well under control in terms of general obedience, and are probably feeling pretty pleased with yourself. If so, then you're in for a shock, as all this discipline is going to go straight out the window, the moment he's let loose on a flock of sheep! Don't despair, though. He won't have forgotten it all, and when he's got over his initial excitement you'll find that the time you put into laying down the foundations was time well spent.

The Training Field

The first mistake that many people make is to assume that it's best to start in a very small field. Nothing could be further from the truth – the bigger the better, within reason. Boundaries are the enemy of the sheepdog trainer trying to start a youngster. Once the sheep get up against the fence there's not much that an inexperienced dog can do about it, and, in frustration, he'll resort to some of the bad habits I mentioned earlier. If the fence is on top of a bank, or low wall, then that's even worse, as the sheep will jump up on there and simply outface the dog from above. The initial aim of the trainer is to get the dog to the far side of the sheep, which can only be achieved if there's sufficient space.

The second mistake is to assume that you need a very small group of sheep. Again, this is not so. The young dog is likely to scatter the sheep to begin with, so you need to be sure that he'll be left with a reasonably sized pack to work with, after the bulk of the flock has fled! Start with 20 or so, and you'll probably find that there is a group of around half a dozen that remain together, which will do for a start. Any less than that and they tend to move about in a rather unpredictable fashion.

Very tame sheep (particularly those that were bottle fed) are hopeless for dog training, but having a few that will follow a bucket can be handy, provided that they still respond to the presence of the dog. Avoid using mature rams, pregnant ewes, or ewes with very young lambs.

Starting On Sheep

Take the dog to the field on a fairly short lead. Again, a piece of baler twine is ideal, and is what most people use, but this time pass the string through the collar and hold both ends, effectively halving the length. You can now keep the dog close to you and under verbal control, and, when it's time to let him go, you simply release one end of the string, which slips from his collar without any fuss. This is much easier than having the dog get all wound up while you struggle to release a snap-hook.

Pause after entering the field, and encourage the dog to look for the sheep (I actually use the command "look"). This also gives the sheep the opportunity to see the dog, and hopefully they'll start bunching together, which makes your job easier. Once you're sure that the dog has spotted all the sheep (not just the one or two which are nearest), you'll need to engineer a situation suitable for letting the dog go. You may need to walk around the field several times with the dog on the lead before a suitable opportunity presents itself. You're aiming to have the sheep in one bunch in the middle of the field, either standing still or moving slowly away from you, with yourself and the dog as close to the sheep as you can get without spooking them. As soon as everything feels fairly settled, turn side-on to the flock, with the dog on the side of you furthest from the sheep. Take a few steps forward and at the same time let go of one end of the string. Don't try to use any command to send the dog off, just give a gentle "Shhhhhh" to send him on his way, and perhaps run with him for a few paces, to keep him on the right course. Now, ideally, as the dog sets off in one direction around the sheep, you turn around and set off the other way to try to stop them bolting (see picture 1 on the next page). Things are going to happen pretty fast though, so, as this is your first attempt, I suggest you just stand still and watch. What you really mustn't do is start shouting at the dog, no matter what's going on. He won't take any notice of you anyway, so you'd effectively be teaching him to ignore commands, which is

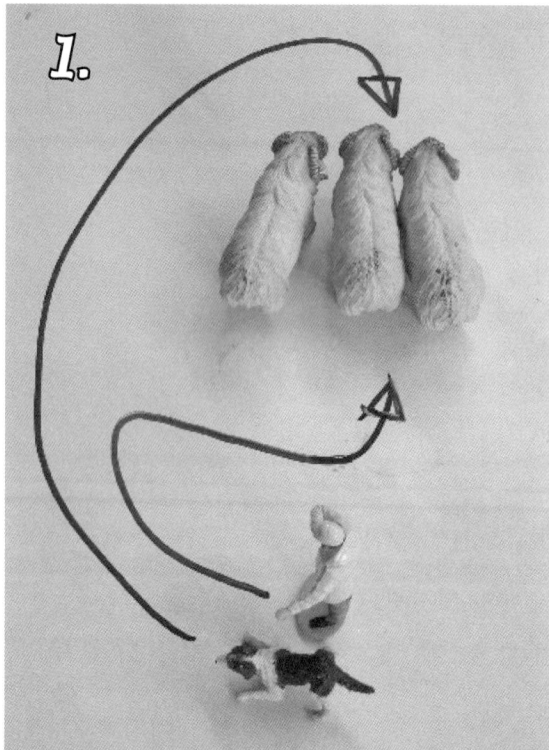

clearly counter productive.

What To Expect, And What To Do About It

Every pup will react differently the first time it's allowed to run sheep, and each must be treated as an individual and trained accordingly. However, we can consider a few broad categories of behaviour within which your youngster might fall:

Sticky

Some youngsters will immediately lie down and stare at the sheep, which is always something of an anti-climax, when you'd prepared yourself for all hell to be let loose! The main thing here is to get the pup excited and moving, so chase the sheep about a bit yourself, if possible running them straight at the pup,

so he's got to move. With a bit of luck, as the sheep break past him he'll turn and follow them. Once the hypnotic spell has been broken, encourage the pup to keep moving, possibly by calling him to you and running after the sheep together. Don't let him stop, and above all don't let the sheep get to the fence or you'll be back to square one, but worse, as there'll be nowhere for the sheep to run.

Natural

Occasionally a young dog will run quite naturally to the other side of the flock, without upsetting them too much, and perhaps even stop at 12 o'clock when he gets there! (The clock face analogy is often used in dog training, with the initial aim being to create a situation where the handler is at 6 o'clock, the sheep are in the centre, and the dog is at 12 o'clock). I admit it looks good, but don't be complacent – it is very easy to turn this style of dog into a sticky one. It's tempting to give a "lie down" command as soon as this type gets to the far side of the sheep, but really you should try to keep things moving at this stage. Back off the sheep, allowing them to move towards you, and hopefully the dog will move forward, effectively fetching the sheep to you.

Free Moving

This type of dog is very keen, and, when released, will probably cut straight across in front of you (despite your efforts to send him on a proper course) and run straight through the middle of the sheep. Utter chaos will reign for a while, as the dog chases first one portion of the flock and then another, gaily scattering them to the four winds! Don't despair! This is the sort that will ultimately make the best worker. As soon as he has let off a bit of steam, you can begin to try to bring things under control. Look out for an opportunity to block the sheep's headlong flight, by effectively placing yourself at the 6 o'clock position (when the dog is at 12 o'clock). You may only succeed in holding things for a few seconds, but make the most of that opportunity to praise and calm the dog. Just saying "steady…steady…" in the right tone of voice will suffice. Don't try to introduce any commands at this stage. A rolled up plastic sack is a handy thing to have, as you can bang it against your leg to attract the dog's attention, wave it in front of the sheep to slow them down, and if you end up hurling it in the dog's general direction, well, it won't do any harm! When the sheep do break past you, you'll find yourself between them and the dog, facing him, and ideally placed to block him as he tries to run after them. To get to the sheep he's now got to go around you, so hopefully he'll go right round the sheep, too (See picture 2). Use the rolled up sack to ward him off and send him out. Turn around to face the sheep and, with a bit of luck you'll once again find yourself in the 6 o'clock position with the dog at 12 o'clock.

By now you're probably beginning to wonder if you'll ever be able to catch the dog, let alone stop him! However, he'll eventually tire and you'll be able to take advantage of a moment of calm to ask him to lie down. Let the sheep drift past you, and when you're between them and the dog (with him still lying down – you'll need to keep repeating the command), crouch down and call him to you. Give him plenty of fuss. Now we come to something really important – having finally caught your dog (which you probably thought you'd never do) you must let him go again! That probably sounds a bit daft, but, if you don't, he'll very soon realise that allowing himself to be caught results in him being taken away from the sheep, and he'll become progressively harder to catch (making you progressively more reluctant to let him go!).

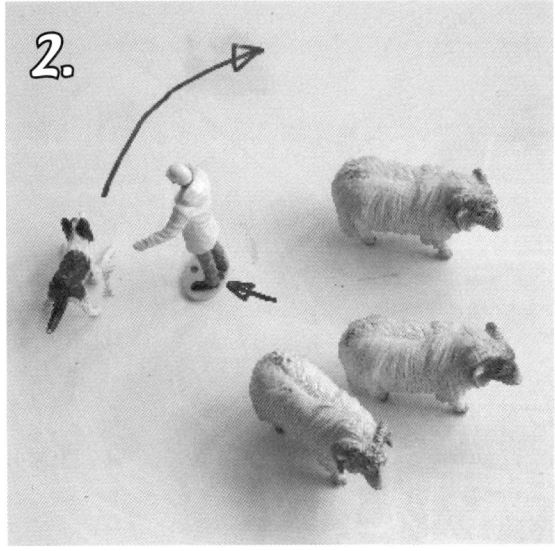

By letting him have a bit more fun on the sheep you're giving him a reward for coming willingly when called. Don't overdo it though, or it'll become a habit. Catch and release a couple of times, setting him up with the sheep properly each time, before putting him back on the lead and walking calmly from the field with the dog at heel.

A Couple Of Guidelines

1. Don't bother giving commands to a young dog unless you think there's a reasonable chance that he'll do as he's told. Shouting "lie down" at the top of your voice when the dog is tearing after the sheep two fields away will get you the canine equivalent of two fingers! Settle the situation, get the dog's attention, and then think about introducing the appropriate commands.
2. Learn to "read" the sheep, and anticipate their (and the dog's) actions. Put yourself in a position ready to deal with what is about to happen, not what has just happened.

Moving On

As I've explained, the initial aim is to get the dog to the 12 o'clock position, resulting in dog, sheep and handler all being in a line. This is called the point of balance, and is brought about by the dog's inherent instinct to get to the head of the flock and fetch the sheep to the pack leader (i.e., you). The exact position of the point of balance can vary slightly, according to the temperament of the dog, the nature of the breed of sheep, and the behaviour of individual sheep within the group, therefore it is something that you feel rather than see. Everything suddenly clicks into place, and momentarily the situation appears to be under some sort of control. You may even notice the dog pause slightly. That's the point at which you introduce a "lie down" command. The timing is critical – too early, and you're wasting your breath, too late, and you miss the moment. Once you've got the dog stopped at the point of balance, you're really making progress.

Fetching And Flanking

With the sheep in front of you, and the dog stopped at the point of balance, begin to walk backwards. This gives the sheep more room, so they move away from the dog, towards you. A quiet "get up" command should encourage the dog to walk on behind the sheep. This is the beginning of the fetch (picture 3). Just carry on going backwards down the field, using the commands "lie down", "get up" and "steady", keeping everything as calm as possible. Now, clearly it's not possible to continue walking backwards indefinitely – sooner or later you're going to come up against the fence, and fences, remember, are to be avoided at this stage. So, before you get to the end of the field you'll need to turn around and go the other way. If you

move around the sheep, the dog will also circle the sheep (picture 4), keeping himself opposite you and maintaining the balance, and you can set off back down the field. If at one end of the field you move clockwise around the sheep, and at the other you move anti-clockwise, you'll follow a figure of eight pattern, and that, really, forms the basis of all sheepdog training. The clockwise and anti-clockwise movements around the sheep are the flanking directions, so you can begin to introduce the appropriate

"come bye" and "away here" commands. Gradually you can make your journeys up and down the field a little more structured, perhaps by placing a couple of traffic cones at either end and aiming to walk the sheep through and around them. And, if you can get your little flock between two cones, it is but a small step to moving sheep from field to field through a gateway, which is probably what you started training the dog for in the first place!

Use Of A Round Pen

In order to speed up the training process a bit you can make use of a round pen to teach the flanking commands. Use 8 – 12 hurdles to create a circular enclosure, and put half-a-dozen (or more) free moving sheep inside. This in itself can be a bit tricky without a trained dog, but if you make the pen reasonably close to the side of the field you will be able to arrange some sort of funnel to run the sheep in. Ideally, though, the pen should have plenty of space all around it. Now, with yourself in the pen with the sheep, and the dog outside, it's a fairly simple matter to get him running around the perimeter as you move the sheep about inside. Introduce the appropriate flanking commands

according to the direction of travel. Most dogs will favour one direction or the other, so make sure you concentrate on the weaker flank.

This type of training is very intense, and if you overdo it you risk destroying the dog's natural sheep sense, so ensure that you intersperse these sessions with plenty of work on loose sheep.

FURTHER TRAINING

Up to this point, all the work you'll have been doing with the dog is what is known as "in-hand", in that you're remaining physically close to him, and helping him control the sheep. In fact, you're probably doing almost as much running about as he is! On many smallholdings, training need not be taken any further than this. Working together with the dog, in close partnership, you'll be able to walk your little flock between fields, and pen them up when required. Gradually, you'll find yourself trusting the dog to do more, so, rather than assisting him in bunching up the sheep prior to moving them from field to field, you'll simply open the gate ready to help guide them through, once he's gathered them together. Without even realising it, you'll have taken the training a step further. This is all very well, but is largely based on the fact that the dog knows the sheep, knows the lie of the land, and above all knows your routine. If you're called away to help a neighbour with his sheep (or if some of yours get out, and end up mixed with someone else's,) you may find that everything goes to pieces, as, in reality, the dog probably isn't obeying your commands as much as you fondly thought. On unfamiliar ground, with different sheep, he may be utterly confused, particularly if the fields are bigger than what he's used to! If your sheepdog is to be a truly versatile asset you'll need to structure his outfield training too. Ultimately, you'll be able to enter any field, anywhere, and be confident that he'll not let you down. There's no buzz quite like working with a dog that's half a mile away, yet still taking every tiny command, such that together you have joint mastery over an awkward packet of sheep. Better still, perhaps, is when the dog is working out of sight, yet pops up right on cue, over the brow of the hill, exactly where you knew he would, with the whole flock trotting on ahead. Call it telepathy, if you like.

The Outrun

Due to the small size of fields characteristic of many smallholdings, teaching a good outrun may be seen as a bit of a problem. Really, though, provided that you don't make the mistake of teaching the dog to follow the fence line, you should be able to achieve results every bit as good as anyone else.

In its simplest form, an outrun consists of sending the dog to fetch the sheep, rather than walking together until you're close to the flock, but there's a bit more to it than that! A young dog, lacking confidence, is likely to run more or less straight at the sheep if asked to run out too far too soon, and, as the sheep bolt in fright, he'll cut in even tighter, maybe gripping the sheep, or crossing over in front of the handler. Therefore, it's important to develop the outrun in small increments, and return to in-hand work if things aren't going particularly well. It's generally a case of "three steps forward, two steps back".

To my mind, the key to developing a nice natural outrun lies in teaching the dog to look for the sheep, and this of course goes back to that very first time you let the dog loose on the flock (see page 273). Hopefully, if you followed my guidelines then, your dog will automatically cast his eye around the field as soon he enters.

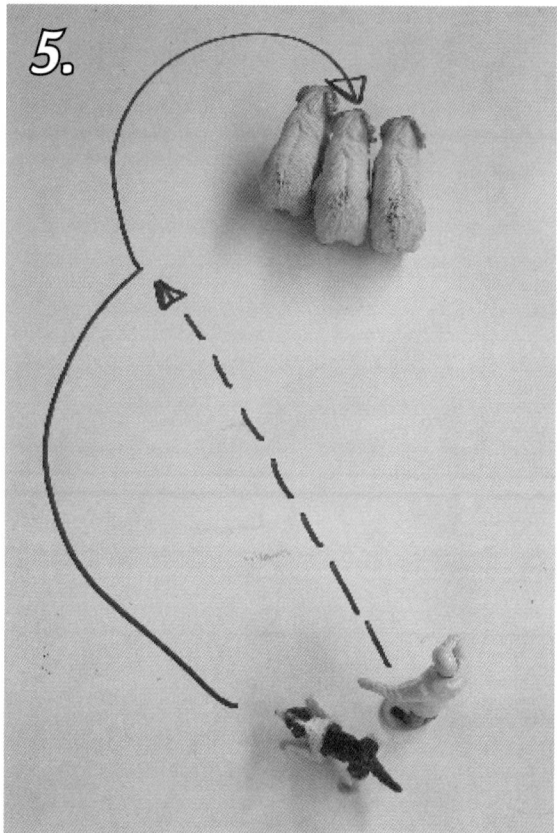

Thus, when he sets off on his outrun he does so with a purpose and a sense of direction, rather than blindly following a set route (or the boundary). What he mustn't do, however, is to run straight at the sheep, so we need to teach him to "cast out", to maintain an appropriate distance between himself and the sheep, and to get to the back of the flock without spooking them unduly. On the other hand, it's vital that the dog isn't forced out too far, or he'll lose contact with the flock. Having a dog that'll run around the perimeter of a 20 acre field isn't much help if you just want to gather 3 stray sheep from the middle! It's important, then, that the cast is in response to the position of the sheep, rather than the size of the field. But of course, the sheep may be distributed widely over the whole area, hence the importance of that first look around.

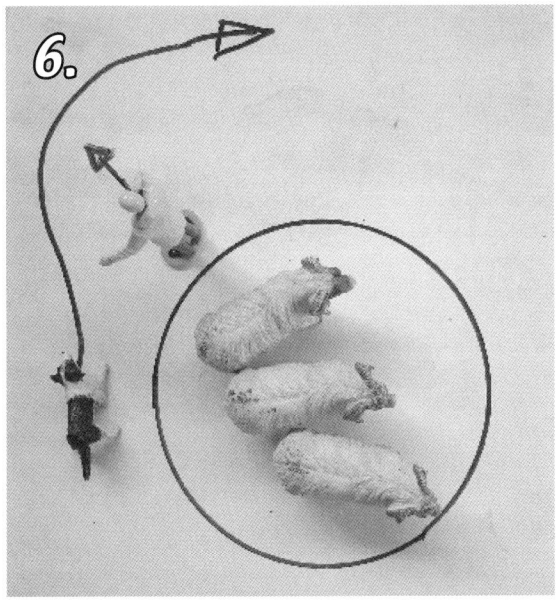

One favoured method of teaching the dog to cast out is by stopping him on his outrun, if he seems to be coming in too tight, then walking up to him and re- sending him at a more appropriate angle (picture 5). This is all very well where fairly tame sheep are being used, but if the sheep are livelier, and have moved away while the dog was stopped, it kind of defeats the object, as you're effectively having to start afresh each time. Also, if this procedure is repeated time and again the dog may get into a habit of slowing up (and perhaps stopping) on his outrun every time. My own preferred method is to teach the outrun in a sloping field, using the lie of the land to ensure that all the sheep are clearly visible to the dog. I find that the contours of the field tend to dictate the way in which the sheep move, making everything a bit more predictable, and, by watching the dog closely as he runs out (helped by the fact that everything is laid out in front of me on the slope) I can give an appropriate "go out" command before things get out of hand. I follow up the "go out" with the correct flanking whistle, given in a long, drawn out,

way. The slope tends to naturally throw the dog out wider anyway. I'll have already taught the "go out" command when using the round pen (page 277), by standing outside the ring of hurdles, between the dog and the sheep, and making him widen out as he begins to circle the pen (picture 6).

With practice and experience it should be possible to bring the dog in or send him out at any point in his outrun without actually stopping him. This can prevent a lot of wasted time and energy as, often, sheep will be visible to you yet out of sight of the dog (remember, he's much closer to the ground than you are!), and you'll be able to direct him to them by the most appropriate route.

Lift And Fetch

Lift

The lift refers to the point at which the dog reaches the end of his outrun, makes contact with the sheep, and begins to move them in the handler's direction. Basically, it's the point of balance that I mentioned earlier. In a strictly controlled situation, such as in training sessions or on the trials field, the dog is generally expected to stop on command at the 12 o'clock position without having put any pressure on the sheep. The appropriate distance between the dog and the sheep at the lift will be dictated by the type of sheep, and only experience can tell you what is correct.

On larger flocks it would be inappropriate for the dog to stop at that position, due to the sheep being more widely scattered. Once again, it's a case of adaptability, and only by gaining as much varied experience as possible will you and your dog learn to read the situation and respond accordingly.

TIP! There appears to be an invisible line drawn on the ground between the dog and the sheep. If each animal remains a respectable distance from the line, on its own side, then all will be well. With different breeds of sheep (and with different types of dog) the line may be drawn closer to the one party or the other. The position of the line also changes if there's young lambs involved. If a sheep crosses the line in the direction of the dog then it's liable to get bitten. On the other hand, if the dog crosses the line in the direction of the sheep, it'll get butted! The sooner you learn to spot that line, the better!

Fetch

You'll already have been teaching the fetch by walking backwards down the field with the dog bringing the sheep to you, but in that situation you were suitably positioned to slow the sheep down and keep everything under control. Now you're expecting the dog to fetch the sheep to you in an equally steady manner, without you being able to check their progress!

Provided that the sheep are moving in a fairly straight line, the fetch appears to be relatively uncomplicated, but it's all too easy to get into a habit of simply stopping and starting the dog all the way down the field as he follows the sheep. In this case, though, he's not actually in control of the sheep, and you'll be hard pressed to get them back on line once they begin to deviate. It's best if you can keep the dog on the move, and in contact with the sheep, the whole time. It's also highly unlikely that the sheep will come strolling along in a straight line, so you'll need regular flanking movements from the dog to maintain the desired direction. This is another area where there's a significant difference between the trials field technique and the more workaday big flock method, but again we're aiming for versatility – the trials field technique is just what's needed on the farm, when you're trying to extricate a couple of escapees from your neighbour's barley field – and it all comes back to the early in-hand training. On small groups of sheep we're looking for small, square, flanking movements from the dog, and on a big numbers he needs to be freer moving, and push in tight at the edges of the flock to keep them bunched up. This is where the idea of having shortened versions of each command

The fetch

Driving

Teaching the dog to drive is the antithesis of all that's gone before. Up 'til now we've depended wholly on the dog's natural herding instinct to bring the sheep to the handler, yet here we are, expecting him to push them away! This appears, to the dog, to be totally contrary to everything he's been taught so far. I think, for driving to be successful, there needs to be confidence and trust between dog and handler. As soon as the dog feels that you're no longer supporting him he'll nip around to the front and turn the sheep back.

I usually teach driving by returning to the business of walking backwards with the dog fetching the sheep to me, but this time we travel along next to a fence (fences shouldn't pose a problem by this stage). When we're going along nicely I move away from the fence, allowing the sheep to pass between me and it. I turn and walk next to the sheep, then gradually I hang back, until I'm alongside the dog, and, eventually, behind him (picture 8). This won't all be achieved in one session, as, initially, as soon as the dog realises I'm no longer in front of the sheep he'll try to get around them to maintain the point of balance. (To a certain extent he's prevented from doing so by the fence, which helps maintain a degree of control). Gradually, by degrees, I get to where I want to be. It's important not to overdo it, and it's also important that you don't always follow the fence in the same direction. Once the dog has accepted the idea of working the sheep with you somewhere other than in front of him, it's time to move away from the fence. With the fence no longer acting as a guide you're going to have to ask the dog to flank from side to side in order to keep the flock moving in the required direction, but the dog is, of course, itching to get around to the front – as soon as he hears a

comes into play, and both versions can be taught on the round pen or in-hand. For the small square movements I use the round pen, with myself on the outside between the sheep and the dog. I get the dog to flank only short distances, perhaps using each hurdle of the pen as a guide, and make sure I turn him off square each time I send him (picture 7). The command given is the shortened form (e.g., "'way", instead of "away here", or a shorter, softer, whistle). When teaching the sharper flanking movements required for big flock work I generally get the dog dashing around a loose bunch of sheep pretty fast, with plenty of rapid changes of direction, and not many stops. The full command is given, but in quite a staccato fashion, and rapidly repeated a couple of times.

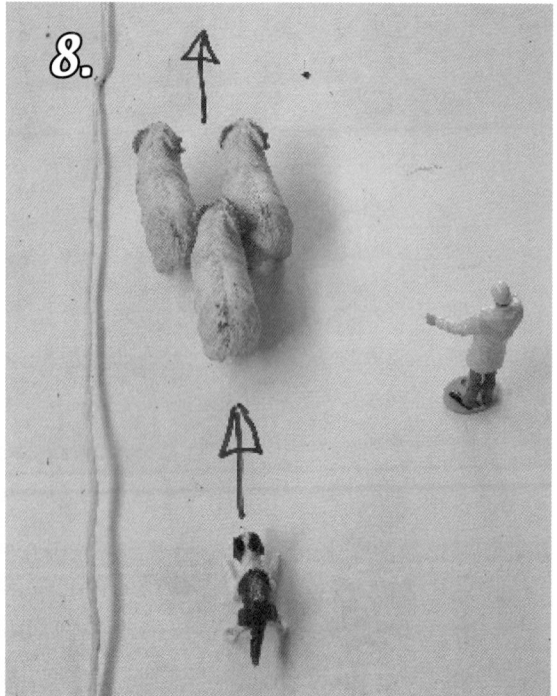

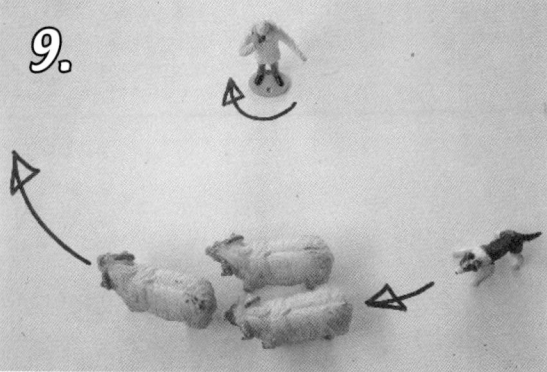

TIP! Even after you've reached the point where you consider your dog to be fully trained, it is well worth going right back to square one from time to time, and, over the course of about a week, running through the whole training program from start to finish.

flanking command he'll be off to the head of the flock! The way to avoid this is to get the dog driving the sheep around you in a circle. Provided you keep turning to face the action, he is, to all intents and purposes, driving the sheep across in front of you the whole time (picture 9). To begin with you'll be able to do this without flanking commands. His instinct will be sending him away from you, in an attempt to head the sheep, and you can use a recall command ("that'll do") to prevent him from doing so. The result will be a series of small square flanking movements, which is exactly what you want! You can re-introduce the proper short flanking commands when he's built up sufficient confidence to understand that the sheep aren't going to get away!

NB: Although I have consistently referred to the sheepdog as "he", there is no implied difference in ability between dogs and bitches. The same applies to handlers!

Tansy, my workmate for more than ten years

"There's a feeling you get at the very end of a busy lambing season, when the last ewe of the year begins to give birth - very glad, slightly worried, and a little sad. I feel exactly like that, having just finished writing this book". Tim Tyne

Appendices

STATUTORY REQUIREMENTS

Holding Number

Whether you keep a commercial flock or just one animal (of a farmed species) as a pet, you need to be registered with Defra. Apply to the Rural Payments Agency / Welsh Assembly for a County Parish Holding (CPH) number for the land where your livestock will be kept. You must register your holding within one month of the time you first keep animals there.

Flock Number

Contact your local Animal Health Divisional Office (AHDO) to obtain a unique flock number, usually referred to as a "UK" number. Currently this is six digits preceded by the letters UK, but from January 2010 an extra digit (0) will be inserted before the current six. The UK number must appear on your official Defra compliant eartags, together with an individual number for each animal (also issued by Defra).

Movement Licences

Animals may be moved onto / off holdings under a general licence, which forms part of the Disease Control Orders (2003). Details of the general licence (including current standstill rules) are obtainable from your AHDO. All animals being moved must be accompanied by the relevant movement document (AML 1 for sheep). Blank forms are available from local authorities (Trading Standards Department). One copy (the white one) of the completed form must be sent to Trading Standards within 3 days of the movement. The yellow copy remains on the premises of departure, and the

pink one is kept at the destination. Sometimes there is also a blue copy, which is for the haulier.

All movements must be recorded in the flock register (an example page of which is included on page 288). Sheep movement records must be retained for 6 years.

A brief summary of the current rules for the identification of sheep (and goats), and new rules due to come into force on 1st January, 2010

• The current "double tagging" requirement applies to all animals born after 12th January 2008, but before 1st January 2010.
• If you do not have a holding number (CPH) or a flock number then contact your nearest Animal Health Divisional Office. You must register your holding within one month of the time you first keep animals there, even if you only keep a few as pets.
• UK tags, printed with flock number and an individual identifier, must be applied at the holding of birth.
Individual identification numbers will be issued by Defra.
• All breeding animals initially require two identical tags (primary and secondary).
A tattoo may be used in place of the secondary tag.
• Tags may be any colour other than red. Red is for replacement ("R") tags. R tags are used in the case of sheep that lose one of their UK tags after leaving their holding of birth. An alternative would be to obtain an identical tag to the one that was lost.
• A tag which is lost while an animal is still on its holding of birth can be replaced with another UK tag bearing the same flock mark,

but not necessarily the same individual number.
- UK tags must be applied:
- within six months of birth, in the case of animals that are housed day and night for most of the year,
- within nine months of birth, where animals are generally kept outdoors,
- before an animal leaves its holding of birth, (whichever is sooner).
• Animals intended for slaughter at less than twelve months of age will only require a single UK tag.
• Animals intended for export must have two identical UK tags applied, regardless of their age.
• For emergency treatment only, an animal may be taken to a vet without official identification. Animals being taken to shows must have both a primary and a secondary means of identification, and the full number must be recorded on the movement licence and in your flock register.

Electronic Identification (E.I.D) – new EU rules as from 1st January 2010

(Please note that the UK sheep industry is strongly opposed to compulsory E.I.D., and is constantly lobbying for amendments to the legislation, or even to overturn it altogether. The situation is therefore subject to change, so do check out the current state of affairs before purchasing tags and equipment).

• Sheep born after 31st December 2009 that are not intended for slaughter at less than 12 months old must have an electronic form of I.D. This could be either an eartag or a bolus. A secondary form of I.D will also be required. This could be either a tag or tattoo. The visually readable number must match the E.I.D number. Tattoos must not be applied to

animals intended for export.
• The existing 6 digit UK flock number will be preceded by a zero.
• Individual numbers will be issued by Defra, and will start from 00001, regardless of what numbers you've used in previous years.
• EID tags must be yellow (except in Scotland). Replacement tags will be red, and need not have the same individual number as the lost tag that they replace.
• If an animal that is double tagged loses one of its tags it will be acceptable to remove the other tag and replace with a new pair of tags, provided that the change is cross referenced in the flock register.
• Animals intended for slaughter at under 12 months old will not require E.I.D. However, as it will not be possible to retrospectively apply electronic tags to sheep that have left their holding of birth, this poses some problems for store lamb buyers and finishers, particularly those that currently retain some female store lambs to sell as breeding ewes.
• All sheep born or first identified after 31st December 2009 must be recorded individually in the holding flock register.
Individual animal identities will need to be recorded on movement licences from 1st January 2011.
• Sheep passing through "critical control points", such as markets and abattoirs, can have tags electronically read at that point, rather than before leaving the holding. Farmers will then be provided with print-out lists for their records.

Animal Transport Regulations (as from 5th January 2007)

• The regulation doesn't apply to the transport of animals not in connection with an economic activity, or transport to or from veterinary practices or clinics.

• Farmers transporting their own animals in their own transport up to 50 km from their holding have limited exemption from parts of the regulation, but must still comply with the general requirements.

• Anyone transporting sheep on journeys over 65 km (but under 8 hours) will need to hold a Type 1 Transporter's Authorisation.

• Journeys over 8 hours require a Type 2 Transporter's Authorisation. The vehicle will need to be inspected and approved.

• Sheep in the last 15 days of gestation cannot be transported unless it is for the purposes of improving the conditions of birth.

• Sheep that have given birth within the last 7 days cannot be transported or presented for sale.

• If ewes "in milk" are transported without their offspring, they must be milked at intervals not exceeding 12 hours.

• Maximum permitted loading ramp angle is 50% (e.g. a rise of 4′ over a distance of 8′).

• Sheep must be transported separately from other species.*

It Is Also Necessary To Separate

• Sheep of significantly different ages or sizes;*
• Sexually mature males from females;
• Sheep with horns from those without;*
• Sheep that are hostile to one another;
• Tied from untied sheep. (i.e. if sheep are haltered and tied for transporting to a show).
* Except where separation would cause distress, for example sheep with dependant lambs, or where animals have been raised in compatible

groups and are accustomed to one another.

• During journeys of more than 8 hours the temperature within the vehicle must not fall below 0°C.

• Lambs under 7 days old must not be transported more than 100km. Any young lamb whose navel is not yet fully dried is not fit to travel.

• Lambs weighing less than 20kg must be provided with a warm, comfortable, bedding material.

Minimum Space Allowances Within Vehicles Are As Follows:

Category	Approximate Weight (Kg)	Area (m²/animal)
Shorn sheep & lambs 26kg and over	<55	0.20 - 0.30
	>55	>0.30
Unshorn sheep	<55	0.30 -0.40
	>55	>0.40
Heavily pregnant ewes	<55	0.40 - 0.50
	>55	>0.50
Source: Defra		

Cleansing And Disinfection

Vehicles used to transport livestock should be cleansed and disinfected at the destination premises, unless the vehicle is returning directly to the place of departure, in which case it must be washed out within 24 hours or before it's next used to carry livestock, whichever is the sooner. At markets and abattoirs you may be required to sign a declaration undertaking to carry out cleansing and disinfection within the required timescale.

Fallen Stock Disposal

Fallen stock – livestock that dies on the farm due to natural causes, or is killed on farm other than for human consumption – is classified as an animal by-product (ABP). Carcasses must not be buried or burnt in the open other than in exceptional circumstances, such as during an outbreak of a notifiable disease. Fallen stock must be taken to, or collected by, an approved knacker, hunt kennel, incinerator or renderer, either by private arrangement or under the National Fallen Stock Scheme.

Medicine Storage

Medicines (and other products such as sheep dips and pour-ons) must be stored in accordance with the label instructions, in a secure lockable store. This could be a container, cupboard, room or separate building. A cupboard within a locked building would be acceptable. Drains must be protected from pollution in the event of leakage. Vaccines that need to be refrigerated should not be stored in fridges used for food. Medicine usage must be recorded, and the record retained for 5 years.

Feed Storage

Feed must be stored separately from any chemicals and other substances prohibited for use in animal feeds. Storage areas and containers must be kept clean and dry, and appropriate pest control measures implemented where necessary. Regular cleansing of feed bins etc. should be carried out, in order to prevent cross-contamination. Medicated and non-medicated feed must be stored correctly to reduce the risk of feeding to non-target animals. Ruminant (sheep, cattle, goats) and non-ruminant (pigs, poultry, etc.) feeds must be stored separately, preferably in different buildings, but, failing that, in different bays or bins. Pet foods should not be stored with other feeds.

Waste Plastic Disposal

It is no longer permissible to burn waste plastic, including feed sacks, silage wrap, baler twine and pesticide containers. Nor is it permissible to place agricultural waste in the bin with domestic rubbish. Waste plastics may only be stored on farm for a limited period before being taken to an approved recycling site or being collected by an approved contractor.

ANNUAL FLOCK RECORD

FLOCK RECORD FOR YEAR BEGINNING JAN 20____

Holding No _____   UK Flock no _____

Tick Here if Continuing from previous page _____

Date	Bought/sold, born/died, lost/found, transferred	Number of Sheep	Identification	Comments (including holding number of buyer/seller)	Ewes	Ewe Lambs retained for breeding	breeding rams	Lambs, wethers, etc	TOTAL
1st Jan	Opening Numbers (must be entered before 31st Jan)								
Examples									
01/04/07	died	1	6345	Lamb died on field - pneumonia?	18	10	1	32	61
05/05/07	transferred	10	534.6-5355	yearlings transferred to breeding stock	28	0	1	32	61
09/05/07	sold	8	UK S tags	cull ewes sold to John Jones 57/001/8000	20	0	1	32	53

RUNNING TOTALS

CPH NO:

LIVESTOCK MEDICINE ADMINISTRATION RECORD								Date withdrawl period ends			
Date of Medicine Purchase	Quantity Purchased	Name of Medicine, Batch Numbers and Where purchased	ID of animal or group treated	Numbe-Treated	Treatment Start Date	Treatment Finish Date	Total Quantity of medicine	Meat	Milk	Other	Medicine administered by

LAMBING DATE CALCULATOR (based on a 145 day gestation length), with additional information for A.I and embryo transfer work. Information kindly supplied by Innovis Ltd.

Synchronise Ewes Insert Sponges or use Teasers	Remove Sponges	Remove Teasers AI or Natural Service	Embryo Flush & Transfer	Recommended Scanning Date	Expected Lambing Date
5th September	17th September	19th September	24th September	18th December	11th February
10th September	22nd September	24th September	29th September	23rd December	16th February
15th September	27th September	29th September	4th October	28th December	21st February
20th September	2nd October	4th October	9th October	2nd January	26th February
25th September	7th October	9th October	16th October	7th January	3rd March
30th September	12th October	14th October	19th October	12th January	8th March
5th October	17th October	19th October	24th October	17th January	13th March
10th October	22nd October	24th October	29th October	22nd January	18th March
15th October	27th October	29th October	3rd November	27th January	23rd March
20th October	1st November	3rd November	8th November	1st February	28th March
25th October	6th November	8th November	13th November	6th February	2nd April
30th October	11th November	13th November	18th November	11th February	7th April
4th November	16th November	18th November	23rd November	16th February	12th April
9th November	21st November	23rd November	28th November	21st February	17th April
14th November	26th November	28th November	3rd December	26th February	22nd April
19th November	1st December	3rd December	8th December	3rd March	27th April
24th November	6th December	8th December	13th December	8th March	2nd May
29th November	11th December	13th December	18th December	13th March	7th May
4th December	16th December	18th December	23rd December	18th March	12th May
9th December	21st December	23rd December	28th December	23rd March	17th May
14th December	26th December	28th December	2nd January	28th March	22nd May
19th December	31st December	2nd January	7th January	2nd April	27th May
24th December	5th January	7th January	12th January	7th April	1st June
29th December	10th January	12th January	17th January	12th April	6th June
3rd January	15th January	17th January	22nd January	17th April	11th June
8th January	20th January	22n January	27th January	22nd April	16th June
13th January	25th January	27th January	1st February	27th April	21st June
18th January	30th January	1st February	6th February	2nd May	26th June
23rd January	4th February	6th February	11th February	7th May	1st July
28th January	9th February	11th February	16th February	12th May	6th July
2nd February	14th February	16th February	21st February	17th May	11th July
7th February	19th February	21st February	26th February	22nd May	16th July
12th February	24th February	26th February	3rd March	27th May	21st July
17th February	1st March	3rd March	8th March	1st June	26th July

22nd February	6th March	8th March	13th March	6th June	31st July
27th February	11th March	13th March	18th March	11th June	5th August
4th March	16th March	18th March	23rd March	16th June	10th August
9th March	21st March	23rd March	28th March	21st June	15th August
14th March	26th March	28th March	2nd April	26th June	20th August
19th March	31st March	2nd April	7th April	1st July	25th August
24th March	5th April	7th April	12th April	6th July	30th August
29th March	10th April	12th April	17th April	11th July	4th September
3rd April	15th April	17th April	22nd April	16th July	9th September
8th April	20th April	22nd April	27th April	21st July	14th September
13th April	25th April	27th April	2nd May	26th July	19th September
18th April	30th April	2nd May	7th May	31st July	24th September
23rd April	5th May	7th May	12th May	5th August	29th September
28th April	10th May	12th May	17th May	10th August	4th October
3rd May	15th May	17th May	22nd May	15th August	9th October
8th May	20th May	22nd May	27th May	20th August	14th October
13th May	25th May	27th May	1st June	25th August	19th October
18th May	30th May	1st June	6th June	30th August	24th October
23rd May	4th June	6th June	11th June	4th September	29th October
28th May	9th June	11th June	16th June	9th September	3rd November
2nd June	14th June	16th June	21st June	14th September	8th November
7th June	19th June	21st June	26th June	19th September	13th November
12th June	24th June	26th June	1st July	24th September	18th November
17th June	29th June	1st July	6th July	29th September	23rd November
22nd June	4th July	6th July	11th July	4th October	28th November
27th June	9th July	11th July	16th July	9th October	3rd December
2nd July	14th July	16th July	21st July	14th October	8th December
7th July	19th July	21st July	26th July	19th October	13th December
12th July	24th July	26th July	31st July	24th October	18th December
17th July	29th July	31st July	5th August	29th October	23rd December
22nd July	3rd August	5th August	10th August	3rd November	28th December
27th July	8th August	10th August	15th August	8th November	2nd January
1st August	13th August	15th August	20th August	13th November	7th January
6th August	18th August	20th August	25th August	18th November	12th January
11th August	23rd August	25th August	30th August	23rd November	17th January
16th August	28th August	30th August	4th September	28th November	22nd January
21st August	2nd September	4th September	9th September	3rd December	27th January
26th August	7th September	9th September	14th September	8th December	1st February
31st August	12th September	14th September	19th September	13th December	6th February

CONVERSION TABLES

LENGTH

inches - cm	x 2.54	cm - inches	x 0.394
inches - mm	x 25.4	mm - inches	x 0.0394
feet - metres	x 0.305	metres - feet	x 3.29
yards - metres	x 0.914	metres - yards	x 1.09
chains - metres	x 20.12	metres - chains	x 0.0497
miles - kilometres	x 1.61	kilometres - miles	x 0.621

AREA

Sq feet - m²	x 0.093	m² - sq feet	x 10.8
Sq yards - m²	x 0.836	m² - sq yards	x 1.20
acres - hectares	x 0.405	hectares - acres	x 2.47

VOLUME

Cubic feet - m³	x 0.0283	m³ - cubic feet	x 35.31
cubic yard - m³	x 0.7646	m³ - cubic yard	x 1.308
pints - litres	x 0.568	litres - pints	x 1.76
gallons - litres	x 4.55	litres - gallons	x 0.22
gallon - m³	x 0.0045	m³ - gallons	x 219.97
fluid ounces - ml	x 28.41	ml - fluid ounces	x 0.0352

WEIGHT

ounces - grams	x 28.3	grams - ounces	x 0.0353
pounds - grams	x 454	grams - pounds	x 0.0022
pounds - kilograms	x 0.454	kilograms - pounds	x 2.2
cwt - kilograms	x 50.8	kilograms - cwt	x 0.020
cwt - tonnes	x 0.0508	tonnes - tons	0.984
tons - kilograms	x 1016		
tons - tonnes	x 1.016		
cwt = hundredweight	112 lbs		

TEMPERATURE

Fahrenheit - Centigrade	(°F-32) x 0.556	Centigrade - Fahrenheit	(°C x 1.8) + 32

RATE OF USE

pounds/acre - kg/hectare	x 1.121	kg/hectare - pounds/acre	x 0.8922
cwt/acre - kg/hectare	x 125.5	kg/hectare - cwt/acre	x 0.000797
ton/acres-kg/hectare	x 2511	kg/hectare - ton/acre	x 0.000398
pounds/gallon - g/litre	x 99.78	g/litre- pounds/gallon	x 0.01
gallons/acre - litres/hectare	x 11.23	litres/hectare - gallons/acre	x 0.08902
fert.units/acre - kg/hectare	x 1.25	kg/hectare-fert.units/acre	x 0.8

NOTIFIABLE DISEASES

Diseases are described as "notifiable" when you are legally required to notify the authorities if you suspect that one may be present in your flock. These are listed under section 88 of the Animal Health Act 1981. There are a number of reasons why a disease may be classified as notifiable. Some, such as foot and mouth disease, would, if left unchecked, cause serious economic loss to the whole livestock industry. Others, such as anthrax, represent a threat to human health, or a perceived threat, as in the case of scrapie. Many of these diseases would have an immediate adverse effect on the UK's export status. From time to time diseases are removed from the list (sheep scab), and new ones are added (bluetongue). The way in which outbreaks are controlled will depend upon epidemiological studies of previous outbreaks, together with an assessment of risk and the likelihood of spread. Usually, the first line of action is to restrict livestock movements, and to set up control and surveillance zones around the suspected outbreak. In many cases compulsory slaughter of all animals on the infected farm is carried out, and occasionally there will be anticipatory culling of contiguous contacts. Vaccination may be carried out on all susceptible livestock in the surrounding area, in order to create a protective barrier that will limit the spread of the disease. However, when blood testing animals to determine the presence of disease, it can be difficult in some cases to determine between healthy, vaccinated individuals and those that have been challenged by the disease itself, as both will carry antibodies. Therefore, vaccination is not generally the preferred option in the UK, except in the case of bluetongue, which, being vector born, is more difficult to contain within a specific area.

Most outbreaks are traced to imported animals or animal products, so tight controls at docks and airports, together with pre-movement testing and quarantine of live animals, are necessary to prevent future epidemics. Routine bio-security measures on farms, at markets and at shows all play an important part in limiting the potential spread of many animal diseases, not just the notifiable ones.

The official UK list of notifiable diseases contains many that have not (as yet) occurred in this country, or have not been seen in the British Isles for a very long time. Here, I'll outline just a few which are currently (or have recently been) present:

Bluetongue

Bluetonge is the most recent exotic disease to rear it's head in Britain, although this has not been entirely unexpected – as climate has changed in recent years, we've seen bluetongue extend its range from North Africa, gradually spreading across Northern Europe. The disease is carried by a species of midge (Culicoides), and, in 2007 it appeared that infected midges had finally made it into the UK, having been blown across from Holland. Further concern was caused by the discovery of infection in imported animals. Thankfully, being late in the year, spread of the disease was curbed by the onset of winter. Due to the method of transmission, traditional control methods (i.e., movement restrictions and slaughter) are of limited value, so, having witnessed the decimation and economic loss within the sheep flocks of mainland Europe, the UK government decided on a policy of vaccination. This has also been the policy in other affected EU countries, where it has been very effective. Here, however, uptake of the voluntary scheme has been rather poor, but conscientious shepherds in the South and East who have vaccinated their stock are effectively safeguarding the rest of the country. The cited reason for the poor uptake of vaccine elsewhere in England and Wales has been concerns over possible side effects, particularly if administered during the breeding season or pregnancy, but these fears have proved unfounded.

The Scottish government has made bluetongue vaccination mandatory. The message to the rest of the UK is: "Don't hesitate; Vaccinate!"

Symptoms

Oedema (swelling) of the face, sometimes extending to the whole head and neck. In severe cases, blood supply to parts of the face may be affected, resulting in areas turning blue. Salivation and nasal discharge. Difficult breathing. Fever. Depression. Lameness. Ulcers in the mouth. Reddening of the mucous membranes. Death.

Ewes infected during pregnancy will abort.

Foot & Mouth Disease

Few of us who were around at the time will ever forget the foot and mouth epidemic of 2001. I was engaged on my third lambing contract of the year when the disease broke out. The second had been on Anglesey, and, looking across the Menai Straits, I could see the smoke of the pyres darkening the sky. All that hard work up in flames! (As it happened, the flock I had worked with escaped the contiguous cull, by the width of a road!) The risk zone around the Anglesey outbreaks was extended to include part of mainland Gwynedd, and driving to work each morning and passing a sign informing me that I was entering an infected area soon became too much to bear.

I 'phoned in to say that they would have to manage without me, and like everyone else we withdrew into our own farm and shut the gate, preparing to sit it out. In all, over 7 million animals were culled in the attempt to control the disease, and the battle, which raged for a year, cost the industry and the tax payer some £8 billion. Rural life ground to a halt, with numerous ancillary businesses forced to close. Many families' lives were changed forever. You can well imagine then, that the whole country cried out in disbelief when, in 2007, the disease was once again confirmed, having apparently escaped from a government laboratory! Thankfully, lessons had been learned during the 2001 outbreak, and this time the disease was quickly brought under control, with minimum disturbance to the rest of the country.

Symptoms

The symptoms of FMD in sheep are often mild, allowing the disease to go undetected in extensive flocks. However, infected sheep will be shedding the virus from ruptured pustules, and, once cattle become infected, the disease is more likely to be identified, as the symptoms are expressed more severely.

Generally, in sheep, the whole affected flock would appear lacklustre, with quite a large proportion being lame and reluctant to move. Where lameness is particularly severe, sheep will adopt a characteristic crouching posture. In-lamb ewes may abort. Closer examination of the lame animals will reveal the presence of blisters around the top of the hoof, and also between the cleats and on the soft heel. There'll be heat in the feet, and the outer horn might have begun to separate from the tissue beneath. Small ulcers may be seen in the mouth, on the gums and tongue, but often these do not occur in sheep.

Incidentally, there are numerous non-specific

causes of similar lesions in sheep's mouths, and, during the 2001 FMD outbreak, these were given the title Ovine Mouth and Gum Other Disorders, or "OMAGOD", which was probably pretty much the first thing said by anyone who came across anything vaguely resembling a blister in a sheep's mouth!

Scrapie

Scrapie is a progressive degenerative nervous disorder of sheep, very similar to BSE (mad cow disease) in cattle. Both are transmissible spongiform encephalopathies (TSEs), though just how they are transmissible remains something of a mystery! The possibility that there may be links between scrapie, BSE and nvCJD in humans prompted a whole raft of legislation and control measures, though subsequently it was shown that scrapie did not present a human health issue. Susceptibility to scrapie appears to be hereditary, so it has been possible, by a national genotyping scheme (NSP), to identify those strains of sheep with most resistance to the disease. As a result of the scheme, incidence of clinical scrapie has fallen to a very low level, but, as it is now considered that there was no threat to human health, financial support for the project has ended. However, scrapie remains a significant sheep health issue, with some breeds appearing to be more susceptible than others. It is to be hoped that breeders will continue to carry out genotyping, despite the lack of government funding. Simple test kits are available (from Innovis) that enable farmers to collect the necessary blood samples themselves, before sending them away for analysis.

Symptoms

Affected sheep appear to be very itchy, and will, quite literally, rub, scratch and scrape

themselves to death! Whilst engaged in scratching they often make compulsive licking and nibbling actions with the head raised. They may also show signs of incoordination, such as an unusual walking action or general wobbliness. The disease only becomes apparent in older animals, as it has a long incubation period.

Anthrax

Anthrax is zoonotic, i.e., it can be transmitted to humans, where it causes a potentially fatal illness. Thankfully human infection is rare nowadays, and we have an armoury of antibiotics with which to treat the disease, but it used to appear from time to time in wool depots, hence its other name – Wool Sorters disease.

When exposed to air, for example by bleeding from infected animals and carcasses, the anthrax bacteria forms spores, which are capable of remaining viable in the environment for a very long time – maybe 60 years or more. It was inhalation of these spores that led to the disease occurring in the dusty atmosphere of the wool grading sheds. At one time, wool shoddy was used as a mulch in hop gardens, so, although hops are no longer grown in the UK on the scale that they once were, the former hop gardens remain a potential source of infection to grazing animals, particularly after localised flooding. The same can be said for land that has, in the past, been dressed with bonemeal fertiliser, or in the vicinity of sites where casualty animals have been buried.

Symptoms (in animals)

Generally characterised by sudden death. The carcass will usually have unclotted blood exuding from the mouth, nose and back end. Occasionally animals will be ill for a day or two before death, in which case they'll run a very high temperature, show signs of severe internal discomfort, and pass diarrhoea consisting almost entirely of blood. Death is preceded by convulsions.

GLOSSARY & ABBREVIATIONS

A.I – Artificial insemination.

Aftermath – Lush growth of grass after taking a crop of hay or silage. Often used for flushing ewes or fattening lambs.

Away – Universally recognized command used to send a sheepdog in an anti-clockwise direction around the flock. (Sometimes Away-to-me or Away-here).

B.S.E – Bovine Spongiform Encephalopathy. (Mad cow disease).

Back – Sheep that have lost condition are said to have "gone back".

Bag – A ewe's udder. (See also Purse).

Barrener – A ewe that is not in-lamb when she should be, ie, she is barren.

Batt – Rectangular mat of carded wool.

Bloom – Yellowish colour in the fleece of sheep in particularly good health and condition. May be artificially applied to show sheep in the form of bloom dips or sprays.

Bone – Thickness of the legs, as in horses. A sheep with thin, spindly legs is said to be "fine boned".

Britch – Coarse wool from on and around the tail, and from the hind legs.

Broker – An old sheep with loose or missing incisors. Broken mouthed.

Brokes – The foreleg area of a fleece.

C.A.P – Common Agricultural Policy.

C.C.N – Cerebro-cortical necrosis. The only disease of sheep that does not have a colloquial name.

C.O.D.D – Contagious Ovine Digital Dermatitis.

Cade – A bottle reared lamb.

Cake – Concentrate feed, generally in pellet form. Also called nuts.

Cast – A sheep that has become stuck on its back. Also a grade of wool.

Cleats – The two main digits of a cloven hoof.

Cleek – See Crook.

Clarts – See Daggings.

Colostrum – The first milk, rich in antibodies. Essential for the new born lamb.

Come-bye – Universally recognized command used to send a sheepdog in a clockwise direction around the flock.

Condition score – A universal method of assessing body condition, on a scale of 0 – 5.

Count – The Bradford Count is the UK measure of wool quality, based on potential spinning capacity. The figure refers to the number of hanks (560 yards) that could be spun from one pound of wool. Thus, the higher the number, the finer the fleece.

Couples – Ewes with lambs at foot. Single couples or double couples, depending on the number of lambs.

Creep – An area accessible only to lambs, where they can be given extra feed away from the ewes. The type of feed used is also called creep.

Crook – The traditional shepherds stick. Decorated, horn topped versions are generally kept for special occasions such as shows, sales and sheepdog trials. Everyday, working models tend to be metal or (nowadays) fibreglass. Used to catch a sheep by the neck or hind leg. The narrower leg crook is called a cleek in some areas.

Crutching – Clipping wool from the tail and around the back end.

Cull – To remove an animal from the flock that is no longer required for breeding, usually due to age or unsoundness.

Daggings, Dags – Lumps of dung adhering to the wool around the back end of the sheep. The task of removing these lumps is called dagging.

Dam – Mother.

Diamond – Wool from the back of the sheep.

Do – Sheep that are thriving are said to be doing well. Sheep will do better on some types of land than others. A sheep that always looks thinner than her flock mates is a poor doer.

Dock – The base of a sheep's tail, where it joins the body. A docked sheep is one that has had its tail cut off short. Docking also means the same as crutching in some areas.

Draft – A ewe (usually 4 years old) sold out of a hill flock for further breeding on a lowland holding; To separate off a sector of the flock using a drafting gate (see also Shed). Also used to describe the action of drawing fibres from the end of a rolag when spinning.

Draw – The characteristic of sheep to follow one another.

Drench – Medication given as liquid by mouth.

E.B.V – Estimated Breeding Value. Individual animal's EBVs for various traits are calculated from performance recording data.

E.I.D – Electronic identification. Compulsory on all breeding sheep born after 31st December 2009.

E.T – Embryo transfer.

Entropion – Turned in eyelids. A hereditary condition.

Ewe – Adult female sheep.

Extra Diamond – Wool from the sides of the sheep.

F.Y.M – Farm yard manure.

Finish – The final degree of fleshing on a fattening lamb that may make all the difference between a quality carcass and a mediocre one.

Flyer – A ewe which, by shearing time, has lost the wool from her belly, throat and crutch, meaning that she can be shorn very fast. Also the part of a spinning wheel that winds yarn onto the bobbin.

Flushing – Moving ewes to better pasture a few weeks before tupping, in order to increase ovulation and conception rates.

Fold – A temporary field, often erected using electric fence (formerly hurdles), which is moved regularly onto fresh grazing. In some areas, sheep are folded over root crops in winter.

Full mouth – A sheep that is full mouthed has all of its adult incisors, and it is no longer possible to determine age by dentition. (4 years old and onwards).

Garget – Mastitis.

Gear – Shearing equipment.

Grade – The grader at a fatstock market checks lambs on arrival to ensure that they reach a minimum standard of quality and finish. Lambs sold on a deadweight basis are graded on the hook.

Gimmer – Young female sheep.

Halfbred – A crossbred breeding ewe sired by a Border Leicester ram.

Heavy – Sheep that are slow, stubborn, and won't move easily away from a sheepdog. Lowland breeds in particular.

Heft – The area of a farm (usually upland) to which a particular sheep is tied by hereditary instinct. On unfenced mountain farms the whole flock is hefted to the hill, so won't stray over the boundary.

Hogget – A sheep from one year old until first shearing. Males and females may be described as Tup Hoggs and Ewe Hoggs respectively.

Hurdle – Small, lightweight gate, used to erect temporary handling systems or individual lambing pens.

In-bye – Sheltered fields close to the farm buildings, often used at lambing time.

In-lamb – Pregnant.

Keel – See Raddle.

Kemp – Short, thick, hairy fibres in a fleece. Usually white, but often red (in Welsh sheep) and occasionally black. Conveys hardiness in hill breeds, but de-values the fleece as kemp cannot be dyed.

Lazy Kate – Holds 2 or more bobbins of spun yarn for plying.

License – Since the Foot & Mouth outbreak in 2001, it has been necessary to report all sheep movements to Trading Standards using form AML 1, generally called a license.

Light – Flighty sheep requiring very careful handling with a sheepdog. Hill breeds in particular.

Masham – A crossbred breeding ewe sired by a Teeswater or Wensleydale ram.

Moiety – Fleece contaminated with vegetable matter (hay seeds, etc).

Moiled – See Polled.

Mordant – A substance used in dyeing in order to fix the colour, otherwise it would simply wash out.

Mule – A crossbred breeding ewe sired by a Blue-faced Leicester ram.

nvCJD – Human new variant Creutzfeldt-Jakob disease. At one time thought to be linked to scrapie in sheep, via BSE in cattle.

N.S.P – National Scrapie Plan.

Niddy Noddy – Used for winding skeins of wool.

Open coated – A sheep with a loose, open fleece, often with a parting down the spine. Not very weather proof

Picklock – Belly wool.

Pitch – Sheep marking fluid. Formerly tar or pitch was used, and the name has stuck.

Pizzle – Penis.

Polled – Naturally hornless.

Prime Britch – Wool from the sheep's haunches.

Purse – Ram's scrotum. The common wild flower, Shepherd's Purse, earns its name from the shape of its seed pods. Rather confusingly, purse is also used to mean udder in some areas.

Quarter – Half of the ewe's udder.

Race – A narrow passage (part of a handling system) down which sheep will draw, and in which they can be held for dosing, vaccinating, etc.

Raddle – Coloured crayon or paste on a ram's chest, in order to show which ewes have been mated. Also called keel in some areas.

Ram – Adult male sheep.

Rejects – Lambs presented at a fatstock market which fail to grade. These may be sold as stores or taken home for further fattening.

Rig – A ram lamb with one testicle undescended, or which has been improperly castrated.

Rolag – A piece of fleece prepared for spinning.

Rooed – Plucked. Some primitive breeds of sheep are rooed rather than shorn.

S.R.M – Specified Risk Material. Certain parts of a carcass that must be removed and disposed of at the point of slaughter, in line with T.S.E regulations.

Sadden – Modifying the tone of a natural dye using iron (ferrous sulphate).

Scurr – Small growth of horn on an animal that should be polled. Often found on ewes of breeds in which the males only are horned, where it is considered a serious fault.

Scour – Diarrhoea.

Shafty – Short, fine neck wool.

Shearling – A sheep after its first shearing, at just over one year old. Age in subsequent years may be denoted as 2-shear, 3-shear, etc.

Shed – Separating off a sector of the flock, either on the field with a well trained dog, or by using a shedding (or drafting) gate, usually situated at the exit end of a race. (Also a term used in weaving, referring to the space between two separated sets of warp threads, through which the shuttle passes).

Sheet – A large sack provided by the wool merchant for packing fleeces. Weighs 65 – 70kg when full.

Shoddy – Waste wool recovered from re-processed fabrics, formerly used for mulching hop plants.

Sire – Father.

Slack coated – see Open Coated.

Slug – see Scurr.

Stamp – A ram is said to "stamp his stock" when his offspring tend to take after himself rather than the ewes.

Standstill – Since the Foot & Mouth outbreak in 2001, any movement of sheep (or cattle) onto a holding triggers a standstill of 6 days, before any stock can be moved off the holding.

Staple – A naturally grown lock of wool, run together at the tips.

Strike – Blowfly attack. A sheep with maggots is said to have been "struck". Struck is also a colloquial name for one of the clostridial diseases (*Clostridium perfringens type C*).

Stones – Testicles. Also urolithiasis, or urinary stones.

Store – Weaned lambs that are held in condition prior to fattening at a later stage.

Super Diamond – Wool from the shoulder of the sheep. The best quality part of a fleece.

T.S.E – Transmissible Spongiform Encephalopathy (eg, Scrapie and BSE).

Tack – Ewe lambs from hill farms that spend their first winter on a lowland holding are said to be "on tack". Also known as wintering.

Teaser – A vasectomised ram used to stimulate oestrus behaviour in ewes prior to tupping.

Teg – Females and wethers from weaning until first shearing.

Terminal sire – A ram used for producing prime lambs for slaughter, ie, the terminal generation. Generally down breeds and continentals.

That'll do – Universally recognized command used to call a working dog off the sheep and back to the handler.

Theave – Young female (as Gimmer).

Two-tooth – As for shearling.

Tup – Ram.

Tupping – Mating.

Wether – Castrated male.

BREED SOCIETIES

Badger Face Welsh Mountain Sheep Society
General Secretary: Peter Weale
Llanarithon, Howey, Llandrindod Wells, LD1 5PP
Tel: 01597 823238
www.badgerfacesheep.co.uk

Balwen Welsh Mountain Sheep Society

Secretary: Mrs Anne Groucott
Swffryd Farm, Hafodyrynys, Crumlin, Gwent,
NP11 5HY 01495 247869
www.balwensheepsociety.com

Beltex Sheep Society

Secretary: Fiona Sloan, Shepherds View, Barras,
Kirkby Stephen, Cumbria, CA17 4ES Tel/Fax: 01768
341124
www.beltex.co.uk

British Berrichon Du Cher Sheep Society

Secretary: Susan Powell
Tregwynt, Three Ashes, Hereford, HR2 8LY
Tel: 01981 770071/07974 360807
www.berrichonsociety.com

Eppynt Hill & Beulah Speckled Face Sheep Society

Secretary: Mr D Jones
The Firs, 63 Garth Road, Builth Wells, Powys,
LD12 3NH Tel: 01982 553726
www.beulahsheep.co.uk

Blackface Sheep Breeders Association

Secretary: Fiona Sloan
Strathearn, Ruthwell, Dumfries, DG1 4NN
Tel: 01387 870653
www.scottish-blackface.co.uk

Black Welsh Mountain Sheep Breeders Asso

Secretary: Lesley Lewin
Lake Villa, Bradworthy, Holsworthy, Devon,
EX22 7SQ Tel: 01409 241579 / Fax: 01409 241579
www.blackwelshmountain.org.uk

Blue Texel Sheep Society

Secretary: Mr Andrew Pajak
The Bungalow, Sheepwash Lane, Steventon,
Abingdon, OX13 6SD Tel: 01235 831573
www.bluetexel.co.uk

Bluefaced Leicester Sheep Breeders Association

Secretary: Helen Carr-Smith
Riverside View, Warwick Road, Carlisle, CA1 2BS
Tel:01228 598022
www.blueleicester.co.uk

UK Mules

Secretary: Helen Carr-Smith
Riverside View, Warwick Road, Carlisle CA1 2BS
Tel: 01228 598022
www.ukmules.co.uk

The Society Of Border Leicester Sheep Breeders

Secretary: Katie Keiley
Carton Croft, Dalbeatie, Kirkcudbrightshire,
DG5 4NH www.borderleicester.co.uk

Boreray Sheep

Rare Breeds Survival Trust
National Agricultural Centre, Stoneleigh, Warwickshire, CV8 2LG Tel: 02476 696551
www.rbst.org.uk

Brecknock Hill Cheviot Sheep Society

Secretary: Peter Francis
13 Lion Street, Brecon, Powys, LD3 7HY
Tel: 01874 622488.

British Bleu Du Maine Sheep Society

Secretary: Mrs Jane Smith
Long Wood Farm, Trostrey, Usk, Gwent, NP15 1LA
Tel: 01291 673816 www.bleudumaine.co.uk

British Coloured Sheep Breeders Association

Secretary: Sarah Stacey Tel: 01873 890712
www.bcsba.org.uk

The British Dorper Sheep Society

Secretary: Nicky Morgan
Holt Valley Farm, Crab Tree Lane, Stoke, Andover,
Hants SP11 0LX Tel: 07710 352793
www.dorpersheepsociety.co.uk

The British Inra 401 Sheep Society
Secretary: Mr G F Burrough
Sheldon Grange, Dunkeswell, Honiton, Devon,
EX14 oRW Tel: 01823 680565.

British Milksheep Society
Secretary: Mr W J Hopkins
St Kenelms, Broad Lane, Tanworth in Arden, Solihull, West Midlands, B94 5HX Tel: 01564 742398
www.britishmilksheep.com

Cambridge Sheep Society
Secretary: Alun Davies
Pharm House, Neston Road, Willaston, Neston,
CH64 2TL Tel: 0151 327 5699
www.cambridge-sheep.org.uk

Castlemilk Moorit Sheep Society
Secretary: Mrs Sheila Cooper
Hillcrest Farm, Coventry Road, Berkswell, Nr. Coventry, CV7 7AZ Tel: 01676 535242
www.castlemilkmooritsociety.co.uk

Charmoise Hill Sheep Society
Secretary: David Trow
Llandinam Hall, Llandinam, Powys, SY17 5DN
Tel: 01686 688234
www.charmoisesheep.co.uk

Charollais Sheep Society
Secretary: Mrs Carroll Barber
Youngmans Road, Wymondham, Norfolk,
NR18 oRR. Tel: 01953 603335 Fax: 01953 607626
www.charollaissheep.co.uk

Cheviot Sheep Society
Secretary: Mrs Pat Douglas
Carlenrig Farm, Teviothead, Hawick, TD9 oLH
Tel: 01450 850218.
www.cheviotsheep.org

Clun Forest Sheep Society
Secretary: Sandra Williams
Tel: 01630 685981
www.clunforestsheep.co.uk

Colbred Sheep Society
Secretary: A. O. Colburn,
Crickley Barrow, Northleach, Cheltenham,
Gloucestershire GL54 3QA Tel: 01451 860330

Cotentin Sheep Society Ltd
Secretary: David Littlehales
Pentervin Farm, Minsterley, Shropshire, SY5 OHD
Tel: 01743 891347

Cotswold Sheep Society
Secretary: Lucinda Foster
Hampton Rise, 1 High Street, Meysey Hampton,
Gloucestershire, GL7 5JW Tel: 01285 851197
www.cotswoldsheepsociety.org.uk

Dalesbred Sheep Breeders Association
Secretary: Mrs J Bradley
Brackenber Lane Farm, Brackenber Lane,
Giggleswick, Settle, North Yorkshire, BD24 oEB
Tel: 01729 822228 www.dalesbredsheep.co.uk

Dartmoor Sheepbreeders Association (Greyface)
Secretary: Patrick Aubrey-Fletcher
Lower Stockadon Farm, St Mellion, Saltash,
Cornwall, PL12 6QF Tel: 01363 85205
www.greyface-dartmoor.org.uk

Derbyshire Gritstone Sheep Breed Society
Secretary: Mrs S Coppack
5 Bridge Close, Waterfoot, Rossendale, Lancashire,
BB4 9SN Tel: 01706 228520
www.derbyshiregritstone.org.uk

Devon Closewool Sheep Breeders Society
Secretary: Mr Terry Squire
Bratton Fleming, Barnstaple, North Devon
www.devonclosewool.co.uk

Devon And Cornwall Longwool Flockbook
Secretary: Mr M Britton
Peckham View, Kentisbeare, Cullompton, Devon
EX15 2EY Tel: 01884 266201
www.devonandcornwalllongwool.co.uk

Dorset Down Sheep Breeders Assocation
Secretary: Carolyn Opie
Havett Farm, Dobwalls, Liskeard, Cornwall.
PL14 6HB Telephone: 01579 320273
www.dorsetdownsheep.org.uk

Dorset Horn & Poll Dorset Sheep Breeders
Secretary: Mrs Margarite Cowley
Agricultural House, Acland Road, Dorchester,
Dorset DT1 1EF Tel: 01305 262126
www.dorsetsheep.org

Welsh Dorset Club
Geraint Jones Tel: 01974 261521

Easy Care Sheep Society
Secretary: R.I. Owen, M.B.E., F.R.Ag.S.
Glantraeth, Bodorgan, Anglesey, LL62 5EU, U.K.
Tel. 01407 840250 www.easycare.com

Est A Laine Merino Sheep Society
Secretary: Mrs C Barber
The Merino Centre, Wymondham, Norfolk,
NR18 0RR Tel: 01953 607860.

Exmoor Horn Sheep Breeders Society
Secretary: Mrs Jan Brown
Cornercott, Oldways End, East Ansty, Tivoton,
Devon, EX16 9JQ Tel: 01398 341372
www.exmoorhornbreeders.co.uk

British Friesland Sheep Society
Secretary: Mr P Baber
Weir Park Farm, Waterwell Lane, Christow, Exeter,
Devon EX6 7PB Tel: 01647 252549

British Gotland Sheep Society
Secretary: Mrs A E Barlow
Whitehall Farm, Luppitt, Honiton, DevonEX14 4TR
Tel: 01404 42141, www,gotlandsheep.com

Hampshire Down Sheep Breeders Association
Secretary: Richard Davis
Rickyard Cottage, Denner Hill, Great Missenden,
Bucks, HP16 0HZ Tel: 01494 488388
www.hampshiredown.org.uk

Hebridean Sheep Society
Secretary: Mrs Helen Brewis
Coney Grey, Gun Lane, Sherington, Newport
Pagnell, MK16 9PE, Tel: 01908 611092
www.hebrideansheep.org.uk

Herdwick Sheep Breeders Association
Secretary: Mrs Amanda Carson
Howe Cottage, Seascale, Cumbria, CA20 1EQ
Tel: 01946 729346, www.herdwick-sheep.com

Hill Radnor Flock Book Society
Secretary: Mr J Lewis
16 Ship Street, Brecon, Powys LD3 9AD
Tel: 01874 623200, www.hillradnor.co.uk

British Icelandic Sheep Breeders Group
Secretary: Mrs Jill Tyrer
Cefn Maen Isaf, Saron, Nr Denbigh, Denbighshire,
LL16 4TH Tel: 01745 550515
www.icelandicsheepbreedersofbritain.co.uk

Ile De France Sheep Society
Secretary: Edward Adamson
6 Fort Road, Kilroot, Carrickfergus, Co. Antrim,
BT38 9BS, Tel: 07711 071290

Jacob Sheep Society Ltd
Secretary: Jean Simms, Camster Fold, Aston, Stafford, ST18 9LJ, Tel: 01785 282818
www.jacobsheepsociety.co.uk

Kerry Hill Flock Book Society
Secretary: Pam Chilman
The Bramleys, Broadheath, Presteigne, Powys
LD8 2HG Tel: 01544 267353
www.kerryhillsheep.net

Leicester Longwool Sheep Breeders Assocation
Secretary: Mrs Ruth Mawer
Meadow View, Kelby, Grantham, Lincolnshire
NG32 3AJ Tel: 01400 230142
www.leicesterlongwoolsheepassociation.co.uk

Lincoln Longwool Sheep Breeders Assocation

Secretary: Ruth Mawer
LLSBA, Lincolnshire Showground, Grange-de-Lings, Lincoln, LN2 2NA. Tel: 01522 568660
www.lincolnlongwools.co.uk

Llanwenog Sheep Society

Secretary: Emily Addis
Preswylfa, Dihewyd, Lampeter, SA48 7PN
Tel: 01570471777
www.llanwenog-sheep.co.uk

Lleyn Sheep Society

Secretary: Gwenda Roberts,
Gwyndy, Bryncroes, Sarn, Pwllheli, Gwynedd,
LL53 8ET Tel: 01758 730366, www.lleynsheep.com

Lonk Sheep Breeders Association

Secretary: Christine Scrivin
Park House Farm, Elslack, North Yorks, BD23 3AT-
Tel: 01282 433047 www.lonk-sheep.org

Manx Loaghtan Sheep Breeders Group

Secretary: Carol Kempson
Cannons, Huntley Road, Tibberton, Gloucester,
GL19 3AB Tel: 01763 838485
www.manxloaghtansheep.org

Masham Sheep Breeders Association

Secretary: Mrs V J Lawson
Oak Bank, Bentham, Lancaster, LA2 7DW
Tel: 01524 261606

Meatlinc Sheep Company

Chairman: H R Fell
Church House, Horkstow, Barton-on-Humber,
North Lincolnshire DN18 6BG
Tel: 01652 618329 Fax: 01652 618447
www.meatlinc.co.uk

Norfolk Horn Breeders Group

Secretary: Suzannah Coke
Woodcarver's Cottage, School Road, East Rudham,
Kings Lynn, PE31 8RF Tel: 07599 566657
www.norfolkhornbreeders.co.uk

North Country Cheviot Sheep Society

Secretary: Alison Brodie
Wallacehall West, Waterbeck, Lockerbie, DG11 3HR
Tel: 01461 600646 www.nc-cheviot.co.uk

North Of England Mule Sheep Association

Secretary: Marion Hope
Albierigg Farm, Canonbie, Dumfriesshire,
DG14 0RY Tel: 01387 371677
www.nemsa.co.uk

North Ronaldsay Sheep

Rare Breeds Survival Trust
National Agricultural Centre, Stoneleigh, Warwickshire, CV8 2LG Tel: 02476 696551
www.rbst.org.uk

Oxford Down Sheep Breeders Association

Secretary: Mrs Ruth Mawer
Meadow View, Kelby, Grantham, Lincolnshire,
NG32 3AJ Tel: 01400 230142
www.oxforddownsheep.org.uk

Portland Sheep Breeders Group

Secretary: Michelle Jones
Hogchester Farm, Wooton Fitzpaine, Nr Charmouth, Dorset, DT6 6BY Tel: 01297 561072
www.portlandsheep.org.uk

Romney Sheep Breeders Society

Secretary: Alan West
2 Woodland Close, West Malling, Kent, ME19 6RR
Tel: 01732 845637 www.romneysheepuk.com

British Rouge De L'ouest Sheep Society Ltd

Secretary: Sue Archer
Marston Mill Farm, Wolston, Coventry, Warwickshire, CV8 3FX Tel: 02476 541766
www.rouge-society.co.uk

Rough Fell Sheep Breeders Association

Secretary: Jayne Knowles
High Borrow Bridge, Selside, Kendal, LA8 9LG
Tel: 01539 823270 www.roughfellsheep.co.uk

Roussin Sheep Society
Secretary: Mrs Andrea Molyneux
Mount Pleasant Farm, Deckport Cross, Hatherleigh,
Oakhampton, Devon, EX20 3LN
Tel: 01837 810006
www.bohdgaya.net/roussinsheepsociety

Ryeland Flock Book Society Ltd
Secretary: Mrs Dot Tyne
Ty'n-y-Mynydd Farm, Boduan, Pwllheli, Gwynedd,
LL53 8PZ Tel: 01758 721898 www.ryelandfbs.com

Scotch Mule Association
Secretary: George Allan
Bogside Cottage, Ochiltree, Cumnock, Ayrshire,
KA18 2QF Tel: 01292 591821
www.scotchmule.co.uk

Shetland Flock Book Trust
Secretary: Jim Nicolson
Lonabrek, Aith, Bixter,Shetland ZE2 9ND
Tel: 01595 810343

Shetland Sheep Society
Secretary: 15 Cross Lane, Braunston, Northants,
NN11 7l ll l Tel: 01788 890865
www.shetland-sheep.org.uk

Shetland-cheviot Marketing Society
Secretary: J Johnson
Fairview, Vidlin, Shetland, ZE2 9QB
Tel: 01806 577227

Shropshire Sheep Breeders Assocation
Secretary: Mrs Jane Wilson
Gibshiel, Tarset, Hexham, Northumberland,
NE48 1RR Tel: 01434 240435
www.shropshire-sheep.co.uk

Soay Sheep Society (Friends of the Soay & Boreray)
Secretary: Julie Suffolk
Black Forest Farm, Swythamly, Rushton Spencer,
Macclesfield, SK11 0RF Tel: 01947 840924
www.soaysheep.org

Southdown Sheep Society
Secretary: Mrs Gail Sprake
Meens Farm, Capps Lane, All Saints, Halesworth,
Suffolk, IP19 0PD Tel: 01986 782251
www.southdownsheepsociety.co.uk

South Wales Mountain Sheep Society
Secretary: Mr G Davies
40 Rhys Road, Blackwood, Caerphilly County Borough, Gwent NP12 3QR Tel: 01443 839234

Suffolk Sheep Society
Secretary: Catherine Fleck
Unit B, Ballymena Livestock Market, 1 Woodside
Park, Ballymena, Co Antrim, Northern Ireland,
BT42 4HG Tel: 02825 632342
www.suffolksheep.org

Suffolk Young Breeders (part of the Suffolk Sheep Society).
Helen Davies, Arddleen, Welshpool
Tel: 01938 590535

Swaledale Sheep Breeders Assocation
Secretary: John Stephenson
The Shooting Lodge, High Shipley, Eggleston, Barnard Castle, Co. Durham DL12 0DP
Tel: 01833 650516 www.swaledale-sheep.com

Talybont Welsh Mountain Sheep Society
Secretary: Mr John A. Lewis,
Montague Harris & Co,
16, Ship Street, Brecon, Powys.LD3 9AD

Teeswater Sheep Breeders Assocation
Secretary: Mrs D Newey
Maedowcroft FArm, Ugthorpe, Whitby, YO21 2BL
Tel: 01947 840924 www.teeswater-sheep.co.uk

British Texel Sheep Society Ltd
N.A.C., Stoneleigh Park, Kenilworth, Warwickshire,
CV8 2LG Tel: 02476 696629 / Fax: 02476 696472
www.texel.co.uk

Derbyshire Texel Club
Secretary: Mrs Janet Young
Littlecote 118, Doles Lane, Findern, Derbyshire,
DE65 6BA Tel: 01332 519494

Dutch Texel Producers Club
Secretary: Mrs Paula Barlow
Mossgiel, Babell, Nr Holywell, Flintshire CH8 8PT
Tel: 01352 720 902 / Fax: 01352 720 902

British Vendeen Sheep Society
Secretary: Mr Andrew John
Darkes House, Conderton, Tewkesbury, Gloucestershire, GL20 7PP Tel: 01386 725229
www.vendeen.co.uk

Welsh Halfbred Sheep Breeders Association
Secretary: Anna Johnson
Corriecravie, Cross Lane, Bignall End, Stoke on
Trent, ST7 8ND
Tel: 01782 721165
www.welshhalfbredsheep.co.uk

Welsh Hill Speckled Face Sheep Society
Secretary: Mr R Griffiths
Nanty Farm, Pantmawr, Llanidloes, Powys
SY18 6SY Tel: 01686 440279

Welsh Mountain Sheep Society - Hill Flock
Secretary: W G M Jones
PO Box 8, Gorseland, North Road, Aberystwyth,
Ceredigion, SY23 2HE.
Tel: 01970 636688 / Fax: 01970 624049.

Welsh Mountain Sheep Society – Pedigree
Secretary: B J Roberts
Llais Afon, Llangwm, Corwen, Clwyd, LL21 0RA
Tel: 01490 420696

Welsh Mule Sheep Breeders
Secretary: Mr W G M Jones
PO Box 8, Gorseland, North Road, Aberystwyth,
Ceredigion, SY23 2HE.
Tel: 01970 636688 / Fax: 01970 624049.
www.welshmules.co.uk

Wensleydale Longwool Sheep Breeders Association
Secretary: Dr D L Clouder,
Coffin Walk, Sheep Dip Lane, Princethorpe, Nr
Rugby, Warwickshire, CV23 9SP
Tel: 01926 633439
www.wensleydale-sheep.com

White Face Dartmoor Sheep Breeders Association
Secretary: Mrs Fiona West
8 East End Terrace, Ashburton, Devon, TQ13 7LD
Tel: 01364 653653
www.whitefacedartmoorsheep.co.uk

White Faced Woodland Sheep Society
Secretary: Rachel Godschalk
Low Thornberry Farm, Bowes, Barnard Castle, Co.
Durham, DL12 9JJ Tel: 01883 628416
www.whitefacedwoodland.co.uk

Wiltshire Horn Sheep Society
Secretary: Mrs Christina Cormack
Little Bache House Farm, Hurleston, Nantwich,
Cheshire CW5 6BU Tel: 0844 8001029
www.wiltshirehorn.org.uk

Zwartbles Sheep Association
Secretary: Christina Cormack
Stickle Heaton Farm, Cornhill on Tweed, TD12 4XG-
Tel: 05603 466931 www.zwartbles.org

RESOURCES

SCOPS – An industry led initiative for the Sustainable Control of Parasites in Sheep (SCOPS) which
offers farmers and vets advice on how to combat
the growing problem of anthelmintic resistance
(AR) in the treatment of worm parasites.
www.defra.gov.uk/animalh/diseases/control/para-
site_control.htm

NSA – The National Sheep Association – The UK
sheep farmers organisation representing the views
and interests of sheep producers throughout the
UK. NSA is funded by its membership of sheep

farmers, and its activities involve it in every aspect of the sheep industry.

The Sheep Centre, Malvern, Worcestershire
United Kingdom , WR13 6PH
Telephone: 01684 892 661 / Fax: 01684 892 663
www.nationalsheep.org.uk

EBLEX – English Beef & Lamb Executive – EBLEX represents the interests of English beef & lamb levy payers; to encourage and facilitate the industry to become more efficient as consumer's needs change.

EBLEX, Winterhill House, Snowdon Drive,
Milton Keynes, MK6 1AX.
Tel: 0870 242 1394
www.eblex.org.uk

HCC – Hybu Cig Cymru / Meat Promotion Wales - the organisation responsible for the development, promotion and marketing of Welsh red meat. Working with all sectors of the Welsh red meat industry, from the farmers through to the retailers, to develop the industry itself as well as developing profitable markets for Welsh Lamb and Welsh Beef.

HCC, Ty Rheidol, Parc Merlin
Glanyrafon Industrial Estate, Llanbadarn Fawr
SY23 3FF Tel: 01970 625050 / Fax: 01970 615148
www.hccmpw.org.uk

QMS – Quality Meat Scotland – QMS's core function is to work with the Scottish red meat industry to improve its efficiency and profitability and to maximise its contribution to Scotland's economy.

Quality Meat Scotland
Rural Centre
West Mains
Ingliston
Newbridge
EH28 8NZ
Tel 0131 472 4040 / Fax: 0131 472 4038
www.qmscotland.co.uk

LMCNI – Livestock & Meat Commission for Northern Ireland – supporting the red meat industry to boost the productivity and sustainability of the entire meat supply chain in Northern Ireland, through improving competitiveness.

LMCNI, Lissue House, 31 Ballinderry Road
Lisburn, BT28 2SL Tel: 028 9263 3000 / Fax: 028 9263 3001 www.lmcni.com

AHDB – The Agriculture and Horticulture Development Board. (inc. Signet Breeding Services.). Formerly the Meat and Livestock Commission. – Their role is to help improve the efficiency and competitiveness of various agriculture and horticulture sectors within the UK.

Agriculture and Horticulture Development Board
Stoneleigh Park, Kenilworth, Warwickshire
CV8 2TL. Tel: 0247 669 2051 www.ahdb.org.uk

Innovis Ltd – Breeding Innovation – Innovis offers the UK livestock industry a wide range of high quality products and services for profitable livestock production and is the leading supplier of sheep breeding technologies.

Peithyll Centre, Capel Dewi
Aberystwyth, Ceredigion, SY23 3HU
Tel: 01970 828236 / Fax: 01970 822018
www.innovis.org.uk

FECPak – (also from Innovis) – Faecal Egg Counting for monitoring of parasitic worm burdens. Contact details as for Innovis

NFU – The National Farmers Union - The NFU represents the farmers and growers of England and Wales. Its central objective is to promote successful and socially responsible agriculture and horticulture, while ensuring the long term viability of rural communities.

Agriculture House, Stonleigh Park, Stonleigh Warwickshire, CV8 2TZTel: 024 7685 8500
Fax: 024 7685 8501 www.nfuonline.com

FUW – The Farmers Union of Wales - An independent union to protect and advance the interests of those whose income is from Welsh agriculture.

Llysamaeth, Plas Gogerddan, Aberystwyth Dyfed Tel: 01970 820820 www.fuw.org.uk

LANTRA – Lantra supports the training and business development needs of employers, employees and volunteers in 17 industries in the environmental and land-based sector.

Lantra House, Stonleigh Park, Stonleigh Coventry, Warwickshire, CV8 2LG Tel: 02476 696996 www.lantra.co.uk

Defra – Department for the Environment, Food and Rural Affairs – The UK government's executive agency primarily responsible the environment, for food and farming, and for rural issues.

Defra, Nobel House 17 Smith Square, London, SW1P 3JR Tel: 08459 335577 / Fax: 0207 238 6951 www.defra.gov.uk

WAG – Welsh Assembly Government – supporting thriving rural communities where people live, work, and enjoy a high quality of life, creating a sustainable environment for the people of Wales and future generations

Department for Rural Affairs, Welsh Assembly Government, Cathays Park, Cardiff CF10 3NQ Tel: 0845 0103300 www.wales.gov.uk/topics/environmentcountrside

SGRPID – Scottish Government Rural Payments and Inspectorates division – Looking after Scotland's farmers.

There is a comprehensive network of local area offices. Telephone : 08457 741741 www.scotland.gov.uk/topics/agriculture

DARDNI – Department for Agriculture and Rural Development in Northern Ireland

Department of Agriculture and Rural Development Dundonald House, Upper Newtownards Road, Belfast, BT4 3SB

RBST – The Rare Breeds Survival Trust – Working to conserve Britain's native livestock heritage.

RBST, Stoneleigh Park, Stoneleigh Warwickshire, CV8 2LG Tel: 02476 696551 / Fax: 024 7669 6706 www.rbst.org.uk

RWAS – The Royal Welsh Agricultural Society – organisers of the annual Royal Welsh Show, the Royal Welsh Winter Fair and Smallholder and Garden Festival.

RWAS, Royal Welsh Showground Llanelwedd, Builth Wells, Powys, LD2 3SY Tel: 01982 553683 / Fax: 01982 553563 www.rwas.co.uk

RASE – The Royal Agricultural Society of England – the RASE aims to play a leading role in the development of British Agriculture and a vibrant economy

RASE, Stoneleigh Park, Warwickshire CV8 2LZ Tel: 024 7669 6969 www.rase.org.uk

RHASS – Royal Highland and Agricultural Society of Scotland – founded in 1784, for people who value the rural areas of Scotland.

RHASS, Royal Highland Centre, Ingliston Edinburgh, EH28 8NF Tel: 0131 335 6200 / Fax: 0131 3335236 www.rhass.org.uk

RUAS – Royal Ulster Agricultural Society – Organisers of the Balmoral Show.

Royal Ulster Agricultural Society, The King's Hall Balmoral, Belfast BT9 6GW
Tel: 02890 665225 / Fax: 02890 661264
www.balmoralshow.co.uk

BWMB – The British Wool Marketing Board – a farmer rum organisation operating a central marketing system for UK fleeces.

Wool House, Roysdale Way, Euroway Trading Estate, Bradford, West Yorks, BD4 6SE
Tel:01274 688666 / Fax: 01274 652233
www.britishwool.org.uk

ISDS – The International Sheepdog Society – UK and World Trials and events information, news and online shop for all aspects of sheep dogs, Border collies and other breeds.

ISDS, Clifton House, 4a Goldington Road Bedford, MK40 3NF
Tel: 01234 352672 / Fax: 01234 348214
www.isds.org.uk

Moredun Research Institute – internationally recognised for its work on infectious diseases of sheep and other ruminants.

Moredun Research Institute, Pentlands Science Park, Bush Loan, Penicuik, Midlothian
EH26 0PZ
Tel: 0131 445 5111 / Fax: 0131 4456111
www.mri.sari.ac.uk

NADIS – National Animal Disease Information Service – NADIS is a network of over 60 veterinary practices and 6 veterinary colleges monitoring diseases in cattle, sheep and pigs in the UK
www.nadis.org.uk

HSA – The Humane Slaughter Association – The only registered charity that works in the UK and internationally, through educational, scientific and technical advances, exclusively towards the highest worldwide standards of welfare for food animals during transport, marketing and slaughter.

HSA, The Old School, Brewhouse Hill Wheathampstead, Herts, AL4 8AN
Tel: 01582 831919 / Fax: 01582 831414
www.has.org.uk

FSA – the Food Standards Agency – Independent UK government department set up to protect the public's health and consumer interests in relation to food.

Tel: 0207 276 8829 www.food.gov.uk

NOAH – National Office of Animal Health – NOAH represents the UK animal medicines industry. Its aim is to promote the benefits of safe, effective, quality medicines for the health and welfare of all animals.

NOAH Ltd., 3 Crossfield Chambers, Gladbeck Way Enfield, EN2 7HF
Tel: 0208 367 3131 / Fax: 020 83631155
www.noah.co.uk

NFSCo – The National Fallen Stock Company – a not for profit, farmer led organisation dedicated to delivering a valued service for the farming community that aims to provide a national service for the collection and disposal of fallen stock that farmers use from choice.

Sallyfield Lane, Stanton, Ashbourne Derbyshire, DE6 2DA Tel: 0845 0548888
www.nfsco.co.uk

Pastoral Alliance and Mutton Renaissance - dedicated to the promotion of mutton, providing detailed production information for farmers, abattoirs and butchers.

Mutton Renaissance, NSA, The Sheep Centre Malvern, Worcestershire WR13 6PH
Tel: 01684 899255 / Fax: 01684 899272
www.muttonrenaissance.org.uk

PSGHS – Premium Sheep & Goat Health Schemes. MV, EAE, CLA and Scrapie monitoring and accreditation.

Premium Sheep and Goat Health Schemes
PO Box 5557, Inverness, IV2 4YT
Tel: 01463 226995 / Fax: 01463 711103
www.sac.ac.uk/consulting/services/i-r/sghs/

Welsh Commons Forum - set up under the auspices of the NSA to look after the interests of Common Land graziers in Wales

Welsh Commons Forum , NSA , The Sheep Centre Malvern, Worcestershire, WR13 6PH
Tel: 01684 892661 / Fax: 01684 892663
www.nationalsheep.org

 SA – Soil Association (Organic certification) – a membership charity campaigning for planet friendly organic food and farming

Soil Association, South Plaza, Marlborough Street Bristol, BS1 3NX
Tel: 0117 314 5000 / Fax: 0117 3145001
www.soilassociation.org

FWAG – Farming and wildlife advisory group – the UK's leading independent and dedicated provider of environmental and conservation advice to farmers.

FWAG, Stoneleigh Park, Kenilworth Warwickshire, CV8 2RX
Tel: 02476 696699 / Fax: 02476 696760
www.fwag.org.uk

Environment Agency – UK government agency mainly concerned with rivers, flooding and pollution.

Environment Agency, National Customer Contact Centre, PO Box 544, Rotherham, S60 1BY
Tel: 08708 506506
www.environment-agency.gov.uk

GAP – The Grazing Animals Project – Bringing together expertise and experience to disseminate conservation grazing advice and awareness.

GAP, Brinkworth House, Brinkworth Nr Chippenham, Wiltshire, SN15 5DF
Tel: 01666 511300
www.grazinganimalsproject.org.uk

PONT – Pori Natur a Threftadaeth – the welsh arm of the grazing animals project.

PONT, PO Box 52, Llangadog Carmarthenshire, SA19 9WZ Tel: 01550 740333
www.grazinganimalsproject.org.uk/pont_home

Small Shepherds Club – providing training and support for people who have, or would like to keep, a small number of sheep. Mainly based in SE England.

Membership Secretary, Broom Cottage, Long Reach, West Horsley, Leatherhead, KT24 6NE
Tel: 02392 412361 www.shepherdsclub.org.uk

BSDA – British Sheep Dairying Association – aims to promote the production, manufacture, marketing and consumption of quality dairy products derived exclusively from sheep's milk in the UK

BSDA Secretary, c/o High Weald Dairy Tremains Farm, Tremains Road, Horsted Keynes West Sussex, RH17 7EA Tel: 01825 791636
www.sheepdairying.co.uk

HAWL – Homeopathy at Wellie Level. Chris Lees, Church Cottage, Alderton Chippenham, Wiltshire, SN14 6NL
Tel: 01666 841213 www.hawl.co.uk

Suppliers Directory

Rappa Fencing Ltd., Steepleton Hill
Stockbridge, Hampshire, SO20 6JE
Tel: 01264 810663 / Fax: 01264 810079
www.rappa.co.uk

Galebreaker Products Gale Breaker House, New
Mills Industrial Estate, Ledbury, Herefordshire, HR8
2SS, Tel: 01531 637900 www.galebreaker.com

Ascott Smallholding Supplies Ltd., Units 21/22,
Whitewalls, Easton Grey, Malmesbury, SN16 0RD
Llanymynech, SY22 6LP Tel: 0845 1306285 or
01666 826931, Fax: 01666 825402 www.ascott.biz

Cotswold Seeds Ltd., The Cotswold Business Village
Moreton-in-Marsh, Gloucestershire, GL56 0JQ
Tel: 0800 252111 / Fax: 01608 652256
www.cotswoldseeds.com

Burgon & Ball, La Plata Works, Holme Lane, Shef-
field, S6 4JY
www.burgonandball.com (hand Shears)

MTL Fabrications Valley Farm Cottage, Fimber,
Driffield, YO25 9LY Tel: 01377 236698

Modulamb,Peter Hall Farm, Coombefields
Coventry. CV2 2DR Tel: 02476 611647
Fax: 02476 613866 www.modulamb.com

The Natural Fibre Company, Unit B, Pipers Court,
Pennygillam Way, Launceston, Cornwall
PL15 7PJ Tel: 01566 777635
www.thenaturalfibre.co.uk

Ashford – New Zealand based manufacturer of
traditional spinning wheels. For a list of UK distributors and stockists see the order info page on their
website. www.ashford.co.nz

Welsh Shearing Equipment Centre Ltd
Salem, Sennybridge, Brecon, Powys, LD3 8RS
Tel: 01874 636455 / Fax: 01874 638959
www.shearing.co.uk

The Woolly Shepherd, Blagdon Hill
Somerset Tel: 01823 421237
www.woollyshepherd.co.uk

Cox Agri Sales, 1 Greencroft Industrial Park
Stanley, County Durham, DH9 7YA
Tel: 0845 6008081 / Fax: 0800 7836655
www.coxagri.com

Fearing International Ltd., Creaton Road
Brixworth, Northampton, NN6 9BW
Tel: 0845 6009070 / Fax: 0800 581606
www.fearing.co.uk

Allflex UK Ltd., Unit 6 Galalaw Business Park
Hawick, Roxburghshire Tel: 01450 364120
Fax: 01450 363121 www.allflex.co.uk

Border Software – Farmit 3000 – fully integrated
farm management software.

Border Software Ltd., Llety Mawr, Llangadfan
Welshpool, Powys, SY21 0PS Tel: 01938 820625
www.bordersoftware.com

ID & SB James, Country Supplies (Mid Wales)
Hundred House, Llandrindod Wells, LD1 5RY
Tel: 01982 570200 / Fax: 01982 570210
www.jamescountrysupplies.net

Tithebarn Ltd., Road Five, Winsford Industrial Estate, Winsford, Cheshire, CW7 3PG
Tel: 01606 595000 / Fax: 01606 595045
www.tithebarn.co.uk

JG Animal Health, Windrush Cottage, Cradley,
Malvern, Worcestershire. WR13 5LF
Tel: 01886 880482 www.animalhealth@aol.com
email: jganimalhealth@aol.com

Wessex Animal Health Ltd., Unit 25/26 Hightown
Industrial Estate, Crow Arch Lane, Ringwood,
Hampshire, BH24 1ND, Tel: 01425 474455
Fax: 01425 474445 www.wessexanimalhealth.co.uk

Osmonds – manufacturers and suppliers of a range
of sheep health products including bactakill 55
www.osmonds.co.uk

Icer Tech Clyweddog Road South, Wrexham Ind. Est.Wrexham, Clwyd, LL13 9XS Tel: 0800 018 1247 www.icertech.co.uk

Weschenfelder Ltd 01642 247524 www.weschenfelder.co.uk

Books (Reference)

This is, quite simply, a list of the titles found on my own bookshelves (and a few from my mother's), that I have dipped into from time to time while working on this book. Everything I've made use of is included, from the most up-to-date veterinary manual on sheep health currently available, down to some older volumes that would not appear out of place in a museum!

Sheep Flock Health. A Planned Approach. N. Sargison. (Blackwell, 2008) ISBN 978-1-4051-6044-5

The Veterinary Book for Sheep Farmers. D. Henderson. (Farming Press, 1990) ISBN 0-85236-189-0

The Modern Shepherd. D. Brown & S. Meadowcroft. (Farming Press, 1989) ISBN 0-85236-188-2

The Showman Shepherd. D. Turner. (Farming Press, 1990) ISBN 0-85236-204-8

Showing Sheep. S. Kendrick. (The Good Life Press, 2008) ISBN 978-1-90487-119-4

Farm Management Pocketbook. J. Nix. (Wye College, University of London) updated annually. SBN 0-86266-151-X

Farm Management Handbook 1993/94. (14th Edition) Editor: L. Chadwick. (The Scottish Agricultural College, 1993)

Compendium of Data Sheets for Veterinary Products 2004. (National Office of Animal Health, 2003) ISBN 0-9526638-9-9

Energy and Protein Requirements of Ruminants. An advisory manual prepared by the AFRC Technical Committee on Responses to Nutrients. (CAB International, 1993) ISBN 0-85198-851-2

Sheepkeeping on a Small Scale. (4th Impression) E. Hart. (Thorsons, 1984) ISBN 0-7225-0516-7

Backyard Sheep Farming. A. Williams. (Prism Press, 1978) ISBN 0-904727-72-6

Profitable Sheep Farming. (5th Impression) M. McG. Cooper & R. J. Thomas. (Farming Press, 1965 and 1991) ISBN 0-85236-117-3

Sheep Ailments. (5th Edition) The TV Vet. (Farming Press, 1986) ISBN 0-85236-161-0

The Production and Management of Sheep. D. H. Goodwin. (Hutchinson Educational, 1971) ISBN 0-09-107311-1

British Sheep & Wool. Editors: J. B. Skinner, D. E. Lord, J. M. Williams. (The British Wool Marketing Board, 1985)

Wool Away – The Art and Technique of Shearing. (2nd Edition) G. Bowen. (Whitcombe and Tombs, 1956) First published 1955

Sheep farming. (7th Edition) A. Fraser. (Crosby Lockwood & Son, 1965) First published 1937

Sheep Farming Today. J. F. H. Thomas. (Faber & Faber, 1966)

Modern British Farming Systems. Editor: F. H. Garner. (Paul Elek, 1972) ISBN 0-236-17730-3

Animal Husbandry. (2nd Edition, 3rd Impression) D. G. M. Thomas & W. I. J. Davies. (Cassell & Co. 1977) ISBN 0-304-93797-5

Sheep Production. An Foras Talúntais (1984) ISBN 0-905442-77-6

The Dorset Horn – a short history of the Dorset horn & poll Dorset sheep breeders association. T. Hearing. Published by the association. ISBN 0-901253-07-3

Good Farming. (4th Impression) V. C. Fishwick. (English Universities Press, 1949) First published 1944

Good Sheep Farming. N. L. Tinley (English Universities Press, 1949)

Good and Healthy Animals. J. D. Paterson. (English Universities Press, 1947)

Good Grassland. D. H. Robinson. (English Universities Press, 1947)

Sheep Farming. (3rd Edition) W. R. Seward. (National Federation of Young Farmers' Clubs, 1947)

Grassland. (5th Edition) J. O. Thomas. (National Federation of Young Farmers' Clubs, 1946)

Sheep Breeding and Management. (3rd edition) Ministry of Agriculture, Fisheries and Food, bulletin number 166. (HMSO, 1964) First published 1956

Agriculture – The Science and Practice of Farming. (11th Edition) J. A. S Watson & J. A. More. (Oliver & Boyd, 1962) First published 1924

British Sheep Breeds. (Shire album 157) E. Henson. (Shire Publications, 1986) ISBN 0-85263-779-9

Rare Breeds. (Shire album 118) Lawrence Alderson.

(Shire Publications, 1984) ISBN 0-85263-577-6
Wool. C. Gorbing-King. (John Baker, 1970) SBN 212-98371-1
Natural Dyes for Spinners & Weavers. Hetty Wickens. (B. T. Batsford, 1983) ISBN 0-7134-2021-9
Colour from Plants – A Simple Dyers Notebook. C. Morton & A. Mortlock. (Abbot Hall Gallery and Museum of Lakeland Life and Industry).
Learn to Spin. (Ashford Handicrafts).
Your Handspinning. Elsie G. Davenport. (Select Books, 1964)
The Women's Institute Book of Country Crafts. W.I. Books Ltd. (Chancellor Press, 1994 edition) ISBN 1-85152-544-0
A Way of Life - Sheepdog Training, Trialling and Handling. H. Glyn Jones. (Farming Press, 1987) ISBN 0-85236-166-1
Working Sheep Dogs – Management and Training. John Templeton. (Crowood Press, 1988) ISBN 1-85223-003-7
Self-Sufficiency. John & Sally Seymour. (Faber & Faber, 1975) ISBN 0-571-09954-8
The New Complete Book of Self-Sufficiency. John Seymour with Will Sutherland. (Dorling Kindersley, 2003) ISBN-13: 978-0-7513-6442-2, ISBN-10: 0-7513-6442-8 (First published 1976)
The Survival Handbook – Self-sufficiency for everyone. Michael Allaby with Marika Hanbury Tenison, John Seymour and Hugh Sharman. (Macmillan, 1975; Pan Books, 1977) ISBN 0-330-24813-8
On Next to Nothing. Thomas & Susan Hinde. (Sphere books, 1977) ISBN 0-7221-4558-6
First Steps in Border Collie Sheepdog Training – from Chaos to Control. DVD Demonstrated by Andy Nickless. (Andy Nickless Media Productions, 2008)

Books (General interest)

The Blue Riband of the Heather – Supreme Champions 1906-1995. (2nd Edition) E. B Carpenter. (Farming Press, 1996) ISBN 0-85236-318-4
A Dog's Life in the Dales. Katy Cropper. (Bantam Books, 1993) ISBN 0-553-40638-8
The Farming Ladder. (3rd Impression) G. Henderson. (Faber and Faber, 1944)
The Shepherd. John Seymour. (Sidgwick & Jackson, 1983) ISBN 0-283-98922-X
Shepherding. John Vince (Sorbus, 1992) ISBN 1-874329-05-2
Rare Breeds in History. Elizabeth Henson. (1982)
Shepherding Tools and Customs. (Shire album 23) Arthur Ingram. (Shire Publications, 1977) ISBN 0-85263-379-3
The Illustrated Shepherd's Life. W. H. Hudson (The Bodley Head, 1987) ISBN 0-370-31101-9 (First published by Methuen & Co. in 1910)
The Downland Shepherds. Barclay Wills. Editors: Shaun Payne & Richard Pailthorpe. (Alan Sutton, 1989) ISBN 0-86299-408-X
Snowdon Shepherd. Keith Bowen. (Pavilion Books, 1991) ISBN 1-85145-425-X
I Bought a Mountain. Thomas Firbank. (Hodder and Staughton, 1990) ISBN 0-450-00049-4 (First published by George G. Harrap & Co, 1940)
101 Ways to Cook a Sheep. John & Barbara Hay. (Sun Books, 1969) Aus 68-1431
Sheep – The remarkable story of the humble animal that built the modern world. Alan Butler. (O Books, 2006) ISBN-13: 978-1-905047-68-0, ISBN-10: 1-905047-68-1
The English Farm. Ralph Whitlock. (J. M Dent & Sons, 1983) ISBN 0-460-04584-9
Hovel in the Hills. Elizabeth West. (John Jones, 1998) ISBN 1-871083-31-1 (First published by Faber & Faber, 1977)
I Went a'shepherding. Richard Perry, Published by Lindsay Drummond (1945)

Reports and Bulletins

SO...you are interested in milking sheep? A bulletin of the Wisconsin Sheep Dairy Cooperative, October 2008. www.sheepmilk.biz

Periodicals

Country Smallholding
Farmers Guardian
Farmers Weekly
Home Farmer
Sheep Farmer
Gwlad
The Smallholder

NATURAL PLANT DYES

COMMON NAME	LATIN NAME	PARTS USED	GENERAL COLOUR GUIDE	SUGGESTED MORDANTS
Agrimony	Agrimonia eupatoria	leaves	gold	alum,chrome
Alder	Alnus spp.	bark	yellow/brown/black	alum,iron,copper sulphate
Alkanet	Anchusa tinctoria	roots	grey	alum,cream of tartar
Apple	Malus spp.	bark	yellow	alum
Barberry	Berberis spp	twigs	yellow	alum
Bilberry	Vaccinium spp.	berries	purple	alum,tin
Blackberry	Rubus spp.	berries, young shoots	pink/purple	alum,tin
Black Crottle (lichen)	Parmelia omphalodes	whole lichen	reddish brown (when boiled)	oak bark
Blackcurrent	Ribes spp.	berries	grey/deep purple	alum,tin
Blackwillow	Salix nigra	bark	red/brown	iron,chrome
Bloodroot	Sanguinaria canan-densis	roots	red	alum,tin
Bluebell	Hyacinthoides non-scripta	flowers	ice blue/bright blue	alum, tin
Bracken	Pteridum aquilinum	young shoots, old tops	yellow/green	alum,chrome
Broom	Cytisus spp.	flowering tops	orange/yellow	chrome,tin
Buckthorn	Rhamnus cathartica	twigs, berries	yellow/brown	alum,cream of tartar, chrome, tin, iron
Cherry	Prunus spp.	bark	pink/yellow/brown	alum
Coffee	Coffea arabica	used grounds	brown/khaki	alum, chrome,none
Coreopsis	Coreopsis tinctoria	flower heads	yellow/orange	chrome, tin
Crab's Eye (lichen)	Orchrolechia parella	whole lichen	orange/red (when fermented in urine then boiled)	alum
Crottle (lichen)	Parmelia saxatilis	whole lichen	yellow/brown (when boiled)	oak bark
Cudbear (lichen)	Orchrolechia tartarea	whole lichen	red/purple (when fermented in urine then boiled)	alum
Cypress	Cypress spp.	cones	tan	alum,chrome
Daffodil	Narcissus pseudonar-cissus	flowers	yellow	alum,tin
Dahlia	Dahlia spp.	petals	yellow/bronze	alum
Damson	Prunum damascenum	fruit skins	dark red/green	alum,ammonia

Day Lily	Hemerocallis spp.	flowers	yellow	alum, tin, copper sulphate
Dandelion	Taraxacum officinale	flower/taproot	yellow/magenta/brown	none, alum
Dog lichen	Peltigera canina	whole lichen	yellow (when boiled)	alum
Dog's Mercury	Mercurialis perennis	whole plants	yellow	alum
Dyer's Broom	Genista tinctoria	flowering tops	yellow	alum, chrome
Dyer's Chamomile	Anthemis tinctoria	flowers	yellow	alum
Elder	Sambucus nigra	leaves, berries, bark	yellow/grey	iron, alum
Fennel	Foeniculum vulgare	whole plant	brown/green	chrome, iron, copper sulphate
French Marigold	Tagetes patula	flower head	yellow/brown/orange	alum, chrome, iron, tin
Golden Rod	Solidago spp.	flowering tops	gold	alum, chrome, iron
Gorse	Ulex europaeus	flower petals	yellow/lime	alum
Heather	Erica spp.	tips	yellow	alum
Horsetail	Equisetum spp.	stalks	green	alum, copper sulphate
Horse Chestnut	Aesculus hippocastanum	conker peels	brown/bronze/grey	alum, tin, iron
Hypogymnia (lichen)	Hypogymnia psychodes	whole lichen	gold/brown	
Ivy	Hedera helix	berries	yellow/green	alum, iron
Ladies Bedstraw	Gallium boreale	roots, tops	yellow/red	alum, chrome, iron
Larch	Larix spp	needles	brown	alum
Lily of the Valley	Convallaria majalis	leaves	gold	lime
Lombardy Poplar	Populus nigra italica	leaves	yellow/gold	alum, chrome
Lungs of Oak (lichen)	Lobaria pulmonaria	whole lichen	orange	none required
Madder	Rubia tinctoria	whole plant	orange/red	alum, tin
Maple	Acer spp.	bark	tan	chrome, copper sulphate
Mahonia	Mahonia aquifolium	roots, berries, whole plant	blue/brown	alum, chrome
Marigold	Calendula spp.	whole plant, flower heads	yellow	alum, chrome
Marjoram	Origannum vulgare	whole flower heads	violet	alum
Meadowsweet	Filipendula ulmaria	roots	yellow/green	alum, iron
Menegussia (lichen)	Menegussia pertussa	whole lichen	pink (when boiled)	washing soda
Nettle	Urtica dioica	fresh tops	yellow/green/grey	alum, iron
Oak	Quercus spp.	inner bark	gold/brown	alum, chrome

Onion	Allium cepa	skins	yellow/orange	alum
Orchil (lichen)	Roccella tinctoria	whole lichen	purple/blue/red	alum, chrome, ammonia
Pansy	Viola spp	yellow petals	lemon/cinnamon/olive/gold	alum,chrome, iron,tin
Pokeweed	Phytolacca americana	berries	red/tan	alum
Privet	Ligustrum vulgare	leaves, berries	yellow/green/red/purple	alum, chrome, tin
Pyracantha	Pyracantha angustifolium	bark	pink/brown/grey	alum, chrome
Ragwort	Senecio	flowers	deep yellow	alum
Rhododendron	Rhododendron ponticum	leaves	yellow/orange/grey	alum, tin, iron
Rhubarb	Rheum rhabarbarum	roots	pale yellow	none required
Saffron	Crocus sativus	stigmata	yellow	alum, chrome, iron, tin
Silver Birch	Betula pendula	leaves, bark	yellow/gold	alum
Sloe (Blackthorn)	Prunus spinosa	berries, bark	red/pink/brown	alum
Snowberry	Symphoricarpus albus	berries	yellow	alum
Spindle-Tree	Euonymus europaeus	seed pods	red	alum
St. John's Wort	Hypericum spp.	flower tops	yellow/grenn/maroon/black	alum,chrome
Sumach	Rhus spp.	berry tops, leaves	tan/brown	alum, chrome
Sunflower	Helianthus annus	petals / seed hulls	yellow/purple	alum
Sweet Pea (purple)	Lathyrus odoratus	petals	deep green/grey	alum,chrome,iron,tin
Sweet Woodruff	Asperula odorata	whole plants	red/pink	cream of tartar
Tansy	Tanacetum vulgare	flowering heads	yellow	alum
Tea	Camelia sinensis	leaves (cheap tea bags)	pink/tan/dark tan	alum,chrome,none
Usnea (lichen)	Usnea barbata	whole lichen	yellow (when boiled)	
	Usnea lirta	whole lichen	purple (when fermented in urine)	
Walnut	Juglans regia	hulls	brown	none required
Weld (wild Mignonette)	Reseda luteula	whole plant	olive green	alum,cream of tartar
Woad	Isatis tinctoria	whole plant	blue	lime
Yellow Flag Iris	Iris pseudaconus	root	grey/black	chrome,tin,iron
Yellow scales (lichen)	Xanthoria parietina	whole lichen	purple/blue	

Much of the information contained in this table first appeared in the original *Home Farm* magazine, and is reproduced here by kind permission of Katie Thear.

Ice-Creams And Sorbets

If you're producing milk at home, then I suggest that an ice-cream maker is a "must have" item!

Lemon Sorbet

¾ pint whole sheep's milk
4 oz caster sugar
Juice of 2 lemons
2 egg whites

Put milk, sugar and lemon juice together in a jug and stir. Whisk egg whites in a bowl, then mix in other ingredients. Put in the ice-cream maker until done, probably about 20 minutes.

Banana Ice-cream

¾ pint whole sheep's milk
3½ oz caster sugar
3 ripe bananas

Puree the bananas in a bowl, then add the milk and sugar. Whisk all together thoroughly, and put it in the ice-cream maker. Will probably take longer than the sorbet.

Yoghurt

Heat milk almost to boiling point, and hold for at least 30 seconds. Allow to cool to 43°C, then stir in a couple of tablespoons of plain natural yoghurt (any shop-bought brand will do, provided that it's live and unflavoured. It needn't be sheep's milk yoghurt). Put the whole lot in a wide-necked thermos flask and leave for twelve hours or so. The result is delicious, thick, creamy yoghurt! Put it in the fridge to cool. Before you eat it all, remember to put some aside as a starter for the next batch. If you want to be purist about it, you can use freeze dried cultures of *lactobacillus bulgaricus* and *streptococcus thermophilus* to start your first batch, but really it's simpler just to go and buy some yoghurt to get things going – you can eat the remainder of the purchased pot while you wait for your own to be ready!

Cheese

A Simple Cheese

Warm 3 gallons of milk to blood heat. Add 16 drops of rennet. Leave in a warm place to coagulate. Cut the curds, and leave in whey for 1 hour. Strain for 3 hours. Crumble, and add salt at a rate of ½ oz / lb. Place in small moulds, and turn at 24 hourly intervals for 5 days.

After removing from the moulds rub salt into the outside and leave for 24 hours. Turn, and leave for another 24 hours.

Spray with *Penicillium Candidum* to obtain a nice velvety brie-like coating.

Skyr

Having referred to the Icelandic sheep, I really must mention skyr. This traditional low fat sheep's milk product, which falls somewhere between soft cheese and yoghurt, is something of a national institution in Iceland, although I suspect it's mostly made with cow's milk these days. Outside Iceland, it's very difficult to make good skyr, as ideally you need some fresh, un-pasteurised, skyr as a starter. In the absence of ready made skyr, sour cream can be used, but the end product is not as good.

1 gallon skimmed sheep's milk (e.g., after separating off the cream for butter making)
½ pint sour cream (or some ready made skyr)
3 drops rennet

Bring the milk to the boil, then cool to blood heat. Mix the cream with a little of the milk until thin and smooth. Pour this mixture into the rest of the milk and stir in the rennet. Leave to stand at room temperature for 24 hours, then strain slowly through several layers of cheesecloth. The skyr, which remains in the cloth, should be whipped to a smooth consistency before eating. Serve with sugar and cream, or fresh fruit.

Mutton Sausages

2lb lean meat scraps
12 oz fat
1 small glass homebrewed beer
8 oz brown breadcrumbs
¼ tsp nutmeg
1 tsp dried sage
½ tsp dried thyme
½ tsp dried marjoram
1½ tsp salt
½ tsp ground black pepper

Thoroughly mince the meat and the fat twice. Add the beer to the minced meat & fat and mix before adding the rest of the ingredients. Mix everything together in a large bowl, ensuring that the herbs and spices are evenly distributed throughout.

Fill the skins, then hang up the strings of linked sausages in a cool airy place for a few days.

Quick Lamb Burgers

8 oz mince
4 oz breadcrumbs
½ onion, finely chopped or grated
1 clove garlic, crushed
Large pinch mixed herbs
1 tsp Bovril

Mix all ingredients together. Divide into burger sized amounts and shape on a floured board. Fry in shallow fat for about 15 mins, or until cooked in the centre, turning once during cooking.

Makes 6 – 8 burgers.

Lamb And Pasta Bake

2 onions, chopped
2 cloves garlic, crushed
1 lb minced lamb
2 tbsp tomato puree
1 tbsp oregano
1 bay leaf
2 tsp cinnamon
1 tsp ginger
¼ pint white wine
14 oz tin tomatoes
5 oz pasta shells
Cheese sauce

Fry the onion, garlic and lamb for about 5 mins. Add the tomato puree, herbs and spices and cook for a further 5 minutes. Add the white wine and tinned tomato and cook for about 30 mins. Cook the pasta shells and mix into the meat sauce. Tip into an oven proof dish, cover with cheese sauce and bake for 35 – 40 mins.

To make the cheese sauce: Melt a good lump of butter in a heavy based pan. Add enough flour to bind to a doughy lump. Cook for 1 min. Add milk, little by little, stirring well, until a thick and creamy consistency is achieved. Season. Add a good handful of grated cheese, preferably mature cheddar, and stir until melted.

Liver Pâté

8 oz liver, cut into small pieces
4 oz butter and some extra to seal pots
1 medium onion, finely chopped
1 clove garlic, finely chopped
Chopped rosemary and thyme
Seasoning
1 tbsp brandy

Soften the onion and garlic in 1 oz of the butter until just turning colour. Add liver, herbs and seasoning and fry together in a shallow lidded pan for about 15 mins, stirring often. Cool, add the rest of the butter and the brandy, then chop in a food processor until reduced to a creamy paste. Fill into small china pots, smooth over the top, and seal with a layer of melted butter.

Durham Lamb Pie

1½ lbs cubed lamb
2 lbs potatoes
1 large onion, sliced or diced.
3 cooking apples, chopped
3 oz butter
1 large tsp muscavado sugar
½ pint stock

Peel and thinly slice the potatoes. Melt half of the butter and fry the lamb until browned. Remove and reserve. Fry apple and onion for about 5 mins. Line the base of an oven proof dish with half of the potato. Add the lamb, apple and onion. Sprinkle the sugar on the top and season. Pour in the stock and cover with the remainder of the sliced potato. Melt the rest of the butter and brush over the top. Cook in oven for 1½ hours at 350°F or until potato is cooked.

Boiled Mutton & Dumplings

This recipe can be used for almost any mutton cut, from best joints right through to scraggy bits. It is most suitable for small joints, coarse cuts, or diced mutton.

A quantity of mutton
Potatoes (leave these out if you prefer to have mashed spud with your meal)
Onions
Carrots
Celery
Any other vegetables that take your fancy
Pearl barley
Dried split peas
Seasoning
Thyme or rosemary
Stock

For The Dumplings

4 oz shredded suet
8 oz plain flour
2 tsp baking powder
Pinch of salt
Freshly ground pepper, to taste
Chopped fresh or dried thyme / parsley
Cold water to mix

Prepare the dried peas by soaking overnight. Prepare and dice all vegetables. Place all the ingredients in a large pan or dish and add enough stock to cover. Put the lid on, and cook in a low oven (150°C / 300°F) for at least 2 hours, until all vegetables are cooked and meat is tender. My mother used to leave it in the bottom oven of the Aga for about 7 hours. About 30 minutes before you want to eat, mix up the dumpling ingredients, divide into 8 pieces, and drop in the dish. Replace the lid and return to heat. (Could be finished off on the hob, if a suitable dish has been used).

INDEX